D0208621

Ecologically Based Municipal Land Use Planning

Ecologically Based Municipal Land Use Planning

William B. Honachefsky, *P.P., P.L.S., Q.E.P.*

Library of Congress Cataloging-in-Publication Data

Honachefsky, William B.
Ecologically based municipal land use planning / William B. Honachefsky
p. cm.
ISBN 1-56670-406-5 (alk. paper)
1. City planning--Environmental aspects--United States. 2. Urban ecology--United States. I. Title.
HT167.H67 1999
333.77--dc219

99-41245
CIP

Direct all inquiries to CRC Press LLC, 2000 N.W. Corporate Blvd., Boca Raton, Florida 33431.

Lewis Publishers is an imprint of CRC Press LLC

No claim to original U.S. Government works
International Standard Book Number 1-56670-406-5
Library of Congress Card Number 99-41245
Printed in the United States of America 3 4 5 6 7 8 9 0
Printed on acid-free paper

Foreword

I admit there was a time in my own early years as a young land surveyor (the pre-Earth Day decade at least) when I too regarded the land simply as a commodity and not part of a larger continuum, intimately linked to the surrounding air, water, vegetation, and wildlife. A change of career roles and further education changed this perspective dramatically. In the three decades that followed the first Earth Day in 1970, an entire generation was similarly enlightened, thanks to the divulgence of the unspeakable insults we had inflicted on our natural environment. With every new revelation, Congress and the states promulgated environmental laws, rules, and regulations — much of it derogatorily described at the time as knee jerk legislation — in unprecedented proportions. The nation's sense of remorse and guilt was so deep that massive state and federal bureaucracies were created, thousands of persons were employed, and billions of dollars were appropriated and spent — all in the name of environmental protection. Being anti-environment was nearly akin to being anti-American and many a political career began or folded on support for, or opposition to, environmental issues alone.

Unfortunately, we were probably 100 years too late. Much to our chagrin, we found that a century or more of abuse could not be legislated away or rectified overnight, in this case, even after three decades of effort; and it is now doubtful that some of the environmental/ecological damage we inflicted can ever be fully remedied. So now we talk of acceptable risks and risk assessments — how many of our neighbors out of a hundred thousand or a million we are willing to sacrifice so that our economy can go on providing the rest of us with all those things that advertising firms insist we absolutely need. In any event, there is much to be learned from this historical American experiment in human ecology. First, of course, is the need to finally acknowledge that a system of natural ecological infrastructure exists everywhere, that it functions with or without our presence, and that when we interfere with its functions, the resulting effects are often cataclysmic — despite our attempts to reconstruct, duplicate, or mend it. Therefore, it is in our best long-term interests to keep as much of this natural infrastructure as healthy and as intact as possible. Second, and perhaps the most important lesson to be learned is we did not fully realize, at least initially, the powerful influence that municipal land use planning has on the quality of our natural environment. As a result, local land planners have been involved only peripherally in the nation's federal- and state-dominated schemes of environmental protection. We may think that federal and state governments have almost total control over protection of the environment. However, we often forget that long before their respective environmental protection agencies sit down to review projects and issue regulatory permits for wide-ranging programs — such as wetlands protection, point source wastewater discharges to surface waterways, air pollution control, floodplain management, and protection of endangered plants and animals — municipal land planners all across the United States will have been out there well ahead of them prescribing, mostly through local zoning ordinances, where residential subdivisions should be placed and what the densities should be. They will also have determined where and how factories and commercial and office buildings are to be located, how much impenetrable parking lot and roadway asphalt are to be allowed, and where their sanitary sewage and storm water collection systems are to be placed, as well as into which waterway they are to be discharged. They will also have determined what water supplies are to be used for all this development and where the new local roadways to accommodate all these new landscape changes are to be placed. By the time state and federal regulators arrive on the scene to fulfill their obligations to protect the environment, the character of the resulting landscape will have already been pretty much determined. The character of these developing landscapes and the resulting impacts, particularly in rural and fringe suburban areas, are making many Americans uneasy, including the very land planners who are helping to create them.

Preface

As previously discussed briefly in the Foreword, local land use planning may be more responsible for the resultant quality of our natural environment than all the state and federal environmental regulations combined. Despite what now seems such an obvious nexus between local land use and environmental quality, local land planners have had only peripheral involvement in the environmental protection strategies devised by federal and state regulators over the past three decades. That, however, is about to change rather dramatically, as state and federal* environmental regulators** intensify their efforts to cultivate much closer partnerships with local governments.

This is not to imply that municipal land planners have not been making some good faith efforts on their own to incorporate environmental protection into their land planning considerations. For instance, many municipalities for decades have required the preparation of environmental impact statements for major land development projects. Some towns have gone even further, compiling comprehensive inventories of some of the major natural resources occurring within their jurisdictional boundaries. Others still, have adopted sophisticated ordinances for storm water control, tree protection, steep slope, and soil and sediment control. Despite all these laudable efforts, an ecologically incongruous pattern of land development (now collectively referred to as *sprawl*) continues to consume more and more of the American landscape, particularly in the nation's rural and suburban environs.*** Even Congress has now taken an interest in sprawl.

Certainly, local land planners have not had an easy time. Decision making during the site plan review process is often a grueling experience with experts on both sides frequently offering diametrically opposed opinions, which are further confounded by exhausting discourses on the results of so-called predictive models that somehow always seem to predict little or no environmental impact. Yet, as we know, postconstruction conditions tell a much different story. Local land planners have been chastised and sanctioned by the courts, who often seem convinced that local planners have a secret agenda, facilitated by large-lot zoning, to thwart the migration of low-income families into their neighborhoods; the courts do not realize that large-lot zoning is often a municipality's last ditch effort to save at least part of its ecological infrastructure intact. Finally, there are the developer's attorneys who, at the slightest suggestion of a density reduction or plan modification to accommodate some environmental concern, threaten to run to the courts to plead for monetary compensation based on a regulatory taking. Add to this the increase in population, (approximately 120 million) since 1950, plus the completion of those development-inducing, pollution-generating interstate highway systems through previously rural and agricultural areas, and it is little wonder that local land planners often feel overwhelmed. Some even argue that land developers and speculators have more to say about how land in their communities is to be used than the local planners or community residents themselves have.

---

* See especially U.S. Environmental Protection Agency (USEPA) publication, Community Based Environmental Protection, EPA 230-B-96-003, September 1997.

** See also Federal Clean Water Action Plan, Federal Multi-Agency Source Water Agreement, November 1998.

*** It is worthwhile noting that sprawl, particularly scattered development in rural areas, poses some serious risks to those living in these more remote subdivisions. A study by the American Farmland Trust and Northern Illinois University[1] found that police response times were as much as 600% longer, on average, than in the adjoining municipality; that ambulance response times were as much as 50% longer, and that fire response times were as much as 33% longer. The economic analysis in this same report was equally unsettling, but generally confirmed the findings of other studies that homes in scatter development sites do not generate enough taxes to educate the children who live there, and that their taxes are generally insufficient to pay for the maintenance of the roads leading to and through these subdivisions.

However, local land planners are not entirely guiltless. In fact, they may be partly or wholly responsible for some of the problems. Take, for example, the preparation of a comprehensive community or municipal master plan (MMP), an acknowledged prerequisite for establishing sensible land use of any kind. Some communities, even at this late date, have yet to prepare, much less adopt such a plan. Even where a plan has actually been crafted, it is often so superficial that it is an MMP in name only. As a consequence, the zoning ordinance, originally intended to support the day-to-day implementation of the recommendations of an MMP, has become the de facto MMP in a majority of the nation's local communities. While zoning has many valid and useful purposes, it is never an adequate substitute for a comprehensive MMP; and if you need further reinforcement of zoning shortcomings in this regard, go to the nearest window and take another look at the landscape that zoning-dominated land planning has produced thus far.

In spite of this bleak assessment, there is reason for considerable optimism. Clearly, the nation is poised on the threshold of a land-planning renaissance that can significantly alter the way in which Americans in general, and local governments in particular, conduct the business of land use planning. This reformation should include a much greater emphasis on environmental and ecological protection, thanks to the public's indignation over the landscape evolving from the present system of land use and the concerted effort by state and federal environmental regulators to recruit and incorporate local governments into their environmental protection agendas. The motto of this reformation simply is, "The quality of our lives is dependent on the quality of our environment, which is largely dependent on the quality of our land use." The fact that the quality of our lives is so dependent on the quality of our natural environment is nothing new. Early American conservationists such as G. P. Marsh[2] have given us plenty of warning in this regard. Likewise for the close relationship between environmental quality and land use, a doctrine eruditely espoused by Aldo Leopold in 1933, an eminent conservationist who would be pleased to know that the "...extension of the social conscience from the people to the land" that he yearned for was about to come to fruition. The more recently published works of McHarg,[3] Arendt,[4] and Lyle[5] continue to demonstrate that there is considerable wisdom in planning our land use around the ecological constraints of the land.

Despite all the compelling evidence and cogent arguments, both recent and historic, favoring the implementation of a more ecologically sensitive approach to land use planning, municipal planners generally remain frustratingly wary. It is apparent that it will take considerably more encouragement and education to win over their hearts and minds completely. This book is an attempt to do exactly that.

Contents

Chapter 1
Why Ecologically Based Land Use? 1

Chapter 2
Environmental Degradation: The Product of Land Use 15

Chapter 3
Land Use and Environmental Protection: Their Origins, Philosophies, and Destinies 21
3.1 Comprehensive Land Use Planning: A Slow Start and Slow Acceptance 21
3.2 Environmental Protection: On a Parallel Course but Destined to Converge with Land Use Planning 23
3.3 The Making of a Land Use Philosophy 26

Chapter 4
Reconciling the Master Plan and Zoning, and Restoring True Home Rule 29

Chapter 5
Additional Science Aids the Process 33
5.1 Beware the Model 34
5.2 Verification, Validation, and Confirmation of Numerical Models in the Earth Sciences: Guest Essay 35

Chapter 6
The Value of Natural Ecosystems and Natural Resources 47
6.1 Setting the Values 47
6.2 Localizing Values 51

Chapter 7
Private Property Rights and Public Trust Resources 53

Chapter 8
Getting Ready 57

Chapter 9
Getting Started 59
9.1 Development of the Community-wide Parcel Base Map 59
9.2 Include Some Historical Perspectives of the Community 61
9.3 Inventory and Importance of Community Resources 61
9.3.1 Ecological Infrastructure and Associated Components 63
9.3.1.1 Water Resources 63
9.3.1.1.1 Introduction 63
9.3.1.1.2 Surface Waterways 63
9.3.1.1.3 Lakes and Ponds 74
9.3.1.1.4 Groundwater 75

9.3.1.2 Wetlands 78
9.3.1.3 Wildlife 79
9.3.1.4 Open Space 85
9.3.1.5 Vegetation 85
9.3.1.6 Soils 89
9.3.1.7 Geology 94
9.3.1.8 Threatened and Endangered Species 96
9.3.1.9 Steep Slopes 96
9.3.1.10 Floodplains 96
9.3.1.11 Unique Habitats 99
9.3.2 Man-made Infrastructure: Agents of Change, Deterioration, and Destruction 99
9.3.2.1 Storm Drainage Systems 99
9.3.2.2 Impervious Surfaces 102
9.3.2.3 Point Source Discharges 103
9.3.2.4 Known Contaminated Sites 107
9.3.2.5 Agricultural Areas 107
9.3.2.6 Potable Surface Water Intakes and Community and Industrial Supply Wells 108

Chapter 10

Assessing Community Health, Analyzing the Data, and Setting Objectives and Strategies for the Municipal Master Plan 109
10.1 Stressor Indicators 114
10.2 Environmental or Exposure Indicators 114
10.3 Response Indicators 117
10.4 Analyzing the Data and Setting the Objectives and Strategies 128
10.4.1 Example Applications 130
10.4.1.1 Surface Waterways 130
10.4.1.2 Groundwater 155
10.4.1.3 Lakes and Ponds 160
10.4.1.4 Wildlife 162
10.4.1.5 Wetlands 165
10.4.1.6 Soils 168
10.4.1.7 Other Applications 171

Chapter 11

New Ideas for a New Millennium 175
11.1 Introduction 175
11.2 One Lot: One Structure 175
11.3 Mountaintops and Towers 176
11.4 Affordable Housing: Proactivity Pays Much Higher Dividends 176

Chapter 12

General Commentary on Best Management Practices 179

Appendices

Appendix A Web Sites Containing Environmental and/or Ecological Information 185

Appendix B Excerpted Tables from the U.S. Environmental Protection Agency Wildlife Exposure Factors Handbook for the American Woodcock (*Scolopax minor*) and the Red Fox (*Vulpes vulpes*) 187

Appendix C Example Data Available from the New Jersey Department of Environmental Protection Natural Heritage Program 203

Appendix D Excerpts from the New Jersey Geological Survey "A Method for Evaluating Ground-Water-Recharge Areas in New Jersey" 219

Appendix E Siltation and Erosion Control Sample Products 243

References 247

Index 251

About the Author

William B. Honachefsky, an environmental scientist, is also licensed as a professional planner, a professional land surveyor, and a health officer; is a Certified Hazard Control Manager, Master Level, and a professional in soil and erosion control. For the past 29 years he has specialized in the fields of environmental protection and land use planning, both in private enterprise and state and federal government. He developed New Jersey's first trace metal analyses protocols and organized and operated that state's first water resources emergency response sampling team. He is the author of two prior books on land use and environmental planning and is a recognized expert in watershed planning and management.

1 Why Ecologically Based Land Use?

> The planet Earth has been the one home for all of its processes and all of its myriad inhabitants since the beginning of time, from hydrogen to men. Only the bathing sunlight changes. Our phenomenal world contains our origins, our history, our milieu; it is our home....

That poignant summation by Ian McHarg in his 1969 tome, *Design with Nature*, should be inscribed over every entranceway to every planning board meeting room in every municipality in the United States. It should likewise be part of, if not the preamble to, every municipal master plan. Perhaps then the nation's land planners would have a permanent reminder as to what land use planning is really all about. It is not about guaranteeing a profit for landowners, investors, or developers. It is not about encouraging a steady improvement in the gross domestic product (GDP) or lowering real estate taxes. It is about protecting our home and standing watch over the air, water, vegetation, and soil on the Earth's crust that makes life possible on what would otherwise be a planet as barren and inhospitable as our moon.

That the nation's local land planners have yet to fully, or as some claim, even partially embrace this stewardship ethos is painfully reflected in the landscapes they are continuing to produce — landscapes that have changed little from three decades ago, when McHarg[3] described them as, "the expression of the inalienable right to create ugliness and disorder for private greed." Beneath the most conspicuous manifestations of this environmentally incongruous pattern of land development, which Gurwitt[6] calls the ills of suburbia — such as traffic congestion*, miles of cookie cutter housing, soul-less commercial strips, and roads that make pedestrians feel like an alien life form — lie even more sinister effects only rarely brought to the public attention. Those narrow ribbons of concrete and asphalt, for example, which we call roadways and which relentlessly carve up the countryside, carrying more and more cars and more and more humans farther and farther into the last remaining open space and farmland, bring hazards much greater than traffic congestion (Figure 1.1 and Table 1.1). Much like your dog or cat constantly sheds hair and dander, so too does the automobile shed; only its by-products are considerably more dangerous. Asbestos and copper particles are scraped from brake pads every time the brake pedal is applied;[7] zinc, lead, and cadmium wear off tire treads; petroleum hydrocarbons and detergents flow from leaking oil pans and crank-cases; and ethylene glycol drips from overheating cooling systems. The amounts are minuscule, but multiplied by the hundreds of millions of vehicles nationwide, a few ounces quickly turn to pounds, and on a national scale the amounts add up to tons per year. Deposited on adjoining soils in the highway corridor or on the roadway surface itself, these by-products await the next rainfall; then they are flushed into the nearest surface waterway or groundwater aquifer to inflict damage on unsuspecting organisms, including unwitting human receptors. If this were not enough, some 12 million tons of deicing salt (principally sodium chloride**) also are applied annually to the nation's roadways, to be carried later into the nearest surface and groundwater, driving up levels of sodium*** in downstream wells and sending fish gasping for breath. Even newer pollutants such

* It is estimated that Americans lose more than 2 billion hours a year to traffic delays (not counting commuting time).[8]

** Chloride concentrations in direct highway runoff have been measured as high as 10,250 ppm in Wisconsin and 11,000 to 25,000 ppm in Chicago. The recommended maximum level of chloride in a potable water supply well is 250 ppm.[9]

*** Sodium concentrations in potable water supplies have a recommended maximum level of 50 ppm. Sodium poses a danger to persons with heart disease and hypertension as well as to those on low-salt diets. Sodium ions also can damage the structure of soils by displacing calcium ions and thereby reducing permeability and water-holding capacity.

FIGURE 1.1 Common highway engineering design practices remain troubling for the nation's waterways. Common techniques shown here, such as channel straightening, channel widening, and application of sterile riprap, combined with an impending increase of storm water runoff, almost assuredly limit or prolong the recovery of this section of trout production waterway.

TABLE 1.1
Highway-Related Pollutants

Constituent	Primary Sources
Particulates	Pavement wear, vehicles, atmosphere, maintenance, snow/ice abrasives, sediment disturbance
Nitrogen, Phosphorus	Atmosphere, roadside fertilizer use, sediments
Lead	Leaded gasoline, tire wear, lubricating oil and grease, bearing wear, fungicides and insecticides
Zinc	Tire wear, motor oil, grease
Iron	Auto body rust, steel highway structures, engine parts
Copper	Metal plating, bearing wear, engine parts, brake lining wear, fungicides and insecticides
Cadmium	Tire wear, insecticide application
Chromium	Metal plating, engine parts, brake lining wear
Nickel	Diesel fuel and gasoline, lubricating oil, metal plating, brake lining wear, asphalt paving
Manganese	Engine parts
Bromide	Exhaust
Cyanide	Anticake compound used to keep de-icing salt granular
Sodium, calcium	De-icing salts, grease
Chloride	De-icing salts
Sulfate	Roadway beds, fuel, de-icing salts
Petroleum	Spills, leaks, blow-by motor lubricants, antifreeze, hydraulic fluids, asphalt surface leachate
PCBs, pesticides	Spraying of highway right of ways, atmospheric deposition, PCB catalyst in synthetic tires
Pathogenic bacteria	Soil litter, bird droppings, trucks hauling livestock/stockyard waste
Rubber	Tire wear
Asbestos[a]	Clutch and brake lining wear

[a] No mineral asbestos has been identified in runoff; however, some breakdown products of asbestos have been measured.

Source: From U.S. Department of Transportation, *Evaluation and Management of Highway Runoff Water Quality*, Publication No. FHWA-96-032, June 1996, as adapted from N. P. Korbinger, 1984.

as the gasoline additive methyl *tert*-butyl ether (MTBE)* are adding to this dangerous cocktail of pollutants. Not only the nation's water resources are being affected. It is estimated that motor vehicle emissions of carbon monoxide, hydrocarbons, sulfur and nitrogen oxides, and carbon particulates in the United States cause up to $93 billion in damage to human health, vegetation, and structures each year.

Assaults on our natural environment, of course, do not come only from our massive system of roadways. Similar insidious negative impacts are regular companions to nearly all our other so-called land improvements. The construction of residential subdivisions, as well as corporate office, commercial, and industrial complexes, generally requires considerable resculpting of the land. In heavily forested areas this usually means the removal of the existing woodland cover completely or its fragmentation into smaller sized parcels, creating in effect oases of woodland in a sea of human development. The most immediate harm, of course, is to the variety of wildlife that once occupied these extensive woodlands and relied on them for uninterrupted passage, feeding, breeding, and refuge. They are suddenly put out on the street, literally. Many other benefits attributable to such forested areas — including filtration, recycling and assimilation of nutrients and pollutants, prevention of erosion, and recharge of groundwater — are likewise diminished, if not eliminated completely. These occur as the wonderfully rich, permeable ground litter typically found on the forest floor is replaced with a combination of impermeable asphalt and concrete driveways, parking lots, sidewalks, and rooftops. To add insult to injury, those areas not macadamized or concreted usually are revegetated with a variety of nonnative grasses that cannot sustain themselves without massive applications of insecticides, herbicides, fungicides, and fertilizers. These substances eventually end up in nearby surface or groundwaters or in the food chains of various organisms including man. It is somewhat difficult, however, to blame completely the nation's local land planners when they merely reflect the attitude of a busy American public, that has little time or interest to contemplate the welfare of the natural systems that keep us alive. If you doubt this widespread disinterest, ask your neighbors where the water in their tap really comes from or into which stream their wastewater is finally discharged after it is flushed down the toilet, kitchen sink, or shower. The public may not be all to blame either because they have been unduly influenced by Madison Avenue merchants and the nation's economists who often glorify self-indulgence, relegate the value of natural resources to inconsequence without any consideration of a sustainable future.

On the surface at least, current municipal land use practices certainly appear to incorporate more opportunities to minimize environmental damage. It is now common practice, for instance, for communities to require the submission of a site-specific environmental impact statement (EIS) for the largest land development projects; and local ordinances now frequently require more on-site control of storm water runoff, soil erosion, vegetation removal, and steep slope construction. While these are certainly well-intentioned additions, there is a dark side to their application. Collectively, they perpetuate a system of segmented reviews, analyses, and mitigation, that is the antithesis of the way natural systems actually function. The scope of inquiry, for example, of all those site-specific EISs rarely extends beyond the boundaries of the project site. This leaves planners to ponder alone (if at all) the off-site, cumulative effects (and there are many) of all these individual projects. Likewise, the mitigation measures embodied in local ordinances and directed toward on-site control of acknowledged negative environmental impacts give a false sense of security that such impacts can be readily ameliorated on-site; and even if they cannot, the effects on the natural environment are alleged to be inconsequential. Figure 1.2 demonstrates the fallacy of this latter assumption.

* MTBE is both an air and a drinking water hazard. Health complaints from airborne MTBE include headaches, dizziness, irritated eyes, burning of the nose and throat, coughing, disorientation, and nausea. The USEPA has tentatively classified MTBE as a possible human carcinogen.

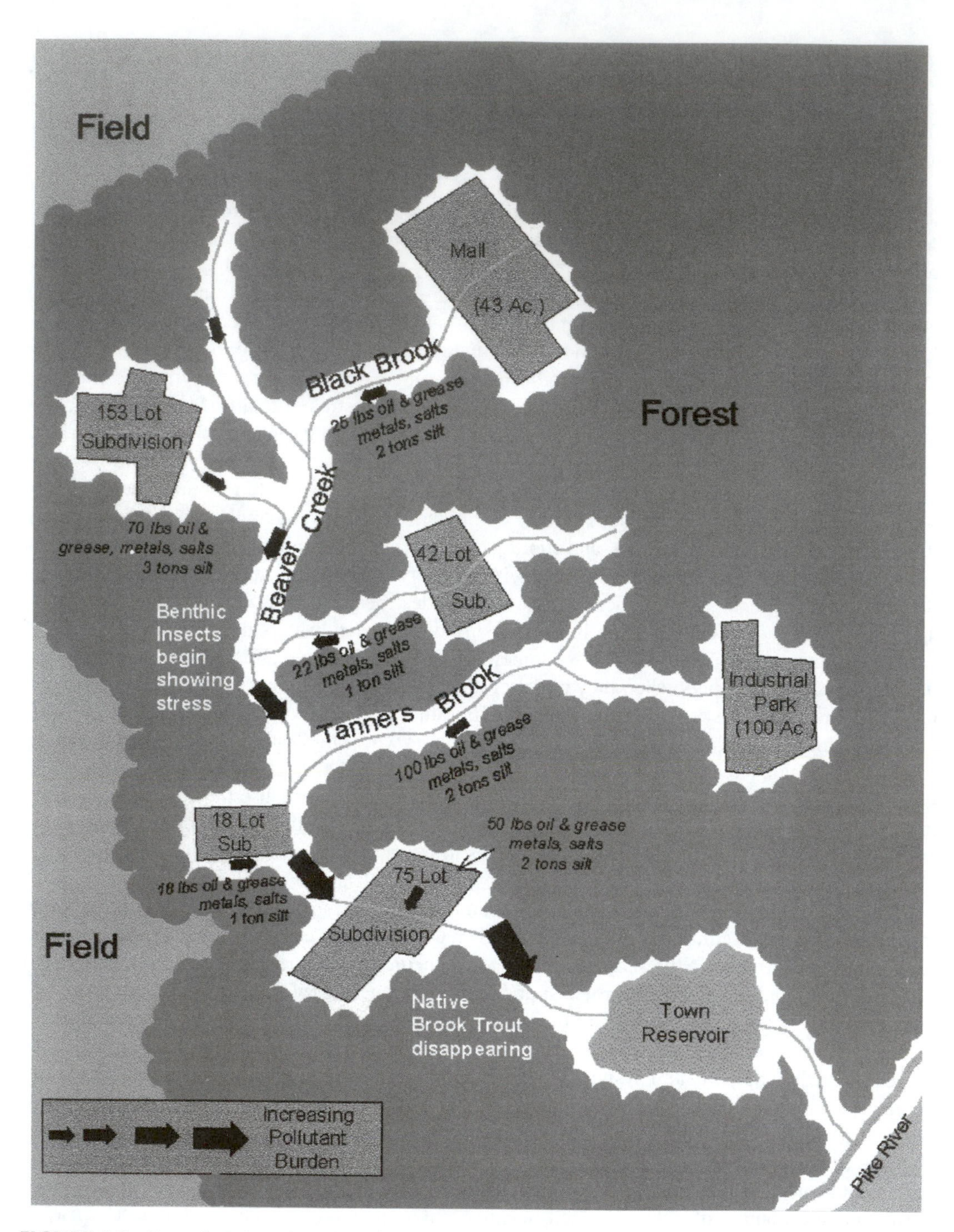

FIGURE 1.2 Even though each project site may contribute only relatively small amounts of pollutants to a stream, when these pollutants are combined with the pollutants discharged from other sites, the burden on the stream grows larger and larger.

While local planners may wring their hands, shake their heads, and complain, there is little they can do; they are, in fact, partly responsible for their own dilemma. They insist on pursuing land-planning practices that should have been discarded years ago. From the outset they preordain themselves to failure, if not at least years of vexing confrontations with landowners, developers, and

FIGURE 1.3 Make no mistake about it; this is more than just a handful of mud. These stream sediments and the disheartening array of pollutants they bear will be our legacy to future generations of Americans, who, I suspect, will judge our role as stewards of Earth's natural resources rather unkindly.

the courts. I am talking about the zoning map, the deadly instrument that is now largely responsible for guiding much of the nation's local land use and for bestowing unearned economic value on private property, primarily through density allocations. Unfortunately, these density allocations are prescribed irrespective of the carrying capacity, health, structure, and functions of the community's natural resources. Zoning rarely benefits the nation's natural resources. It does, however, increase the economic expectations of private landowners, and once granted, such expectations are rarely relinquished voluntarily thereafter. An owner of a 100-acre parcel of land located in a 1-acre zone fully expects to be able to subdivide his or her parcel into 100 lots. Such density allocations rarely are accompanied by a caveat that indicates the densities are conditioned on the carrying capacity or health of the natural resources existing on the property. Land planners should, therefore, not fret if later on during the site plan review process, the landowner is reluctant to give up that density, even if it is for the good of the community's natural resources. Density, once granted, is rarely surrendered. James Kunstler[10] perhaps sums up best what is a growing consensus concerning current zoning practices: "…if you want to make your community better, begin at once by throwing out your zoning laws. Don't revise them — get rid of them. Set them on fire if possible and make a public ceremony of it; public ceremony is a great way to announce the birth of a new consensus."

While I believe that our local land planners have a strong commitment to environmental protection, they, much like the rest of the nation's citizenry, remain appallingly uninformed about the structure, functions, and overall health of the environment they want so strongly to protect. This pronounced lack of knowledge greatly weakens their ability to function as advocates for protecting the community's natural resource systems and puts them constantly at the mercy of self-serving hired experts whose scientific inquiries, as we have just seen, rarely track the impacts of their client's proposed land form changes beyond the borders of a site's property boundaries. Yet often the cumulative effect of these off-site impacts have contributed to the nation's current environmental malaise.

As custodian of Earth's resources, which include those creatures that share the planet with us, we have a rather dismal track record; and unless we quickly change the way we currently prescribe our land use, we are destined to burden the generations to follow, much as we were betrayed and burdened by our predecessors who left us with what now appear to be everlasting legacies of contaminated sediments (Figure 1.3), soil, water, air, and a long list of extirpated species. We should act like this is the only planet we have because it is.

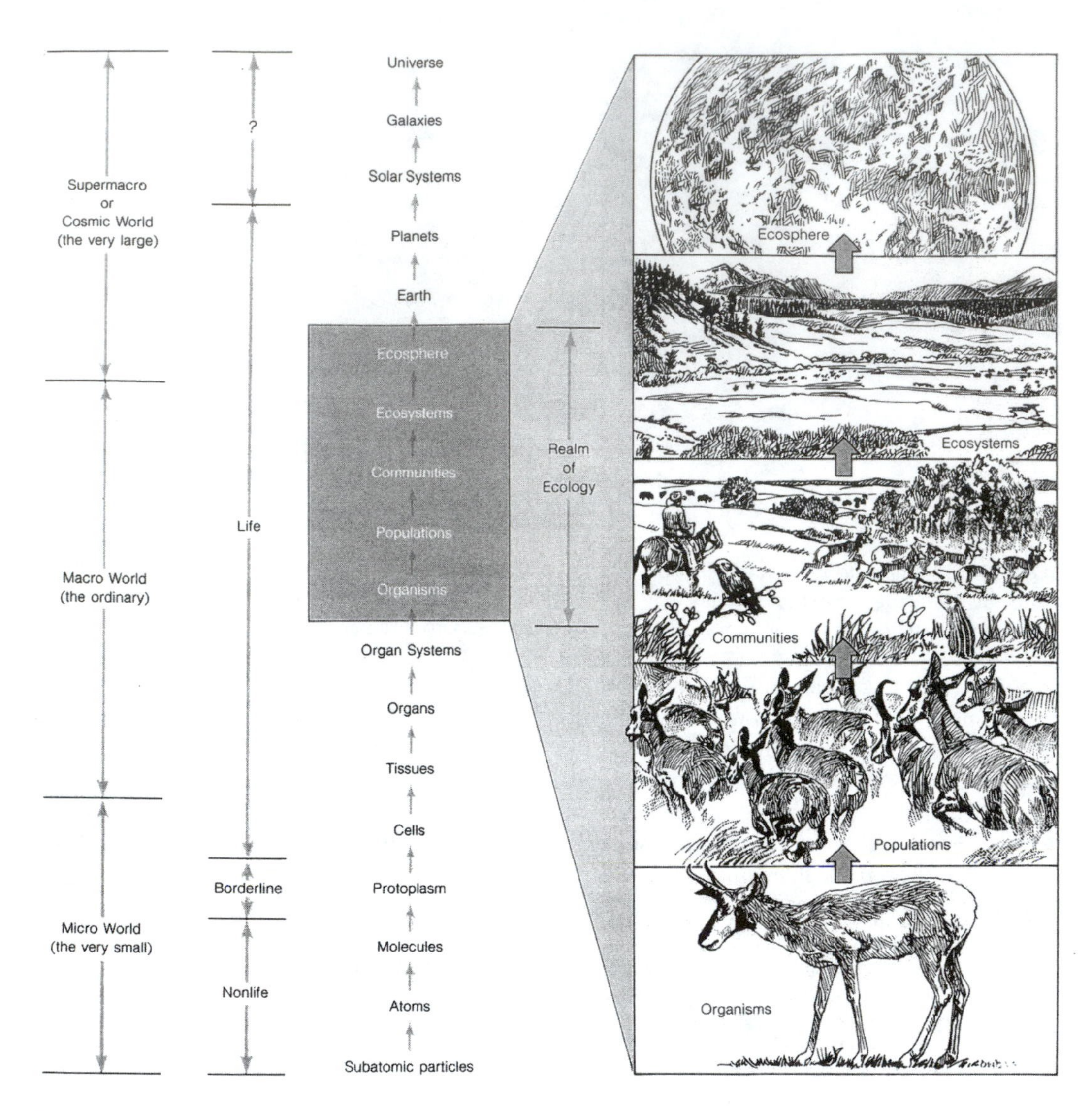

FIGURE 1.4 Levels of organization of matter, according to size and function. (From Miller, G. T., *Living in the Environment*, 6th ed., Wadsworth, Belmont, CA, 1990. With permission.)

How then do we protect Earth and avoid the further construction of landscapes that are inherently environmentally hostile, unsustainable, and often inhumane, and that we sometimes can hardly bear to look at? The answer, of course, lies in the adoption of an ecologically based system of municipal land use planning. This certainly is not a new idea but one that has gained considerably in popularity especially in the past decade as symptoms of an out of control system of land use have grown more ominous. What is ecology anyway? The term, whose derivation is credited to German biologist Ernst Haeckel, comes from a combination of the Greek word for home, "oikos," with the Greek word, "logos," meaning study of. Literally interpreted, it means the study of one's home — certainly a fitting definition. Scientists have enlarged that definition slightly, and ecology is now most often described as the study of the interactions between organisms and their living (biotic) and nonliving (abiotic) environments — a definition whose simplicity belies the enormous number and complexity of those interactions.

The focus of ecology can be global and generalized (Figure 1.4), or local and very specific (Figures 1.5 and 1.6), but in either case, the common unit of study is called an *ecosystem.* Whether

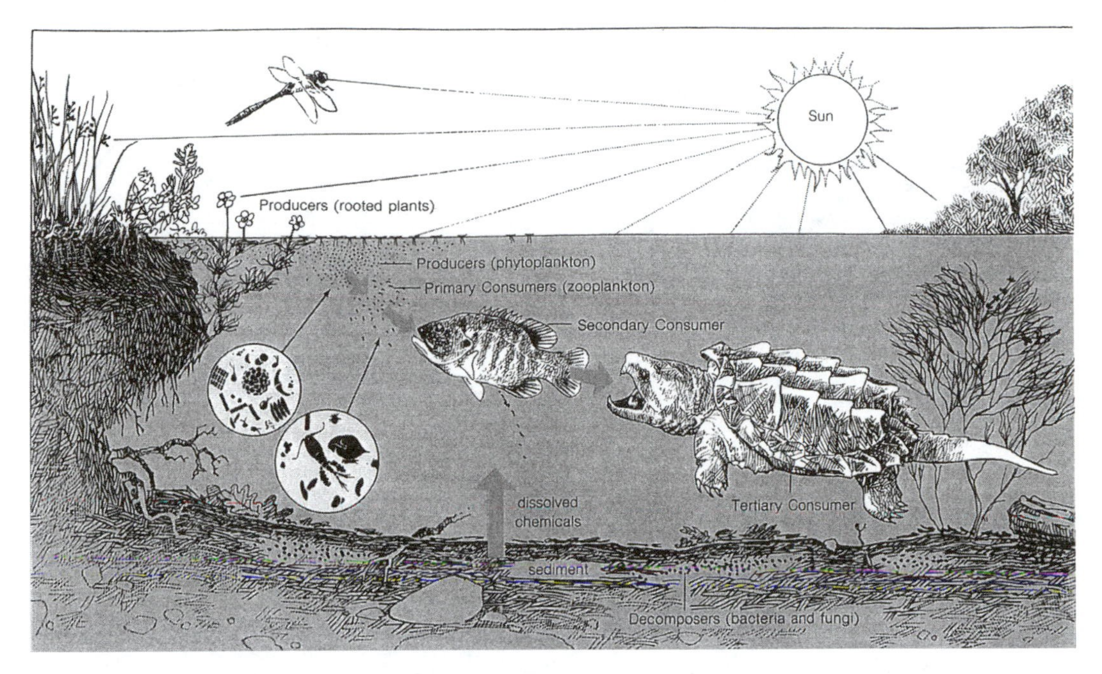

FIGURE 1.5 The major components of a freshwater pond ecosystem. (From Miller, G. T., *Living in the Environment*, 6th ed., Wadsworth, Belmont, CA, 1990. With permission.)

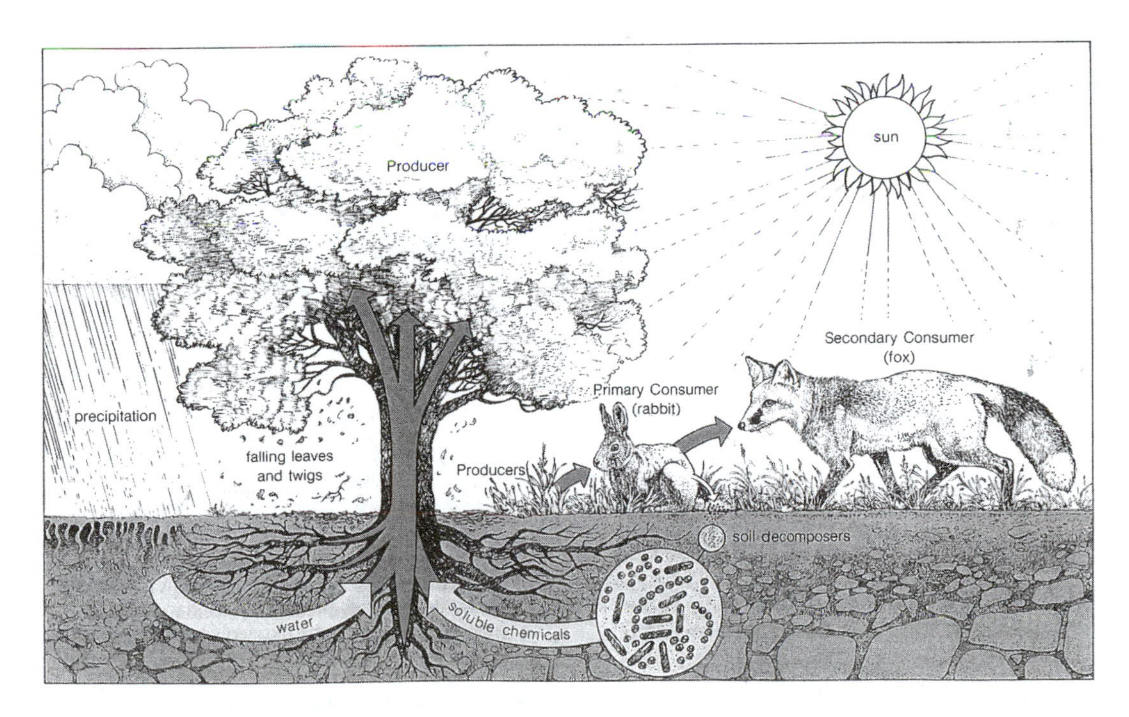

FIGURE 1.6 The major components of an ecosystem in a field. (From Miller, G. T., *Living in the Environment*, 6th ed., Wadsworth, Belmont, CA, 1990. With permission.)

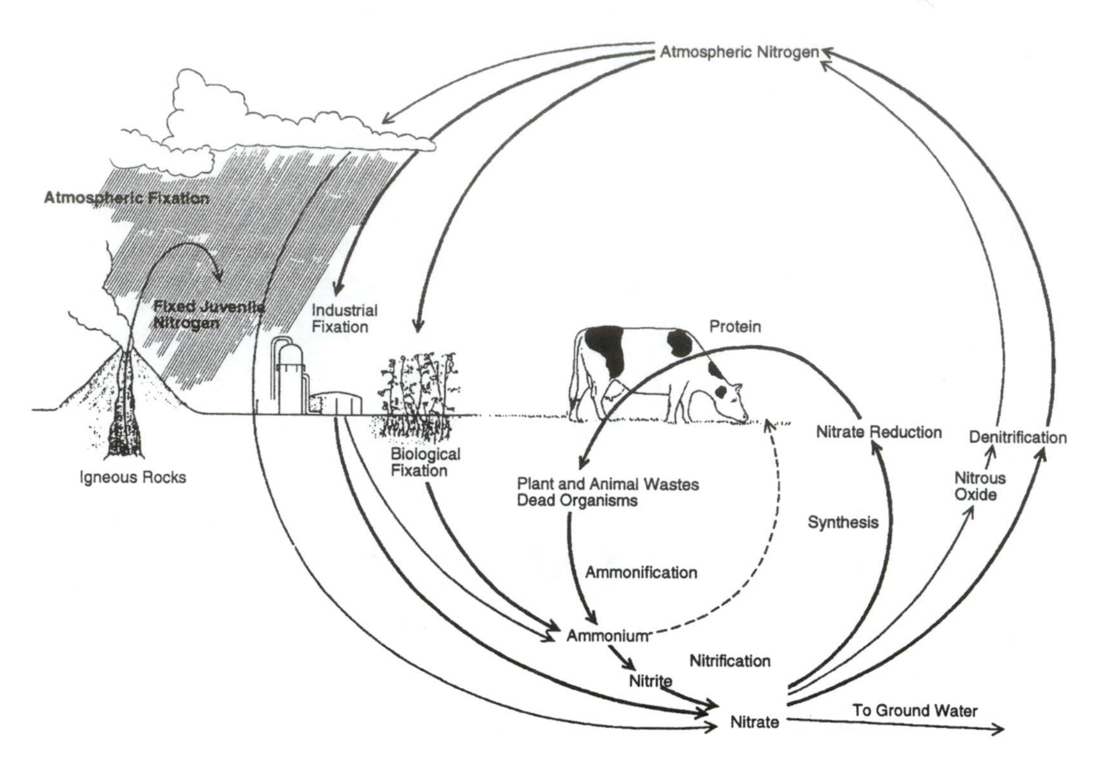

FIGURE 1.7 The nitrogen cycle. (From U.S. Environmental Protection Agency, Manual of Nitrogen Control, EPA/625/R-93/010, September 1993.)

large or small, ecosystems all have common characteristics that are of critical importance to local land planners. First, all ecosystems rely on two fundamental processes — the flow of energy from the sun and the recycling of essential chemicals and nutrients required for living organisms. Elements of significance that are continuously recycled include carbon, nitrogen (Figure 1.7), phosphorus, water (Figure 1.8), and oxygen.

Second, the biotic component of each ecosystem is generally organized into three levels of organisms, often referred to as *trophic* levels. Occupying the first level are the *producers* or autotrophs, those organisms that manufacture their own organic compounds that they then use to satisfy their own needs for energy and nutrients. Most producers are green plants, including algae, which utilize photosynthesis to manufacture carbohydrates, sugars, starches, and celluloses. *Consumers*, or heterotrophs, the second level of organisms, on the other hand, get the nutrients and energy they need by feeding either directly or indirectly on producers. The last group, called *decomposers*, are those organisms of decay, primarily fungi and single-celled bacteria, that release simple substances back to the environment; these substances are to be reused once again by the producers, thus completing a continuing flow of energy and materials through what is collectively called a food chain. Because there are so many food chains operating simultaneously, ecologists have coined the word *food web* (Figure 1.9). Finally, whether large or small, ecosystems do not normally have distinct boundaries. Instead, they operate as continuums, with one ecosystem blending into the next through transition areas called *ecotones.* Ecotones characteristically contain many of the plants and animals found in each adjoining ecosystem, but often have species not found in either.

Despite the importance of these ecological continuums — food chains, food webs, and endless cycling and recycling of nutrients, energy, and other materials — I would be willing to wager that no more than a handful of municipalities (if any at all) have acknowledged, much less incorporated, such considerations into their zoning ordinances and maps, municipal or comprehensive master plans, or in their day-to-day site plan deliberations. Add to this the fact that our present consumption

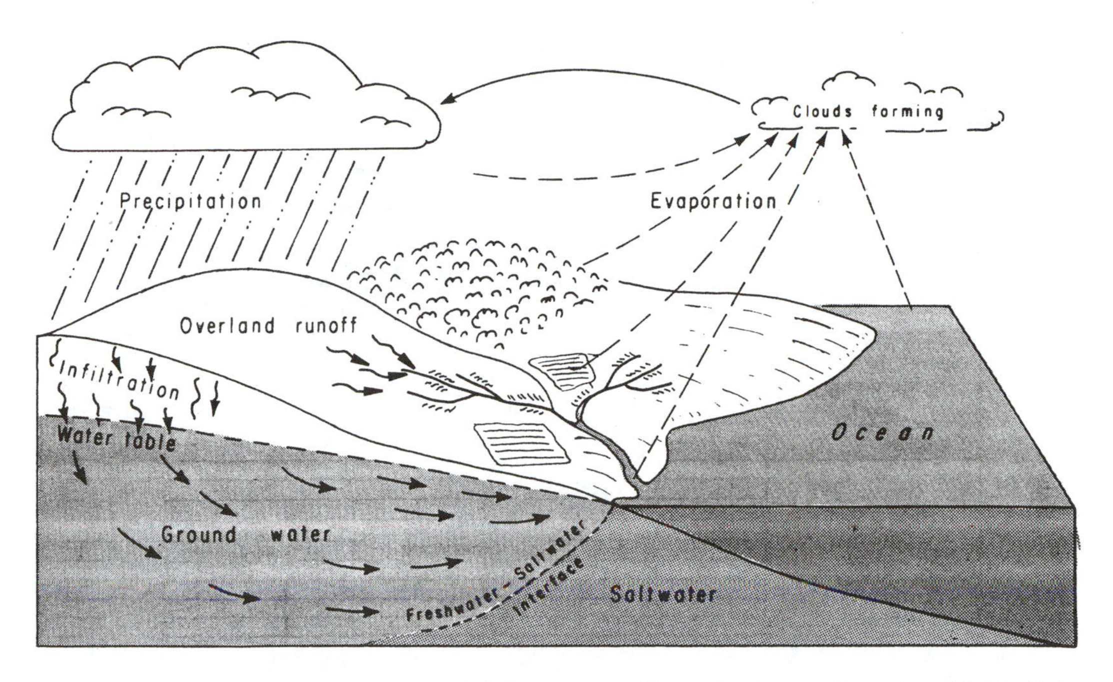

FIGURE 1.8 The hydrologic cycle. (From U.S. Environmental Protection Agency, Protection of Public Water Supplies from Ground Water Contamination, EPA/625/4-85/016, September 1985.)

and use of natural resources simply is not sustainable. We are taking materials from Earth at rates much faster than they can be replaced by natural cycles. Groundwater, for example, is being extracted from our underground aquifers in such great proportions nationwide that normal rainfall is simply unable to replenish it fast enough to keep water tables from dropping. As a result, ground surfaces are subsiding, sometimes by as much as several feet, and we are being forced to drill wells deeper and deeper or find alternate aquifers. Ironically, a good deal of that water is being used to subsidize an agricultural industry in landscapes that were never more than marginal, at best, for the production of crops. Wackernagel et al.[82] argue rather persuasively that we need to begin thinking in terms of *ecological footprints*, which they describe as the total area of ecologically productive land and water occupied exclusively to produce all the resources (food, fuel, and fiber) consumed and to assimilate all the wastes generated by a given population, be it a household, community, or country. Wackernagel and colleagues have calculated the ecological footprint for 52 countries, with results given in hectares per capita (Table 1.2). For example, a country with a 5 ha per capita footprint would mean that 5 ha of biologically productive space* are in constant production to support the average individual of that country. In contrast to what space is being used by each country, Wackernagel et al. also have calculated what the average available ecological footprint is for the entire world population. They argue this is only 1.7 ha per capita.** Thus unless a country has sufficient ecological capacity (i.e., substantial biologically productive space), its citizens must rob space and resources from other nations. Of all the 52 nations evaluated, the United States has the largest footprint, some 10.3 ha per capita, with an available biocapacity of 6.7 ha per capita, resulting in an ecological capacity deficit of 3.6 ha per capita. Thus, as a nation, we are living far beyond our ecological capacity.

* Earth has a surface area of 51 billion ha, of which 36.3 billion are sea and 14.7 billion are land. Only 8.3 billion ha of land area are biologically productive. The remaining 6.4 billion ha are marginally productive or unproductive for human use.[82]

** The actual amount is about 2.0 ha per capita; however, the World Commission on Environment and Development indicates that at least 12% of the world's ecological capacity should be preserved for biodiversity of Earth's 25 million other species. Thus, the reduction to 1.7 ha is explained.

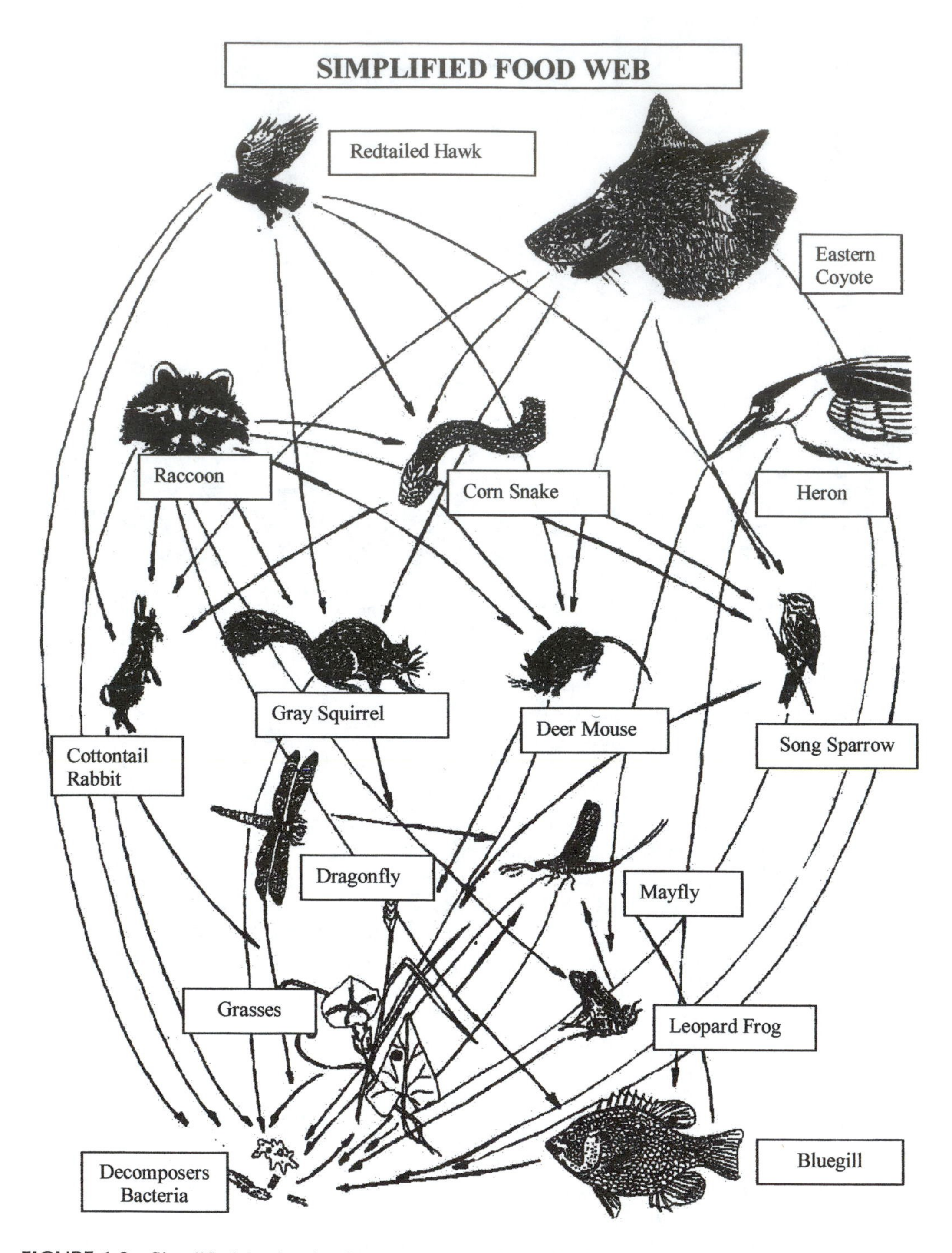

FIGURE 1.9 Simplified food web. (Graphics by Greg Honachefsky.)

There are a number of local land planners, I am sure, who have some trepidation and reservations about adopting a more vigorous ecological approach to municipal land use planning, perhaps because they may believe they are alone in this endeavor, and therefore vulnerable. Nothing could be further from the truth. This is a renaissance of national and global proportions. Support is coming from many groups; one group, in particular, is especially but pleasantly surprising, American

TABLE 1.2
Ecological Footprints of Nations

	Population in 1997	Ecological Footprint (in ha/capita)[a]	Available Biocapacity (in ha/capita)[a]	Ecological Deficit (If Negative) (in ha/capita)[a]
Argentina	35,405,000	3.9	4.6	0.7
Australia	18,550,000	9.0	14.0	5.0
Austria	8,053,000	4.1	3.1	–1.0
Bangladesh	125,898,000	0.5	0.3	–0.2
Belgium	10,174,000	5.0	1.2	–3.8
Brazil	167,046,000	3.1	6.7	3.6
Canada	30,101,000	7.7	9.6	1.9
Chile	14,691,000	2.5	2.1	0.7
China	1,247,315,000	1.2	0.8	–0.4
Colombia	36,200,000	2.0	4.1	2.1
Costa Rica	3,575,000	2.5	2.5	0.0
Czech Republic	10,311,000	4.5	4.0	–0.5
Denmark	5,194,000	5.9	5.2	–0.7
Egypt	65,445,000	1.2	0.2	–1.0
Ethiopia	58,414,000	0.8	0.5	–0.3
Finland	5,149,000	6.0	8.6	2.6
France	58,433,000	4.1	4.2	0.1
Germany	81,845,000	5.3	1.9	–3.4
Greece	10,512,000	4.1	1.5	–2.6
Hong Kong	5,913,000	5.1	0.0	–5.1
Hungary	10,037,000	3.1	2.1	–1.0
Iceland	274,000	7.4	21.7	14.3
India	970,230,000	0.8	0.5	–0.3
Indonesia	203,631,000	1.4	2.6	1.2
Ireland	3,577,000	5.9	6.5	0.6
Israel	5,854,000	3.4	0.3	–3.1
Italy	57,247,000	4.2	1.3	–2.9
Japan	125,672,000	4.3	0.9	–3.4
Jordan	5,849,000	1.9	0.1	–1.8
Korea, Republic	45,864,000	3.4	0.5	–2.9
Malaysia	21,018,000	3.3	3.7	0.4
Mexico	97,245,000	2.6	1.4	–1.2
Netherlands	15,697,000	5.3	1.7	–3.6
New Zealand	3,654,000	7.6	20.4	12.8
Nigeria	118,369,000	1.5	0.6	–0.9
Norway	4,375,000	6.2	6.3	0.1
Pakistan	148,686,000	0.8	0.5	–0.3
Peru	24,691,000	1.6	7.7	6.1
Philippines	70,375,000	1.5	0.9	–0.6
Poland, Republic	38,521,000	4.1	2.0	–2.1
Portugal	9,814,000	3.8	2.9	–0.9
Russian Federation	146,381,000	6.0	3.7	–2.3
Singapore	2,899,000	6.9	0.1	–6.8
South Africa	43,325,000	3.2	1.3	–1.9
Spain	39,729,000	3.8	2.2	–1.6
Sweden	8,862,000	5.9	7.0	1.1
Switzerland	7,332,000	5.0	1.8	–3.2
Thailand	60,046,000	2.8	1.2	–1.6

TABLE 1.2 (continued)
Ecological Footprints of Nations

	Population in 1997	Ecological Footprint (in ha/capita)[a]	Available Biocapacity (in ha/capita)[a]	Ecological Deficit (If Negative) (in ha/capita)[a]
Turkey	64,293,000	2.1	1.3	–0.8
United Kingdom	58,587,000	5.2	1.7	–3.5
United States	268,189,000	10.3	6.7	–3.6
Venezuela	22,777,000	3.8	2.7	–1.1
World	5,892,480,000	2.8	2.1 (2.0 for 1997)	–0.7 (–0.8 for 1997)

[a] Expressed in area with world average productivity, 1993 data.

Source: Adapted from Wackernagel, M. et al., *Ecological Footprints of Nations: How Much Nature Do They Use? How Much Nature Do They Have?* Commissioned by the Earth Council for the Rio+5 forum, International Council for Local Environmental Initiatives, 1997. With permission.

industrialists, where some of the captains of industry are already declaring the first industrial revolution a serious mistake environmentally. Ray Anderson, CEO and chairman of the board of the billion-dollar Interface Corp., for example, recently stunned a packed audience of architects, designers, planners, and builders at a conference in Miami, Florida, when he declared:

> …I'm part of an endemic process that is going on at a frightening, still-accelerating rate, worldwide, to rob our children, their children, theirs, and theirs, and theirs of their futures. There is not an industrial company on Earth, and I feel pretty safe in saying not a company or institution or firm of any kind, not even an architectural firm or an interior design practice, that is sustainable in the sense of meeting its needs without some measure depriving future generations of the means of meeting their needs. When Earth runs out of exhaustible resources, when ecosystems collapse, our descendants will be left holding the empty bag. And some day people like me may be put in jail.[11]

These are very powerful words indeed from a representative of a group once considered indifferent to the plight of our natural environment. His co-keynote speaker, Paul Hawken (author of *The Ecology of Commerce*), was equally as emphatic when he proclaimed:

> This is a reformation, as profound and as important as the earlier Reformation, but there is no template, there is no map, really there is no one to lead us, except us. This reformation is entirely grassroots from ten thousand groups from hundreds of millions of people around the world.[11]

I want to further reassure local land planners, and anyone else who is interested, that ecologically based municipal land use planning does not equate to no growth. To the contrary, ecologically based municipal land use recognizes that providing for the needs of a growing national population (expected to grow by 80 million by the year 2020 and by 254 million by 2050), is a continuing obligation. At the same time, the growth along with accompanying land use needs to be directed to areas that are intrinsically best suited for it. It should not in the process, however, sacrifice the very ecological infrastructure that assures our sustainability if not survivability. Having said that, the reality is that landowners and developers are not going to be able to build wherever or whatever they want. Land restricted by some paramount ecological concern, nevertheless, would be an appropriate candidate for inclusion on a community's "wish list" of open space acquisitions, through either a fee simple purchase or a conservation easement. Linking local open space acquisitions to the maintenance of ecological infrastructure is a much better way of doing business than presently exists for acquiring open space, where parcels are acquired intermittently, are spatially scattered, and only occasionally are designed specifically to preserve the integrity of a community's ecological infrastructure.

The attractiveness of ecologically based municipal land use planning is that it can be accommodated within the existing land use planning tools and practices already in place, such as the municipal master plan and zoning ordinances, with some modification, of course. We need not reinvent the wheel in this case. We have known for decades, if not centuries and millennia, either by intuition or by experience, that certain lands are inherently dangerous or inappropriate for human occupation. We also have learned that our intrusion into relatively undisturbed landscapes produces considerable environmental damage, which we have spent much of the last 30 years trying to characterize in greater and greater detail. To their credit, state, federal, and academic scientists have compiled an enormous amount of valuable research data, with which we are now better able to describe the ecological responses to these intrusions. To say that we can now describe them all would be misleading. We cannot, and research will continue. In the meantime, there are plenty of useful scientific data waiting for local land planners to assimilate and to incorporate into their decision-making processes, bearing in mind that ecological proofs need not meet a standard of beyond a reasonable doubt, but instead a preponderance of the evidence.

The late Winston Churchill once observed, "Americans usually do the right thing, but only after exhausting all their alternatives." It is time to do the right thing. If we do not, our successors may very well curse the day that this generation walked on Earth.

2 Environmental Degradation: The Product of Land Use

On a hot summer day in September 1995, the residents of a small community located in the western highlands of New Jersey, watched in horror as one of the town's favorite waterways rolling past their municipal building, abruptly ceased flowing. While the small trout and other resident fish life sought refuge in what few deep pools remained, the town's leaders began to search for clues to this unprecedented event. The state had been experiencing a severe shortage of rainfall but certainly not a record shortage. By nightfall, however, much to the surprise of the town's residents, the stream suddenly began flowing again. Then on the following morning, the stream again stopped flowing. This alternating pattern of cessation and flow continued for several days, until an astute town employee recognized the correlation of this erratic stream flow with the commencement and cessation of irrigation on an upstream golf course that directly abutted the affected waterway. The golf course had been permitted to withdraw up to 100,000 gallons per day (GPD) of groundwater to keep its fairways green and healthy. While no scientific study was ever undertaken to prove a direct cause and effect relationship, when the golf course was requested to halt its irrigation operations for several days, the stream flow remained continuous. The problem was eventually corrected with the return of more normal rainfall patterns. However, the incident clearly demonstrated the complex interrelationships between various components of a waterway ecosystem and our continued failure to recognize such interrelationships and to provide an adequate level of protection for the maintenance of their integrity. In all likelihood, the golf course was probably just one of many final straws for this water body. For the prior quarter century, new residential subdivisions had been constructed in both the headwaters and primary recharge areas of the small waterway; as a result of the placement of new and impervious roadways, roofs, and driveways, along with the installation of interceptor and storm drains and the withdrawal of groundwater through a myriad of individual wells, less and less recharge of the groundwater was taking place. Because stored and later discharging groundwater is often the primary source of water to a surface water body when there is no rainfall, the underlying aquifer had diminishing amounts of water to give up to the surface stream, especially during stressful drought periods. The system had lost its normal resiliency. Whether the irrigation pumps actually prevented flowing groundwater from reaching and recharging the diminished stream flow or was conversely sucking water back from the streambed itself, can be determined only by further scientific studies — none of which have been proposed at this time.

A similar occurrence in the state several years earlier also demonstrated the precariousness of water supplies taken from areas where groundwater storage is accommodated primarily within the cracks and fissures of underlying geologic bedrock formations. In this particular instance, a residential subdivision of approximately 25 homes was approved for a site located in the state's Piedmont region underlain by a dense but slightly fractured Triassic shale. All homes were to be served by individual wells and septic systems. The discontinuous fractures in this underlying bedrock allowed very little capture and storage of infiltrating rainfall, and potable water wells already installed in this general area for individual households were noted for their low productivity, on the order of 0.5 to 1 gal/min. As is common in such developments, the subdivision was being built in sections. In this case, ten homes were constructed in an initial section that occupied an area located at the base of a hilly prominence. Once this first section was completed, construction began on the remaining 15 homes that were to be placed on the top of this same prominence. As this final group of homes was sold and occupied, the downhill residences constructed in the first phase of the project began to experience problems with their water supply wells. At first there were complaints of turbid, muddy water followed by complaints of insufficient supply and eventually a total loss of all water. The homes in the uphill section had intercepted, through their own water supply wells, the limited amount of groundwater that was flowing from the hilly prominence, consequently leaving their downhill neighbors high and dry.

On a much grander scale, New York and New Jersey have been wrestling with the dilemma of what to do with approximately 7 million cubic yards of sediments contaminated with toxic metals and organic chemicals that need to be dredged from the bottom of major shipping channels and docking facilities in the New York–New Jersey harbor complex. Unless done fairly soon, ship traffic to these areas could be in serious jeopardy. Because open ocean dumping of such contaminated material is now prohibited by the federal government, there are heated debates over possible solutions; and the projected costs, no matter which solution scenario is eventually selected, will be enormous. In all the debate that has taken place so far, however, no one has questioned how these sediments became so contaminated in the first place. That perhaps is the real issue, at least for those concerned with land use and land planning. The answer, of course, lies largely in the upstream watersheds tributary to the New York–New Jersey harbor complex,* where an ounce of prevention, or more correctly an ounce of environmentally sensitive planning, centuries (even decades) earlier might have prevented what is now a very expensive pound of cure.

We could go on for pages and volumes with similar anecdotal stories of "environmental problems" ranging from minor to catastrophic. In the analysis of their genesis we would find them all to be intimately linked to the ways in which we have used or perhaps more correctly, abused the land. After 30 years of investigating nearly every conceivable form of environmental transgression, I am comfortable in suggesting that all environmental problems are in reality land use and land planning problems. The nation's land planners, primarily at the local or municipal level, zone the land to prescribe where residential subdivisions are to be placed and what the density should be. The planners also determine where and how factories and commercial and office buildings are to be constructed, where sanitary sewage and storm drain collection systems are to be placed and into what stream they are to be discharged, what type of water supplies are to be used, and where roads to accommodate these new land use changes are to be placed; all this, when implemented, can seriously alter existing ecosystems that may have been in equilibrium for centuries or even millennia. This disruption of ecological equilibria is largely responsible for the mind-boggling environmental problems that now confront us.

Finally, one of the major premises of ecological theory is that everything is linked to everything else and that in reality there is only a single global ecosystem of which we are all a part. That concept was dramatically reinforced by some unexpected results from a study of near coastal and estuarine water currents in the New York–New Jersey harbor estuary complex in late 1987 and early 1988. New Jersey was greatly concerned about the amount of floatable debris that was washing up on its coastal recreational beaches (Figure 2.1) threatening a multibillion dollar tourism industry. Bathers were being bombarded with such things as nail-studded pieces of lumber, greaseballs, syringes, blood vials (Figure 2.2), wooden pilings, plastic tampon applicators, glass bottles, and in some cases human and animal body parts. Determined to find the potential sources of this obnoxious, sometimes gruesome debris, the state undertook a "drifter" study. This involved deploying specially designed 500-ml plastic bottles (Figure 2.3) in various tributaries of the New York–New Jersey harbor complex, allowing them to drift freely with the existing water currents, and hoping to track their waterborne routes and final destinations. Inside each drifter bottle was a postcard size information packet that the finder could mail back to the state, indicating the location where the drifter bottle had beached itself. In lieu of this, the finder could call a toll free number to provide the same

* The fact is supported by Newman and McIntosh[12] who assert, "Coastal marine and estuarine metal contamination is typically a collection of local problems at the subregional scale on the North Atlantic coast of North America."

FIGURE 2.1 Some of the vast amounts of floatable debris temporarily stranded by the tides and waiting for eventual transport by subsequent tides and storms to New Jersey's bathing beaches in 1988.

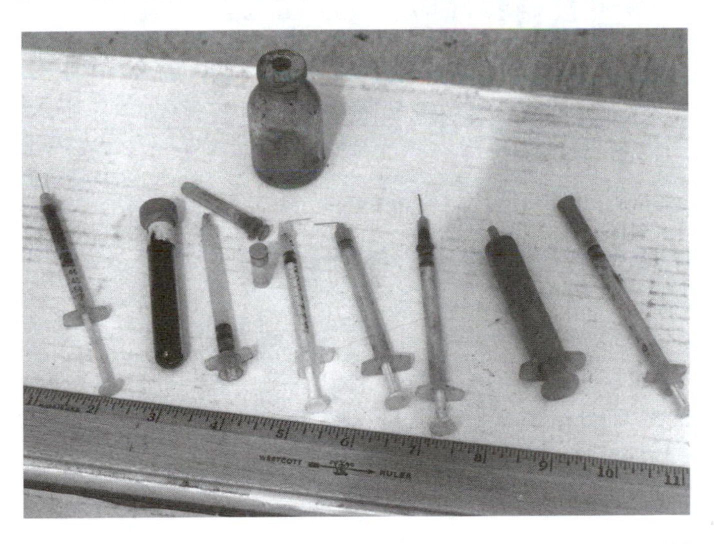

FIGURE 2.2 Typical medical waste found washing up on New Jersey beaches in 1988 much to the horror of summer bathers.

information. Figure 2.4 illustrates the routes and final destinations of some of the drifter bottles. Many of the locations where the drifters had landed were anticipated by the project designers; however, there was some surprise when landings were reported 1 to 6 months later from the coastline of Rhode Island, and Martha's Vineyard and Nantucket Island. However, the biggest surprises came a year later when drifter bottles were found by bathers in the Azores off the coast of Portugal and

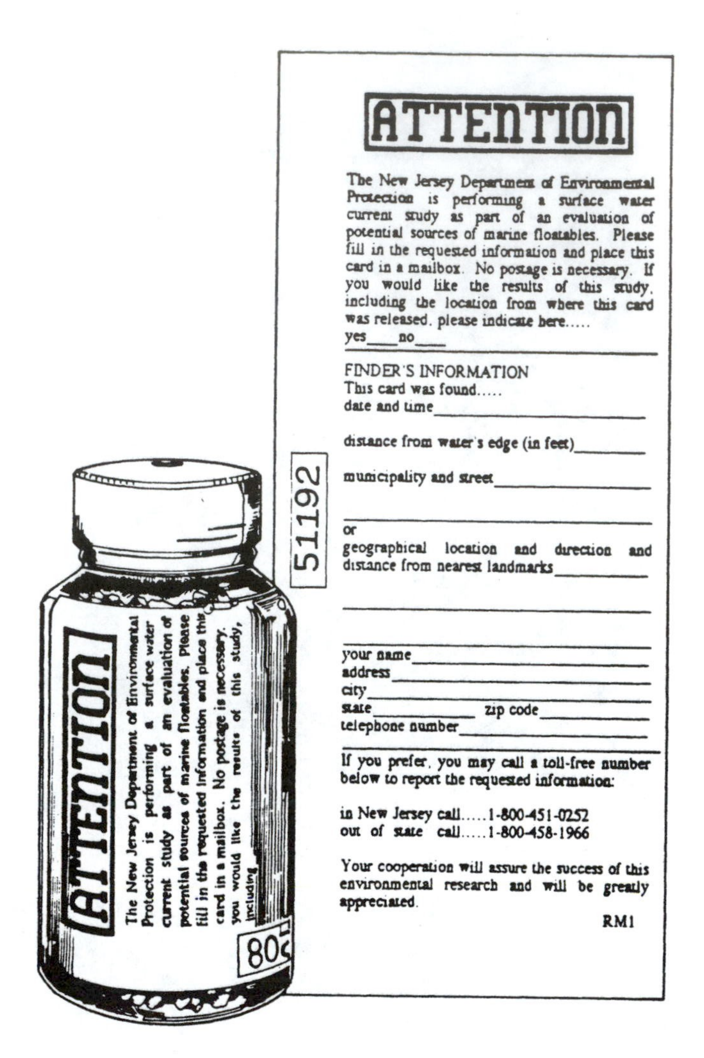

ATTENTION

The New Jersey Department of Environmental Protection is performing a surface water current study as part of an evaluation of potential sources of marine floatables. Please fill in the requested information and place this card in a mailbox. No postage is necessary. If you would like the results of this study, including the location from where this card was released, please indicate here.....
yes____ no____

FINDER'S INFORMATION
This card was found.....
date and time________________

distance from water's edge (in feet)________

municipality and street________________

or
geographical location and direction and distance from nearest landmarks________

your name________________
address________________
city________________
state________ zip code________
telephone number________________

If you prefer, you may call a toll-free number below to report the requested information:

in New Jersey call.....1-800-451-0252
out of state call.....1-800-458-1966

Your cooperation will assure the success of this environmental research and will be greatly appreciated.

RM1

51192

FIGURE 2.3 Diagram of drifter bottle and return form. (From New Jersey Department of Environmental Protection, New Jersey Floatables Study: Drifter Study Results, Science Applications International Corp., August 1988.)

on the beaches of Bermuda, Ireland, Scotland, and Great Britain. One of those bottles had been released near the Statue of Liberty. Some months later one bottle was found stranded on a beach in Norway.

Parts of the planet that once seemed far removed from one other had suddenly taken a quantum leap closer together, and the often quoted contemporary slogan "think globally, act locally" now had significantly greater meaning.

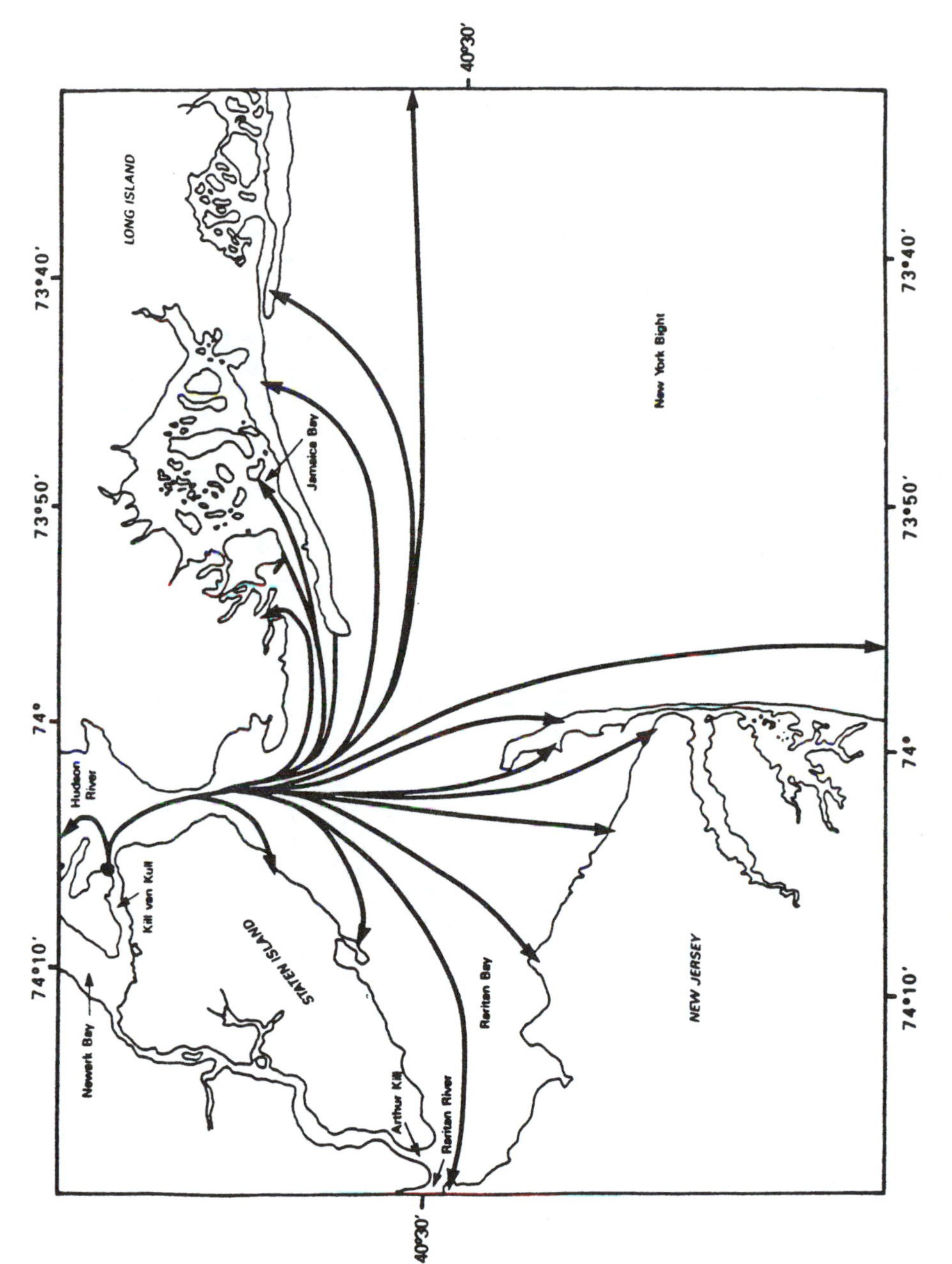

FIGURE 2.4 Trajectories and locations of some of the found drifter bottles. (From New Jersey Department of Environmental Protection, New Jersey Floatables Study: Drifter Study Results, Science Applications International Corp., August 1988.)

3 Land Use and Environmental Protection: Their Origins, Philosophies, and Destinies

3.1 COMPREHENSIVE LAND USE PLANNING: A SLOW START AND A SLOW ACCEPTANCE

It should not come as a surprise to anyone that comprehensive land use planning has never been a integral part of the American culture despite some early seminal town planning done for colonial Philadelphia and Washington, D.C. Such planning has always been regarded as a serious threat to one of the nation's founding principles — unfettered private enterprise. Our deep-rooted aversion to restrict or injure free enterprise, however, brought about environmental conditions in our earliest towns and cities that proved to be both intolerable and costly to remediate.

America's 19th century towns, as described by Hagman and Juergensmeyer[13] were, "...largely unattractive, muddy, cluttered clusters of buildings...characterized by filth, stench and stagnant water in the streets, backyard privies, dampness and the absence of sunlight in residential areas." Their further assessment that these unappealing and sometimes hazardous conditions were largely the result of, "unrestrained, private enterprise," is one that today's local land planners would do well to commit to memory for subsequent recall during their nightly, often chafing, deliberations with prospective land developers and speculators.

American towns were growing rapidly during the 19th century due to highly profitable land speculation as well as the proliferation of new factory construction in existing town centers. The net result of both of these factors was the rapid, almost explosive centralization of population that is clearly shown between the 1840 and 1880 censuses. In 1840, only 12 American cities had populations of over 25,000, of which 3 had populations that exceeded 100,000. By 1880, 75 cities had populations of over 25,000 and 20 of those had more than 100,000.

Despite the horrific conditions produced by this centralization of population, a comprehensive strategy for localized land use planning was slow to materialize. The incidences of several major epidemics including an 1878 yellow fever outbreak in Memphis, Tennessee, in which 5000 people died, may have been an inadvertent catalyst for the promotion of some rudimentary level of land use planning. In response to this particular outbreak, for instance, Congress created a National Board of Health to advise governments and regulate quarantines. Most importantly, however, the board also completed an unprecedented study of the physical and structural conditions of the city of Memphis, which made 12,000 recommendations for remedial improvements to property in the city. While most of the recommendations dealt with nuisance abatements, there were also proposals for a new sanitary public water supply, a sewerage system, the destruction of substandard buildings, the enactment of a sanitary code for the entire city, the repaving of many of the streets, the ventilation of all city houses, and the appointment of a city sanitation officer.[13]

A further catalyst may have been the initiation of the *City Beautiful Movement*, which some claim had its roots at the 1893 Chicago World's Fair. The City Beautiful Movement was a "grass roots" movement resulting in the formation of numerous town improvement associations whose goal was to rectify the lack of order and cleanliness in American cities. By 1901, there were over 1000 such associations across the United States. These civic associations promoted the creation of citizen advisory land use commissions. These advisory commissions, in some instances, became organs of municipal government, but lacked any powers conferred by statute or ordinance. Any plans, therefore, that they might have produced were submitted to the municipal governing body solely as recommendations. The first community in the United States to create such an advisory city planning commission was Hartford, Connecticut in 1907. Milwaukee, Wisconsin followed suit in 1908, as did Chicago, Illinois in 1909.

While the deep-seated aversion to restrict free and open competition remained strong among the nation's early entrepreneurs, the local citizenry was less than enthused with the results of such unrestricted enterprise. Slaughterhouses and smoke-belching factories proved to be highly incompatible neighbors to residential neighborhoods, resulting in numerous citizen complaints that were referred most often to the judiciary for resolution. The judicial decisions (case law) resulting from these referrals made a significant contribution to the implementation of land use planning on a broad scale. First, they recognized the hierarchical primacy of residential land uses, which required a greater degree of safety and quietness than other land uses required. Second, they acknowledged that to protect these family living areas, more intensive land uses had to be segregated from them. Thus was born the concept of *zoning*, whereby land use districts were created to segregate residential, commercial, industrial, and sometimes agricultural uses.

At first blush, such a concept would have seemed to be highly objectionable to free enterprisers. Surprisingly, however, such zoning became a popular, national trend for two principal reasons. First, such zoning validated (protected) pre-existing land uses; and second, it provided the opportunity to *over zone* for the more profitable uses of business and industry. Comprehensive zoning, however, was not comprehensive land use planning. Sewer systems, for example, especially during the era of the sanitary reform movement, continued to be designed and laid out without reference to future locations and densities of different land uses to be served by them. Similarly, highways continued to be constructed without reference to any long-range plan for the types of land uses they would eventually serve. Such an approach to problem solving produced a phenomenon known as *disjointed incrementalism*, in which successive municipal problems such as storm drainage, sewage treatment, or traffic circulation were incrementally solved without consideration for any long-term impacts or integration.

While local zoning maps and their concomitant ordinances rose in popularity, the more visionary, future-oriented, comprehensive land use planning process languished. Zoning ordinances, to the further detriment of comprehensive land use planning, received additional encouragement at the national level when the U.S. Department of Commerce promulgated its 1924 (later revised in 1926) Standard State Zoning Enabling Act. The act provided a ready-made model for state legislatures to follow in delegating police power to municipalities to prepare, adopt, and administer zoning codes. By 1926, 564 cities and towns had adopted zoning ordinances and several state courts had upheld zoning as a valid exercise of police powers delegated by states to their municipalities. That same year, the United States Supreme Court, in a landmark case, *Village of Euclid vs. Ambler Realty Co.*, upheld the use of police power to zone. While the act required that zoning regulations and zoning decisions be made "in accordance with a comprehensive plan," it failed to detail what exactly a comprehensive plan was.

Not until 1928 did the U.S. Department of Commerce finally produce the Standard City Planning Enabling Act (in contrast to the *zoning* act). This act envisioned a more comprehensive approach to regulating land uses and specified:

> It shall be the function and duty of the [planning] commission to make and adopt a *master plan* for the physical development of the municipality…[showing] the commission's recommendations for the development of said territory, including, among other things, the general location, character and extent of streets, viaducts, subways, bridges, waterways, water fronts, boulevards, parkways, playgrounds, squares, parks, aviation fields and other public ways…[and] the removal, relocation, widening, narrowing, vacating, abandonment, change of use or extension of any of the foregoing…as well as a *zoning plan* for the control of the height, area, bulk, location and use of buildings and premises….

The act further provided:

> In the preparation of such plan the commission shall make careful and comprehensive surveys and studies of present conditions and future growth of the municipality…The plan shall be made with the general purpose of guiding and accomplishing a coordinated, adjusted and harmonious development of the municipality…as well as efficiency and economy in the process of development.

Unfortunately, under this act planning was optional and not mandatory. In addition, the act did not specifically state that any proposed zoning should be enacted in accordance with a comprehensive plan. These shortcomings increased the growing confusion between comprehensive land use planning and zoning, especially since the prior State Zoning Enabling Act had expressly stated that zoning should be enacted in accordance with a comprehensive plan. That one act called for a comprehensive plan (not defined) and the other called for a master plan were further discrepancies that have not gone unnoticed by the courts in their deliberations of land use cases. Unfortunately, they have held in some instances, much to the detriment of comprehensive land use planning, that the plan with which zoning must be in accord with could be found in the entirety of the zoning ordinance itself.

Despite these adverse judicial rulings, the federal government and a significant number of states continue to encourage the preparation of a comprehensive land use plan or master plan for their local governments. The state of New Jersey, for example, the most densely populated state in the nation, has passed a Municipal Land Use Law (New Jersey S.A. 40:55D-1 et seq.) which generally calls for "…the establishment of appropriate population densities and concentrations that will contribute to the well-being of persons, neighborhoods, communities and regions, and preservation of the environment."

While the law provides the opportunity for New Jersey municipalities to prepare a municipal master plan (MMP), it does not specifically make it mandatory; this indicates instead that a planning board, "…*may* [emphasis added] adopt…a master plan." Despite this, many New Jersey municipalities concerned that their zoning ordinances might be invalidated without one have opted to prepare an MMP. Of particular importance is the fact that these municipalities have included as an essential element a *conservation plan*, which is to provide

> …for the preservation, conservation, and utilization of natural resources, including, to the extent appropriate, energy, open space, water supply, forests, soil, marshes, wetlands, harbors, rivers and other waters, fisheries, endangered and threatened species, wildlife and other resources and which systematically analyzes the impact of each other component and element of the master plan on the present and future preservation, conservation, and utilization of those resources.

This conservation plan portion of the MMP, if adequately addressed and implemented to the extent it was intended, has at least the potential even as a stand-alone document to totally revolutionize the ways in which New Jerseyans currently utilize their portion of Earth's landscape.

3.2 ENVIRONMENTAL PROTECTION: ON A PARALLEL COURSE BUT DESTINED TO CONVERGE WITH LAND USE PLANNING

Much like our sluggish acceptance of land use regulations, the adoption of a sense of stewardship toward our natural environment and its component systems and resources was similarly slow and perhaps of even longer duration. The current American philosophy of environmental protection that is now an important part of our national agenda was not easily assimilated into American society. In fact, the process took place over a period of nearly 200 years. Why it took so long for Americans to become serious about their roles as stewards of the nation's natural resources has been examined at great length by earlier authors and historians. There is no need to repeat those detailed examinations in this text.

The development of the American environmental protection (formerly conservation) ethic occurred generally in three phases. The first covered the time period from the end of the Revolutionary War to the beginning of the 20th century and can be described as the *formulation phase*. Abundance was the key word of this period — an abundance of natural resources and human greed. In spite of this, during this time period a small but diverse group of citizens — including philosophers, teachers, historians, public servants, and politicians — began to catalog the nation's natural resources and to observe and critique the performance of their fellow Americans and the human race, in general, toward their natural environment. These early naturalists would formulate the relatively inviolate values of the American conservation ethic. These values were best summed up by George Perkins Marsh who is often considered the nation's first ecologist. In Marsh's[2] 1864 book, *Man and Nature*, he states:

> The ravages committed by man subvert the relations and destroy the balance which nature has established...; and she avenges herself upon the intruder by letting loose her destructive energies.... When the forest is gone, the great reservoir of moisture stored up in its vegetable mould is evaporated....The well-wooded and humid hills are turned to ridges of dry rock...and...the whole earth, unless rescued by human art from the physical degradation to which it tends, becomes an assemblage of bald mountains, of barren turfless hills, and of swampy and malarious plains. There are parts of Asia Minor, of Northern Africa, of Greece and even Alpine Europe, where the operation of causes set in action by man, has brought the face of the earth to a desolation almost as complete as that of the moon.... The Earth is fast becoming an unfit home for its noblest inhabitant, and another era of equal human crime and human improvidence...would reduce it to such a condition of impoverished productiveness, of shattered surface, of climate excess, as to threaten the depravation, barbarism, and perhaps even extinction of the species.

It is not surprising that Marsh's concerns and observations differ very little from those offered by contemporary environmentalists. As we look at the world today, I believe he would be both pleased and terribly saddened to find that his predictions are very close to fulfillment.

The second phase, called the *period of heightened concern*, lasted from around 1900 through 1960. This period was marked by the continued subservience of the nation's natural resources to the personal empires of a wealthy few. The conservation effort peaked and waned during this phase, interrupted by an economic depression and two world wars. Two notable peaks were the presidencies of Theodore Roosevelt (1903 to 1909) and Franklin D. Roosevelt (1933 to 1945). The pugnacious "T.R." is remembered mostly for his attempts to focus the attention of the nation on conservation, for retrieving over 150 million acres of American forest land from corrupt, vested interests, and for beginning an inventory of the nation's natural resources. The black period just prior to Franklin Roosevelt's presidency is described in Stuart Udall's book,[14] *The Quiet Crisis*:

> The economic bankruptcy that gnawed at our country's vitals after 1929 was closely related to a bankruptcy of land stewardship. The buzzards of the raiders had, at last, come back to roost, and for each bank failure, there were land failures by the hundreds. In a sense the Great Depression was a bill collector sent by nature, and the dark tidings were borne on every silt laden stream and every dust cloud that darkened the horizon.

Franklin Roosevelt made conservation, in this case, land restoration, an integral part of his war against depression. His greatest contributions were the creation of the Civilian Conservation Corps (CCC), the Tennessee Valley Authority (TVA), and the Soil Conservation Service (SCS). The active participation of the general populace in his conservation programs established a new sense of personal responsibility to the environment for millions of Americans. Unfortunately, the 15 years that followed his presidency and that ended the second phase were relatively uneventful for the conservation movement.

The third phase of the environmental movement is best described as the *period of panic*, and commenced around 1960 and continues to the present. Aldo Leopold, one of the nation's premier conservationists, had complained in 1933:[15]

> There is as yet no other dealing with man's relation to land and to the animals and plants which grow upon it. The lands' relation is still strictly economic, entailing privileges but not obligations.... Obligations have no meaning without conscience, and the problem we face is the extension of the social conscience from people to the land.

In 1961, the extension of the national conscience to the land that Leopold spoke of would begin in earnest. Why did it begin at this period? Perhaps it was the election of a vibrant and environmentally sensitive president, the worsening condition of the nation's natural resources, or a healthy economy that allowed citizens to focus on issues other than an empty wallet or empty stomach. In all likelihood, it was all these. However, there was also another milestone in the conservation movement about to occur. An adroit, highly principled scientist, laboring away in Maryland was already concluding what would probably be her magnum opus on environmental degradation. Armed with new technologies and a communication network previously unavailable to earlier naturalists and scientists, Rachel Carson jolted the national conscience with her 1962 book, *Silent Spring*. The success of her book was due largely to her ability to distill the incomprehensible into a few hundred pages of easily understood prose. Citizens who may have had difficulty comprehending the magnitude of the nation's misuse of pesticides and chemicals found Carson's book on the topic frighteningly clear. Equally as important was the fact that *Silent Spring* also stimulated a personal and national inquiry into the ramifications of man's other transgressions on the environment.

The old values formulated by the likes of Marsh, Thoreau, Emerson, Muir, Powell, and Leopold were suddenly resurrected by new legions of advocates. By the latter part of the 1960s, environmental protection had risen to the top of the national agenda, culminating in the passage of the National Environmental Policy Act in 1969. One year later, a nation filled with remorse would celebrate the world's first Earth Day.

In the decade that followed, the federal and state governments produced environmental protection laws, rules, and regulations by the ton. Municipal governments, compelled by state mandates or citizen initiatives, responded in a similar fashion and began to accommodate or renew environmental goals in their subdivision ordinances and master plans. Some municipalities even went so far as to prepare maps, albeit sometimes crude, of their natural resources. Being anti-environment nearly equated to being anti-American, so deep was the guilt and remorse of the nation.

Despite the widespread popularity of environmental protection and antipollution legislation, little of it passed through Congress or state legislatures with ease. Political action committees, many newly formed and representing a broad cross section of corporate America, lobbied very effectively for key modifications. Confrontations were common and often bitter. *Compromise* would become a common, but distasteful word in the vocabulary of modern American conservationists, now generally referred to as environmentalists. This large-scale production of environmental protection legislation had negative side effects other than the most obvious one of widening the schism between corporate America and American environmentalists. It also promoted the false theory of the *legislative cure*, the hypothesis of which is that the mere passage of legislation would somehow cure the problem. The most significant negative impact of this mass production effort, however, was the fact that it promoted the diminution of individual participation in the environmental protection movement. Abetted by burnout from over a decade of concerted effort to produce legislation that often had unpalatable compromises, many Americans found it easier to assume that the massive environmental protection bureaucracies they had helped set in place would diligently continue the battle for them. It was a false presumption.

It was not until 1984 that some earnest inquiry began into what all these laws, rules, regulations, and hundreds of billions of dollars we had spent had produced. At that time Congressional hearings on the National Environmental Monitoring Improvement Act concluded, "that, despite considerable expenditures on monitoring, federal agencies could assess neither the status of ecological resources nor the overall progress toward legally-mandated goals of mitigating or preventing adverse ecological effects."[16]

There was no doubt that some substantial progress had been made. Rivers no longer caught on fire or flowed with all different colors of the spectrum. Gross releases of sulfuric acid mists from paper mills and black sooty plumes from coal fired power plants no longer coated residences, ate the paint off cars, or forced residents indoors so they could breathe. We had stripped away some of the grossest forms of pollution; however, having done so, we found that environmental degradation in many instances did not disappear. Thousands of miles of our streams, rivers, and lakes, for example, still failed to meet the long sought goals of being fishable and swimmable thanks to sediments laden with such substances as toxic heavy metals, petroleum hydrocarbons, pesticides, and polychlorinated biphenyls (PCBs). If this were not enough, those domestic sewage treatment plants we had paid to upgrade to secondary treatment levels were now spewing out tons of nitrates a day, which were flowing to our estuaries and coastal waters; there they were inducing massive blooms of unsightly, sometimes toxic algae. Our lakes were similarly continuing to fill up with tons and tons of sediment laden with the same toxic heavy metals, nitrates, pesticides, and bacteria; as a result, fish kills remained a common occurrence as did the excessive growth of emergent weeds and algae, producing the most unsightly conditions.

While the nation's attention had been focused on the grossest point source discharges of pollutants, it had missed the more subtle implications of failing to deal with environmental protection on an ecological scale. The same disjointed incrementalism we had utilized in our approach to land use regulation was, not surprisingly, the same choice here. In fairness, however, some attempts had been made under Sections 208 and 303(e) of the federal Clean Water Act and subsequent amendments that followed to consider and plan for water quality related issues on a regional scale. Unfortunately, after considerable work was done, funds ran out, the programs languished, and many of the recommendations they had produced were left on the shelf to gather dust.

The 1984 inquiry by Congress prompted the federal environmental bureaucracy to take a new look at the way it was doing business. It is now encouraging its state counterparts to think in more holistic terms such as watersheds and ecosystems, and most importantly, of partnerships, not only between state and federal governments but also with municipal governments. The U.S. Environmental Protection Agency's (USEPA) recommended use of *place-based management* is an attempt to recognize the importance of local governments in environmental protection. While the EPA has not yet explicitly identified them as a key component of this newest initiative, local land planners will eventually have a prominent role in such localized management. However, such ecologically based initiatives are doomed to failure if we think that our present approaches to land use planning, primarily through zoning ordinances and subdivision regulations, are sufficient. They are not and any such efforts are futile.

3.3 THE MAKING OF A LAND USE PHILOSOPHY

Why our relationship with the natural environment is so incredibly adversarial has piqued the interest of planners and citizens alike, and the roots of this relationship go back several centuries. Lyle[5] offers some insight in this regard:

> Much of our difficulty in dealing with resource and environmental issues is brought on by the fact that the human landscape...was shaped according to a concept of nature that grew out of the Renaissance notion that humans are the measure of all things.

This concept helped foster a subsequent view of the world as little more than a giant machine whose parts could be systematically dissected and analyzed, thereby providing information that would enable us to gain power over nature and thus control her processes. This *dominion over nature* philosophy has been the foundation of the way in which we view and use the land and its associated resources. Supplemented with the mechanical capabilities provided by the Industrial Revolution, we have clearly demonstrated our potency to overpower nature by our ability to resculpt Earth's landscapes on a catastrophic level. Controlling Earth's processes, however, is another story. Much like the fictitious fairy tale character, Humpty-Dumpty, we have had significant difficulty, after we have dissected the parts, in putting them back together again (if we really ever tried), so that we might restore the integrated whole and subsequently the prior level of diversity produced by thousands, if not millions of years of evolution. We have chosen instead to replace what Lyle[5] describes as "…an ever varying, endlessly complex network of unique places adapted to local conditions…," with a system of relatively simple forms and processes repeated with consistent, often boring regularity, across the face of Earth, leaving in its wake a disaggregated world.

4 Reconciling the Master Plan and Zoning, and Restoring True Home Rule

Chapter 1 indicates that the zoning ordinance and the comprehensive community master plan with some modification could easily accommodate any proposed adaptations for an ecologically based system of land use. However, before any such adaptation can take place, we need to first reconcile the anomalous hierarchical relationship that generally exists between these two documents — a relationship where comprehensive planning, normally first in order of priority, has been largely preempted and subordinated by zoning.

The original mission of zoning, if you recall, was to preserve the sanctity of family living areas, separating them from the obnoxious and sometimes harmful character and by-products of commercial and industrial enterprise. Thanks to some historical quirks, that singular mission evolved into an amalgam of other objectives and goals, to the extent that it intruded into areas more appropriately within the domain of a comprehensive community master plan. That intrusion has come at some considerable cost, not the least of which is the widespread abrogation of comprehensive planning in many of the nation's municipalities, in deference to a system of hybridized zoning. Zoning, however, regardless of its packaging, is *not* planning, a point reinforced by Solnit and others:[17]

> Zoning is only part of the process called "planning." Zoning separates a municipality into districts and regulates, on various bases, building and structures, but planning has a much broader focus — it concentrates on development in relation to the community's current and future well being…a practical distinction is that planning measures, such as adopted plans, goals and so forth are official policy for the future whereas zoning lists the permissible uses for specific properties right now.

It is hard to imagine under what circumstances zoning would be fully capable of successfully addressing all the requisite issues (Table 4.1) normally considered by a legitimate, comprehensive master plan; and if that is the case, which I believe it is, we need to stop trying to force zoning to do something it was never intended to do.

It is clear that zoning was expected to be subordinate to a comprehensive master plan* facilitating the implementation of the plan's provisions, instead of usurping its authority and jurisdiction as is currently practiced. Consequently, in those communities where land use is currently being dictated by zoning ordinances masquerading as comprehensive planning, local land planners and the governing body need to reinstitute legitimate comprehensive planning before inaugurating a program of ecologically based municipal land use.

---

* The New York State Court of Appeals noted in *Udell vs. Haas*, 21 NY 2d 463, 235 N.E. 2d 897, 288 NYS 2d 888 (1968), "…the comprehensive plan is the essence of zoning. Without it, there can be no rational allocation of land use."
See also *Riggs vs. Long Beach Township*, 109 NJ 601, 619–622 (1988), where Supreme Court Justice Alan Handler noted the significance of the master plan; "…The envisioned master plan is a much more detailed, rigorous and systematic exercise in planning than that which sufficed under the old Planning Act…. It is thus clear that the focus of the Municipal Land Use Law is on the enhanced role of planning, that it strengthens the planning process itself…. NJSA 40:55D-28 (b) (2) serves the end of heightening the role of planning as a condition of proper zoning."

TABLE 4.1
Minimum Contents of a Comprehensive Municipal Master Plan

1. A statement of objectives, principles, assumptions, policies, and standards on which the constituent proposals for the physical, economic, and social development of the municipality are based
2. **A land use plan element taking into account topography, soil conditions, water supply, drainage, floodplain areas, marshes and woodlands**
3. A housing plan element including, but not limited too, residential standards for the construction and improvement of housing
4. A circulation plan element showing the location and types of facilities for all modes of transportation
5. A utility services plan element analyzing the need for and showing the future general location of water supply and distribution facilities, drainage and flood control facilities, sewerage and waste treatment and solid waste disposal
6. A community facilities plan element showing existing and proposed location and type of educational or cultural facilities, historic sites, libraries, hospitals, firehouses, and police stations
7. A recreation plan element showing a comprehensive system of areas and public sites for recreation
8. **A conservation plan element that provides for the preservation, conservation and utilization of natural resources, including energy, open space, water supply, forests, soil, marshes, wetlands, harbors, rivers and other waters, fisheries, endangered or threatened species of wildlife and other resources, and that systematically analyzes the impact of each other component and element of the master plan on the present and future preservation, conservation, and utilization of those resources**
9. An economic plan element
10. A historic preservation plan element
11. A recycling plan element
12. An element indicating the relationship of the master plan to the master plan of contiguous municipalities, the master plan of the county in which the municipality is located, and any state master plan

Note: Elements in boldface type will be the building blocks for establishing an ecologically based system of land use.

Source: Adapted from NJSA 40:55 D-1 et seq.

I do not wish to infer, however, that municipal master plans, even where they have not been preempted by zoning, have been any more productive in protecting and preserving community ecological infrastructure than current zoning ordinances have. Both of these documents, as they are presently constructed, have failed to adequately define and incorporate the essential components of community ecological infrastructure and to provide for their preservation and protection. Furthermore, neither contrivance has generally incorporated any substantive, *holistic* evaluation of the health of these same components, relying instead of fragmented, site-by-site diagnoses provided by experts, hired by land speculators, and developers. Yannacone[18] insists these deficiencies are a fatal flaw:

> Any so-called master plan, be it village, town, city, state or region, which fails to evaluate *fully* the effects of any proposed land use on the *overall integrity* [emphasis added] of the system is an inadequate plan at best and is ultimately doomed to become a costly and deadly hoax on the community. Any zoning law — local, state or federal — based upon such inadequate evaluation must fail. It should fail as legislation, and it will fail in the courts; just as every attempt to ignore the natural limitations imposed on man's use of his natural resources must fail.

Our main purpose for strengthening the stature and content, particularly the ecological content, of the master plan, is to restore the ability of community residents (we mean, all landowners, stakeholders, community business leaders, environmental groups, and any other parties of interest) to collectively meet and decide by consensus, well before individual site plans are submitted, what is and is not vital to the long-term interests and future well-being of the community, including its sustainability. These are clearly considerations that should not be argued during individual site plan reviews, where interests are extremely myopic and positions are more polarized and confrontational. Are there, for instance, certain lands in the community that have been determined to be the primary groundwater recharge areas for a headwater stream or an underground aquifer that is the source of the community's potable water supply? These could and probably should be specifically delineated in the master plan along with a prohibition on their disturbance or overlay with any kind of impervious surface. Similarly, is a trout production waterway containing native brook trout something worth protecting? If it is, do we need streamside buffers, and if so, how wide must they be? Do we also need a prohibition on the direct discharge of all storm water outfalls into this same stream? Conversely, are there other natural resources or features of the terrain that the community is willing to sacrifice because collectively the community has determined they are not worth protecting or preserving? All these considerations are now to be decided up front by the community as a whole, and to become an integral part of the comprehensive master plan. Prospective developers therefore will know well ahead of time by a simple trip to town hall and a quick look at the community's master plan [which hopefully by then should be in a Geographic Information System (GIS) format] what vital interests the community holds close and inviolate and that it intends to protect. This knowledge allows the developer to avoid the costs of holding options and of reducing preliminary survey costs, and also to quickly decide that his or her money and time might be better spent on another parcel.

Landowners, farmers, other community residents, and local business leaders will all have been involved in the process of deciding the community's future and this is the way true "grassroots" decision making and home rule were meant to work.

Information discussed in previous chapters provides some valuable hindsight perspectives that are worthy of carrying forward as major considerations for developing the municipal master plan (MMP) for the 21st century and are listed here:

1. Society must end its practice of making future generations pay the costs of environmental degradation.
2. If we accept the claim that society should compensate landowners for the income they forfeit for society's benefit, it is just as logical to claim that the public should be compensated if the landowner degrades or destroys natural processes that yield public benefits.
3. Many of the institutions of modern technological societies are inappropriate for dealing with the cumulative, adverse, environmental impacts produced by human intrusions into our natural ecological infrastructure because they were established when economic activity was small relative to the magnitude of Earth's physical and biological processes. Thus, these institutions are based on the belief that nature's bounty is so vast that it can be tapped relentlessly without adverse consequences. Consequently, we have been slow to incorporate environmental costs into the prices charged consumers; and as long as environmental goods and services are not priced at their true value, market prices cannot provide adequate signals of growing scarcity or cumulative environmental damage.
4. An owner's expectancy for his or her property is to be measured at the time the owner takes title to the property.
5. A requisite degree of connection (nexus) and proportionality must exist between the exactions imposed by a regulatory agency on private property and the projected impact of the proposed development. That is to say, any remedy called for by a government agency has to be directly related to the negative impact the government believes will result from a particular land use, and the remedy called for is in proportion to the predicted impact.
6. There must be greater reliance on preservation of complete ecosystems and less reliance on attempts to mitigate the damage we have done to them.
7. Environmental protection goals and objectives vary for "greenfield" areas (i.e., rural and developing suburban areas) and "brownfield" areas (i.e., areas already heavily urbanized).

8. The scientific community needs to articulate more clearly for local decision makers the underlying ecological processes and the consequences resulting from interference or truncation of those processes.
9. We must conduct our planning with a transgenerational mindset.
10. We need to treat nature as a mentor and a model, not as an obstacle to continually overcome.

5 Additional Science Aids the Process

In Chapter 3 we discuss the enormous volume of environmental laws, rules, and regulations that were produced by Congress and the state legislatures in the decades following the first Earth Day in 1970. That impressive, fecund productivity has been matched, if not exceeded by the scientific community, who in the same 29-year period has collected scientific data and produced reports in such vast quantities that the volume is very nearly incomprehensible. Much has been deduced from the prolific research that is pertinent to local land use planning, including the confirmation that land transformations (land improvements?) undertaken with minimal regard for the integrity of existing ecological infrastructure can have disastrous consequences. That same research has also confirmed that our ability to fully mitigate deleterious environmental impacts or to successfully rehabilitate portions of damaged ecosystems is limited. As a consequence, an increasing number of scientists are advocating a much greater reliance on *preventative* land use planning practices. In addition to these significant conclusions, there has also been recognition that too little of the scientific data has made its way into the hands of local community land planners where it could provide the greatest benefit for the public good. While large numbers of completed environmental studies have been published, many others have simply been accumulating on back shelves, in record storage areas, in basements, and most recently on computer hard drives of federal and state environmental agencies and college campuses all across the nation. Whether municipal planners have been unaware of the existence of such data or simply lacked access to it, is unclear. In any case, the results have been the same — local land planners have had to rely almost exclusively on the scientific documentation presented by experts hired by individual developers at the time they submit their projects for local review and approval. Despite Weaver's characterization of these scientific experts as, "...those priestlike practitioners who bring us truth,"[85] the veracity of their testimony has been called into question with unsettling regularity, especially by planning board members who have been around long enough to contrast the accuracy of expert-predicted outcomes with actual postdevelopment conditions.

Why postdevelopment conditions persistently differ from expert predictions is a cause of great concern for local planners; and while its cause is attributable to a variety of factors, much of the blame can be placed on the scope of the scientific documentation provided, and on the manner in which it is collected and presented by the developer's experts. For example, the site-specific, scientific documentation provided to planning boards during the site plan approval process generally fails to recognize the *continuum* of ecological infrastructure; this in turn results in very spatially constrained scientific inquiries, often limited to activities within the boundaries of the site and a very narrow peripheral band around them, even though the impacts usually occur on a much broader scale. Furthermore, such site-by-site scientific investigations do not effectively reveal the cumulative impacts of this disaggregated type of land development. Take, for example, the situation where three separately designed storm water detention basins are collecting runoff from three separate and unrelated sites and discharging it into the same stream at different locations. Rarely are calculations ever done to determine if there is a potential for the peak discharges from each of the three basins to reach the stream in a sequence that would be hydrologically additive, thus creating the potential to wreak havoc on the stability of the stream channel, accelerate erosion, and at the same time destroy stream bottom habitat and the organisms that occupy it. Furthermore, how many environmental impact statements describe the displacement of wildlife populations that occur from the development activities at a specific site, but never quite identify where those same populations are expected to go once their habitat has been usurped and converted for other uses? Wildlife cannot always keep running to neighboring properties as so many impact statements conveniently continue to claim. It is little wonder that many of our wildlife populations are completely out of balance with their habitat in many parts of the nation — a classic example of which is the white-tailed deer (*Odocoileus virginianus*) whose populations have reached nuisance and sometimes life-threatening proportions in many east coast suburban American communities. Additionally, both the content and the format of the scientific information supplied to local planners can be intimidating, especially for lay planners with minimal experience in the environmental sciences. This has led to a condition that is thankfully short-lived, and which is often referred to as *analysis paralysis*. In other words, we become so intent on trying to understand the complex algorithms, terminology, and acronyms of Gaussian air pollution diffusion models and vehicle trip statistics that we forget to ask the commonsense questions such as, will the rural roadway that now services the vehicles from two farms have more traffic on it, and will air pollution be increased when 200 homes are built on these same farms? The answer to both, of course, is yes, despite attempts to obfuscate the obvious with self-serving scientific data. When we combine all these factors with an acknowledged disaggregated approval process, it is little wonder that municipal planners are perplexed with the results of the existing land use planning process.

How then do we overcome these shortcomings? First, we need to buttress the local decision-making process with more independent scientific evaluation. Second, we need to expand the spatial scope of our scientific inquiries. Third, we especially need to increase the number of environmental matrices evaluated, examined, or measured, by utilizing as much as possible all of the already published scientific studies and results until now left languishing on the back shelves of many institutions. All these improvements in the process, however, need to be accomplished long before individual site plans are submitted for review and approval; and as later chapters will show, the appropriate time to do this is during the formulation of the community's ecologically based municipal master plan (MMP). Among the additional matrices that should be included are *environmental indicators*, often categorized under the following general headings of condition and response indicators. These latter items are especially important because they comport with the *weight or preponderance of the evidence* approach now being recommended by the scientific community.

5.1 BEWARE THE MODEL

Experts and their supposedly predictive models have mesmerized local land planners and even the courts for decades. Do these models, particularly environmental models, deserve the extreme level of reverence they now are afforded? Models are tools and nothing more that provide us with some ability to compress time so that we can *estimate* what the results of a given action or approach might be, years and sometimes even decades into the future. They do not, however, tell us what it will actually be.

One of the greatest shortcomings of expert testimony before local planning boards, and perhaps this is by design, is the failure to disclose what the limits of accuracy are in the model's estimates. This is a problem also identified by the Center for Watershed Protection.[19] This center stresses that

> The staff managing the watershed planning process should be able to understand the input and assumptions of any model used to develop the plan. Since these staff will ultimately make management decisions, *they need to know what portions of the model may be less accurate than others* [emphasis added]. Otherwise, the model acts as a "black box," isolating managers from an important part of plan development.

If I were to tell you that any environmental model has only a 50/50 chance of being somewhat accurate in its estimations, would that influence the final decision that you, as a local land planner, might make? Because natural biological and chemical equlibria can shift dramatically on changes often measured at the parts per billion level, a 50/50 chance seems particularly risky. Therefore, I am recommending that you begin to pose tougher questions to hired experts or even to your own representatives concerning the accuracies of their so-called predictive models. Delve deeper into the input data of such models, especially if the data are extrapolated from sites far removed from the site in question. Finally, carefully read the Guest Essay that follows by Dr. Naomi Oreskes et al.[20] who argue quite convincingly that the value of environmental or earth science models is strictly heuristic. You also will gain further understanding of the terms *verification*, *validation*, and *calibration* — terms frequently used by experts in their presentation to local planning boards.

5.2 VERIFICATION, VALIDATION, AND CONFIRMATION OF NUMERICAL MODELS IN THE EARTH SCIENCES: GUEST ESSAY*

In recent years, there as been a dramatic increase in the use of numerical simulation models in the earth sciences as a means to evaluate large-scale or complex physical processes. In some cases, the predictions generated by these models are considered as a basis for public policy decisions: global circulation models are being used to predict the behavior of Earth's climate in response to increased CO_2 concentrations; resource estimation models are being used to predict petroleum reserves in ecologically sensitive areas; and hydrological and geochemical models are being used to predict the behavior of toxic and radioactive contaminants in proposed waste disposal sites. Government regulations and agencies may be required by law to establish the trustworthiness of models used to determine policy or to attest to public safety;[1,2] scientists may wish to test the veracity of models used in their investigations. As a result, the notion has emerged that numerical models can be verified or validated, and techniques have been developed for this purpose.[1,3–5] Claims about verification and validation of model results are now routinely found in published literature.[6]

Are claims of validity and verity of numerical models legitimate?[2,7] In this article, we examine the philosophical basis of the terms *verification* and *validation* as applied to numerical simulation models in the earth sciences, using examples from hydrology and geochemistry. Because demand for the assessment of accuracy in numerical modeling is most evident at the interface between public policy and scientific usage, we focus on examples relevant to policy.[8] The principles illustrated, however, are generic.

VERIFICATION: THE PROBLEM OF "TRUTH"

The word "verify" (from Latin, verus, meaning true) means an assertion or establishment of truth.[9] To say that a model is verified is to say that its truth has been demonstrated, which implies its reliability as a basis for decision making. However, it is impossible to demonstrate the truth of any proposition except in a closed system. This conclusion derives directly from the laws of symbolic logic. Given a proposition of the form "p" entails "q," we know that if "p" is true, then "q" is true if and only if the system that this formalism represents is closed.

For example, I say, "If it rains tomorrow, I will stay home and revise this paper." The next day it rains, but you find that I am not home. Your verification has failed. You conclude that my original statement was false. But in fact, it was my intention to stay home and work on my paper. The formulation was a true statement of my intent. Later, you find that I left the house because my mother died, and you realize that my original formulation was not false, but incomplete. It did not allow for the possibility of extenuating circumstances.[10] Your attempt at verification failed because the system was not closed.

---

* From Oreskes, N., Schraeder-Frecette, K., and Belitz, K., Verification, validation, and confirmation of numerical models in the earths sciences, *Science Magazine*, 263, February 4, 1994. All references and notes pertaining to the essay are cited at the end of this chapter.

This example is trivial, but even an apparently trivial proposition can be part of a complex open system. Indeed, it is difficult to come up with verbal examples of closed systems because only purely formal logical structures, such as proofs in symbolic logic and mathematics, can be shown to represent closed systems. Purely formal structures are verifiable because they can be proved by symbolic manipulations, and the meaning of these symbols is fixed and not contingent on empirically based input parameters.[11]

Numerical models may contain closed mathematical components that may be verifiable, just as an algorithm within a computer program may be verifiable.[12] Mathematical components are subject to verification because they are part of closed systems that include claims that are always true as a function of the meanings assigned to the specific symbols used to express them.[13] However, the models that use these components are never closed systems. One reason they are never closed is that models require input parameters that are incompletely known. For example, hydrogeological models require distributed parameters such as hydraulic conductivity, porosity, storage coefficient, and dispersivity, which are always characterized by incomplete data sets. Geochemical models require thermodynamic and kinetic data that are incompletely or only approximately known. Incompleteness is also introduced when continuum theory is used to represent natural systems. Continuum mechanics necessarily entails a loss of information at the scale lower than the averaging scale. For example, the Darcian velocity of a porous medium is never equal to the velocity structure at the pore scale. Finer scale structure and process are lost from consideration, a loss that is inherent in the continuum mechanics approach.

Another problem arises from the scaling-up of nonadditive properties. The construction of a numerical simulation model of a ground flow system involves the specification of input parameters at some chosen scale. Typically, the scale of model elements is on the order of meters, tens of meters, or kilometers. In contrast, the scale on which input parameters are measured is typically much smaller, and the relation between those measurements and larger scale model parameters is always uncertain and generally unknown. In some cases, it is possible to obtain input data at the scale chosen by the modeler for the model elements (for example, pump tests) but this is not often done, for practical reasons. Even when such measurements are available, they are never available for all model elements.

Another reason hydrogeological and geochemical models are never closed systems is that the observation and measurement of both independent and dependent variables are laden with inferences and assumptions. For example, a common assumption in many geochemical models of water–rock interaction is that observable mineral assemblages achieve equilibrium with a modeled fluid phase. Because relevant kinetic data are frequently unavailable, kinetic effects are assumed to be negligible.[15] But many rocks contain evidence of disequilibrium on some scale, and the degree of disequilibrium and its relation to kinetic controls can rarely, if ever, be quantified. To attempt to do so would necessarily involve further inferences and assumptions. Similarly, the absence of complete thermodynamic data for mineral solid solutions commonly forces modelers to treat minerals as ideal end-members, even when this assumption is known to be erroneous on some level. Measurement of the chemical composition of a mineral phase to estimate the activities of chemical components within it requires instrumentation with built-in assumptions about such factors as inference effects and matrix connections. What we call data are inference-laden signifiers of natural phenomena to which we have incomplete access.[16] Many inferences and assumptions can be justified on the basis of experience (and sometimes uncertainties can be estimated), but the degree to which our assumptions hold in any new study can never be established a priori. The imbedded assumptions thus render the system open.

The additional assumptions, inferences, and input parameters required to make a model work are known as "auxiliary hypotheses."[17] The problem of deductive verification is that if verification fails, there is often no simple way to know where the principal hypotheses or some auxiliary hypothesis is at fault. If we compare a result predicted by a model with observational data and the comparison is unfavorable, then we know something is wrong, and we may or may not be able to determine what it is.[17] Typically, we continue to work on the model until we achieve a fit.[19] But if a match between the model result and observational data is obtained, then we have, ironically, a worse dilemma. More than one model construction can produce the same output. The situation is referred to by scientists as nonuniqueness and by philosophers as underdetermination.[20,21] Model results are always underdetermined by the available data. Two or more constructions that produce the same results may be said to be empirically equivalent. Then there is no way to choose between them other than to invoke extraevidential considerations like symmetry, simplicity, and elegance; or personal, political, or metaphysical preferences.[19,23–25]

A subset of the problem of nonuniqueness is that two or more errors in auxiliary hypotheses may cancel each other out. Whether our assumptions are reasonable is not the issue at stake. The issue is that often there is no way to know that this cancellation has occurred. A faulty model may appear to be correct. Hence, verification is only possible in closed systems in which all the components of the system are established independently and are known to be correct. In its application to models of natural systems, the term "verification" is highly misleading. It suggests a demonstration of proof that is simply not accessible.[26]

Validation

In contrast to the term "verification," the term "validation" does not necessarily denote an establishment of truth (although truth is not precluded). Rather, it denotes the establishment of legitimacy, typically given in terms of contacts, arguments, and methods.[27] A valid contact is one that has not been nullified by action or inaction. A valid argument is one that does not contain obvious errors of logic. By analogy, a model that does not contain known or detectable flaws and is internally consistent can be said to be valid. Therefore, the term valid might be useful for assertions about a genetic computer code but is clearly misleading if used to refer to actual model results in any particular realization.[28] Model results may or may not be valid, depending on the quality or quantity of the input parameters and the accuracy of the auxiliary hypotheses.

Common practice is not consistent with this restricted sense of the term. Konikow and Bredehoeft[2] have shown that the term "validation" is commonly used in at least two different senses, both erroneous. In some cases, validation is used interchangeably with verification to indicate that model predictions are consistent with observational data. Thus, modelers misleadingly imply that verification and validation are synonymous, and that validation establishes the veracity of the model. In other cases, the term validation is used even more misleadingly to suggest that the model is an accurate representation of physical reality. The implication is that validated models tell us how the world really is. For example, the U.S. Department of Energy defines validation as the determination "that the code or model indeed reflects the behavior of the real world."[29] Similarly, the International Atomic Energy Agency has defined a validated model as one that provides "a good representation of the actual processes occurring in a real system."[30] For all the reasons discussed above, the establishment that a model accurately represents the "actual processes occurring in a real system" is not even a theoretical possibility.

How have scientists attempted to demonstrate that a model reflects the behavior of the real world? In the Performance Assessment Plan for the proposed high-level nuclear waste repository at Yucca Mountain, Nevada, Davis and co-workers[1] suggest that "[t]he most common method of validation involves a comparison of the measured response from the *in situ* testing, lab testing, or natural analogs with the results of computational models that embody the model assumptions that are being tested."[31] But the agreement between any of these measures and numerical output in no way demonstrates that the model that produced the output is an accurate representation of the real system. Validation in this context signifies consistency within a system or between systems. Such consistency entails nothing about the reliability of the system in representing natural phenomena.

"Verification" of Numerical Solutions

Some workers would take as a starting point for their definition of terminology the analytical solution to a boundary value or initial value problem. In this context, they may compare a numerical solution with an analytical one to demonstrate that the two match over a particular range of conditions under consideration. This practice is often referred to as verification.[4,pp.7–8;32]

The comparison of numerical with analytical solutions is a critical step in code development; the failure of a numerical code to reproduce an analytical solution may certainly be cause for concern. However, the congruence between a numerical and an analytical solution entails nothing about the correspondence of either one to material reality. Furthermore, even if a numerical solution can be said to be verified in the realm of the analytical solution, in the extension of the numerical solution beyond the range and realm of the analytical solution (for example, time, space, and parameter distribution), the numerical code would no longer be verified. Indeed, the raison d'être of numerical modeling is to go beyond the range of available analytical solutions. Therefore, in application, numerical models cannot be verified. The practice of comparing numerical and analytical solutions is best referred to as bench-marking. The advantage of this term — with its cultural association with geodetic practice — is that it denotes a reference to an accepted standard whose absolute value can never be known.[33]

Calibration of Numerical Models

In the earth sciences, the modeler is commonly faced with the inverse problem: the distribution of the dependent variable (for example, the hydraulic head) is the most well-known aspect of the system; the distribution of the independent variable is the least well known. The process of tuning the model — that is, the manipulation of the independent variables to obtain a match between the observed and simulated distribution or distributions of a dependent variable or variables — is known as calibration.

Some hydrogeologists have suggested a two-step calibration scheme in which the available dependent data set is divided into two parts. In the first step, the independent parameters of the model are adjusted to reproduce the first part of the data. Then in the second step the model is run and the results are compared with the second part of the data. In this scheme, the first step is labeled "calibration," and the second step is labeled "verification." If the comparison is favorable, then the model is said to be "verified."[3,p.110;4,p.253] The use of the term "verification" in this context is highly misleading, for all the reasons given above. A match between predicted and obtained does not verify an open system. Furthermore, models almost invariably need additional tuning during the so-called verification phase.[3,p.110] That is, the comparison is typically unfavorable, and further adjustments to the independent parameters have to be made. This limitation indicates that the so-called verification is a failure. The second step is merely a part of the calibration.

Given the fundamental problems of verification, van Fraassen[22] has argued that the goal of scientific theories is not truth (because that is unobtainable) but empirical adequacy. Using van Fraassen's terminology, one could say a calibrated model is empirically adequate. However, the admission that calibrated models invariably need "additional refinements"[3,p.110] suggests that the empirical adequacy of numerical models is forced. The availability of more data requires more adjustments. This necessity has serious consequences for the use of any calibrated model (or group of models) for predictive purposes, such as to justify the long-term safety of a proposed nuclear or toxic waste disposal site. Consider the difference between stating that a model is verified and stating that it has "forced empirical adequacy."[34]

Finally, even if a model result is consistent with present and past observational data, there is no guarantee that the model will perform at an equal level when used to predict the future. First, there may be small errors in input data that do not impact the fit of the model under the time frame for which historical data are available, but which, when extrapolated over much larger time frames, do generate significant deviations. Second, a match between model results and present observations is no guarantee that future conditions will be similar, because natural systems are dynamic and may change in unanticipated ways.[35]

Confirmation

If the predicted distribution of dependent data in a numerical model matches observational data, either in the field or laboratory, then the modeler may be tempted to claim that the model was verified. To do so would be to commit a logical fallacy, the fallacy of "affirming the consequent." Recall our proposition, "If it rains tomorrow, I will stay home and revise this paper." This time you will find that I am home and busily working on my paper. Therefore, you conclude that it is raining. Clearly, this is an example of faulty logic. The weather might be glorious, but I decided this paper was important enough to work on in spite of the beautiful weather. To claim that a preposition (or model) is verified because empirical data match a predicted outcome is to commit the fallacy of affirming the consequent. If a model fails to reproduce observed data, then we know the model is faulty in some way, but the reverse is never the case.[36]

This conclusion, which derives strictly from logic, may seem troubling given how difficult it can be to make a model or develop a hypothesis that reproduces observed data. To account for this discrepancy, philosophers have developed a theory of confirmation, founded on a notion of science as a hypothetico-deductive activity. In this view, science requires that empirical observations be framed as deductive consequences of a general theory or scientific law.[37] If these observations can be shown to be true, then the theory or law is "confirmed" by those observations and remains in contention for truth.[17] The greater the number and diversity of confirming observations, the more probable it is that the conceptualization embodied in the model is not flawed.[38] But confirming observations do not demonstrate the veracity of a model or hypothesis; they only support its probability.[39,40]

Laboratory tests, *in situ* tests, and the analysis of natural analogs are all forms of model confirmation. But no matter how many confirming observations we have, any conclusion drawn from them is still an example of the fallacy of affirming the consequent. Therefore, no general empirical proposition about the natural world can ever be certain. No matter how much data we have there will always be the possibility that more than one theory can explain the available observations.[41] And there will always remain the prospect that future observations may call the theory into question.[42] We are left with the conclusion that we can never verify a scientific hypothesis of any kind. The more complex the hypothesis, the more obvious this conclusion becomes. Numerical models are a form of highly complex scientific hypothesis. Confirmation theory requires us to support numerical simulation results with other kinds of scientific observations and to realize that verification is impossible.

Numerical Models and Public Policy

Testing hypotheses is normal scientific practice, but model evaluation takes on an added dimension when public policy is at stake. Numerical models are increasingly being used in the public arena, in some cases to justify highly controversial decisions. Therefore, the implication of truth is a serious matter.[43] The terms verification and validation are now being used by scientists in ways that are contradictory and misleading. In the earth sciences — hydrogeology, geochemistry, meteorology, and oceanography — numerical models always represent complex open systems in which operative processes are incompletely understood and the required empirical input data are incompletely known. Such models can never be verified. No doubt the same may be said of many biological, economic, and artificial intelligence models.

What typically passes for validation and verification is at best confirmation, with all the limitation that this term suggests. Confirmation is only possible to the extent that we have access to the natural phenomena, but complete access is never possible, not in the present and certainly not in the future. If it were, it would obviate the need for modeling. The central problem in the language of validation and verification is that it implies an either–or situation. In practice, few (if any) models are entirely confirmed by observational data, and few are entirely refuted. Typically, some data do agree with predictions and some do not. Confirmation is a matter of degree. It is always inherently partial. Furthermore, both verify and validate are affirmative terms: they encourage the modeler to claim a positive result.[44] And in many cases the positive result is presupposed. For example, the first step of validation has been defined by one group of scientists as developing "a strategy for demonstrating [regulatory] compliance."[1,45] Such affirmative language is a roadblock to further scrutiny.

A neutral language is needed for the evaluation of model performance. A model can certainly perform well with respect to observational data, in which case one can speak of precision and accuracy of the fit. Judgmental terms such as excellent, good, fair, and poor are useful because they invite rather than discourage, contextual definition. Legitimately, all we can talk about is the relative performance of a model with respect to observational data, other models of the same site, and our own expectations based on theoretical preconceptions and experience of modeling other sites. None of these things can be discussed in absolute terms.

Then What Are Good Models?

Models can corroborate a hypothesis by offering evidence to strengthen what may already be established through other means. Models can elucidate discrepancies in other models. Models can also be used for sensitivity analysis — for exploring "what if" questions — thereby illuminating which aspects of the system are most in need of further study, and where more empirical data are most needed. The primary value of models is heuristic: models are representations, useful for guiding further study but not susceptible to proof.

The idea of model as representation has led the philosopher Nancy Cartwright to claim that models are "a work of fiction."[46] In her words, "some properties ascribed to objects in the model will be genuine properties of the objects modeled, but others will merely be properties of convenience." Her account, which is no doubt deliberately provocative, will strike many scientists as absurd, perhaps offensive. While not necessarily accepting her viewpoint, we might ponder this aspect of it: a model, like a novel, might resonate with nature, but it is not a "real" thing. Like a model, a model might be convincing — it may "ring true" if it is consistent with our experience of the natural world. But just as we may wonder how much the characters in a novel are drawn from real life and how much is artifice, we might ask the same of a model: how much is based on observation and measurement of accessible phenomena, how much is based on informed judgment, and how much is convenience? Fundamentally, the reason for modeling is a lack of full access, either in time or space, to the phenomena of interest. In areas where public policy and public safety are at stake, the burden is on the modeler to demonstrate the degree of correspondence between the model and the material world it seeks to represent and to delineate the limits of that correspondence.

Finally, we must admit that a model might confirm our biases and support incorrect intuitions. Therefore, models are most useful when they are used to challenge existing formulations, rather than to validate or verify them. Any scientist that is asked to use a model to verify or validate a predetermined result should be suspicious.

REFERENCES AND NOTES

1. Davis, P. A., Olague, N. E., and Goodrich, M. T., Approaches for the Validation of Models Used for Performance Assessment of High-Level Nuclear Waste Repositories, SAND90-0575/NUREG CR-5537, Sandia National Laboratories, Albuquerque, NM, 1991. These workers cite the case of *Ohio vs. EPA*, in which a federal appeals court found the state responsible for testing computer models used to set emission limits on electric power plants (*U.S. Law Week*, 54, 2494, 1986). The court found the government liable because it had made no effort to determine the reliability of the model used. Given that the legal necessity of model testing has been established, what claims are justified on the basis of such tests?

2. Konikow, L. F., and Bredehoeft, J. D., *Adv. Water Resour.*, 15, 75, 1992.
3. Wang, H. F., and Anderson, M. P., *Introduction to Groundwater Modeling: Finite Difference and Finite Element Methods*, Freeman, San Francisco, 1982.
4. Anderson, M. P., and Woessner, W. M., *Applied Groundwater Modeling: Simulation of Flow and Advective Transport*, Academic Press, New York, 1992.
5. The volume of *Adv. Water Resour.*, 15, 1992 is entirely dedicated to the discussion of validation and verification of computer models.
6. In our experience, such claims are particularly abundant in cases in which an obvious public policy interest is at stake, such as in work surrounding the proposed high-level nuclear repository at Yucca Mountain, NV. Examples include Hayden, N., Benchmarking NNMSI Flow and Transport Codes: Cove 1 Results, Sandia National Laboratories, Albuquerque, NM, 1985; Stephens, K. et al., "Methodologies for Assessing Long-Term Performance of High-Level Radioactive Waste Packages, NUREG CR-4477, ATR-85 (5810-01) 1 ND, U.S. Nuclear Regulatory Commission, Washington, D.C., 1986; Brikowski, T. et al., Yucca Mountain Program Summary of Research, Site Monitoring, and Technical Review Activities, State of Nevada, Agency for Projects-Nuclear Waste Project Office, Carson City, NV, 1988; Costin, L., and Bauer, S., Thermal and Mechanical Codes First Benchmark Exercise, Part 1; Thermal Analysis, SAND88-1221 UC814 (Sandia National Laboratory, Albuquerque, NM, 1990; Barnard, R. and Dockery, H., Nominal Configuration, Hydrological Parameters and Calculation Results, Vol. 1 of Technical Summary of the Performance Assessment Calculational Exercises for 1990 (PACE-90), SAND90-2727, Sandia National Laboratories, Albuquerque, NM, 1991.
7. For recent critiques of verification and validation in hydrology, see Reference 2; Bredehoeft, J. D., and Konikow, L. F., *Groundwater*, 31, 178, 1993. For a similar critique in geochemistry, see Nordstrom, D. K., *Eos* 74, 326, 1993; Proc. 5th CEC Nat. Analogue Working Group Meet. Alligator Rivers Analogue Project Final Workshop, Toledo, Spain, October 5–9, 1992; von Maravic, H., and Smellie, J. Eds., EUR 15176 EN, Commission of the European Community, Brussels, 1994.
8. Two recent editorials dealing with the interface of modeling and public policy at Yucca Mountain are Malone, C. R., *Environ. Sci. Technol.*, 23, 1452, 1989; and Winograd, I. J., *Environ. Sci. Technol.*, 24, 1291, 1990.
9. For example, the *Random House Unabridged Dictionary, New York, 1973* gives the first definition of verify as, "to prove the truth of." Dictionary definitions of verify, validate, and confirm reveal the circularity present in common use, thus highlighting the imperative for consistent scientific usage.
10. This is the same as saying that there was an implicit *ceteris paribus* clause.
11. Godel questioned the possibility of verification even in closed systems (see Nagel, E., and Newman, J. R., *Godel's Proof*, New York University Press, New York, 1958).
12. On the problem of verification in computer programming, see Fetzer, J. H., *Commun. ACM*, 31, 1048, 1988; *North Am. Math. Soc.*, 36, 1352, 1989; *Minds Mach.*, 1, 197, 1991.
13. This is equivalent to A. J. Ayer's classic definition of an analytic statement as one that is "true solely in virtue of the meaning of its constituent symbols, and cannot, therefore, be confirmed or refuted by any fact or experience" (Ayer, A. J., *Language, Truth and Logic*, Dover, New York, 1946, [reprinted, 1952] 16). Analytic statements are verifiable because, "they do not make any assertion about the empirical world, but simply record our determination to use symbols in a certain fashion" (Ayer, A. J., *Language, Truth, and Logic*, Dover, New York, 1946 [reprinted 1952], 31). Also see Ayer, A. J., Ed. *Logical Positivism*, Free Press, New York, 1959.
14. If it were technically and economically possible to undertake exhaustive sampling on the scale of model elements, then we would run the risk of modifying the continuum properties we are trying to measure. The insertion of closely spaced drill holes into a porous medium may change the hydraulic properties of that medium. Furthermore, the dependent variables of the system — hydraulic head, solute concentration, and mineral assemblages — cannot be obtained at the model element scale. To know these parameters perfectly would be to mine out the region being modeled. This point is also made by Tsang, C. F., *Groundwater*, 29, 825, 1991.
15. Recently, geochemists have made considerable progress on the kinetics of mineral reactions, but the point remains the same: in the absence of adequate data, many modelers assume that kinetics can be ignored. Similarly, in the absence of complete thermodynamic data, modelers necessarily extend available data beyond the range of laboratory information. To call this bad modeling is to miss the point: data are never complete, inferences are always required, and we can never be certain which inferences are good and which ones are not as good.

16. An obvious example for atmospheric modeling is the notion of the mean global temperature. How do we measure the average temperature of Earth? Our most basic data can be deeply layered.
17. Hempel, C. G., and Oppenheim, P., *Philos. Sci.*, 15, 135, 1948; Hempel, C. G., *Aspects of Scientific Explanation*, Free Press, New York, 1965; *Philosophy of Natural Science*, Prentice-Hall, Englewood Cliffs, NJ, 1966.
18. For this reason, C. F. Tsang,[14] proposes that model evaluation should always be a step-by-step procedure.
19. This perspective refutes a simple Popperian account of falsification, where we are expected to throw out any models whose predictions fail to match empirical data. As many philosophers have emphasized, especially Imre Lakatos and Thomas Kuhn, scientists routinely modify their models to fit recalcitrant data. The question is, at what point do scientists decide that further modifications are no longer acceptable? Philosophers are still debating this question. (Kuhn, T. S., *The Structure of Scientific Revolution*, University of Chicago Press, Chicago, 1970; *The Essential Tension: Selected Studies in Scientific Tradition and Change*, University of Chicago Press, Chicago, 1977; Lakatos, I., in *Criticism and the Growth of Knowledge*, Lakatos, I., and Musgrave, A., Eds., Cambridge University Press, Cambridge, 1970, pp. 91–196; Popper, K. R., *The Logic of Scientific Discovery*, Basic Books, New York, 1959; *Conjectures and Refutations: The Growth of Scientific Knowledge*, Basic Books, New York, 1963.)
20. Nonuniqueness may arise on a variety of levels: Konikow and Bredehoeft[2] have emphasized the heterogenity of the natural world; C. Bethke (*Geochim. Cosmochimi. Acta*, 56, 4315, 1992) has emphasized the possibility of multiple roots to governing equations. Also see Plummer, L. N., *Practical applications of groundwater geochemistry*, 1st Canadian-American Conf. Hydrogeology, National Water Well Association, Worthington, OH, 1964, 149; Plummer, L. N., Prestemon, E. C., and Parkhurst, D. L., *U.S. Geol. Surv. Water Resour. Invest. Rep.*, 91-4078, 1991, 1.
21. This also is referred to as the Duhem-Quine thesis, after Pierre Duhem, who emphasized the nonuniqueness of scientific explanation, and W. V. O. Quine, who emphasized the holistic nature of scientific theory. Both perspectives refute any simple account of the relation between theory and observation. The classic essays on underdetermination have been reprinted and critiqued in S. Harding, Ed., *Can Theories Be Refuted? Essays on the Duhem-Quine Thesis*, Reidel, Dordecht, the Netherlands, 1976.
22. van Fraassen, B., *The Scientific Image*, Oxford University Press, New York, 1976.
23. H. E. Longino (*Science as Social Knowledge*, Princeton University Press, Princeton, NJ, 1990) examines the role of personal and political preference in generating sex bias in scientific reasoning. Her point is that extraevidential considerations are not restricted to bad science but are characteristic of all science, thus making differentiation between "legitimate" and "illegitimate" preferences difficult.
24. For a counterpart, see Glymor, C., in *The Philosophy of Science*, Boyd, R., Gasper, P., and Trout, J. D., Eds., Massachusetts Institute of Technology Press, Cambridge, MA, 1991, 485.
25. Ockham's razor is perhaps the most widely accepted example of an extraevidential consideration. Many scientists accept and apply the principle in their work, even though it is an entirely metaphysical assumption. There is scant empirical evidence that the world is actually simple or that simple accounts are more likely than complex ones to be true. Our commitment to simplicity is largely an inheritance of 17th century theology.
26. In the early 1920s a group of philosophers and scientists known as the Vienna Circle attempted to create a logically verifiable structure for science. Led by the philosopher Rudolf Camap, the "logical positivists" wished to create a theoretically neutral observation language that would form a basis for purely logical structures, free of auxiliary assumptions, for all of science. Such logical constructions would be verifiable (Camap, R., reprinted in *Logical Positivism*, Ayer, A. J., Ed., Free Press, New York, 1959, 62. Also see Ayer, A. J., in Reference 13. For a historical perspective on Camap and logical positivism, see Hacking, I., *Representing and Intervening*, Cambridge University Press, New York, 1983; Creath, R., Ed., *Dear Camap, Dear Quine: The Quine–Camap Correspondence and Related Work*, University of California Press, Berkeley, 1990; and Boyd, R., in Reference 24, 3. The influence of the Vienna Circle on philosophy of science was profound; W. V. O. Quine has called Camap "the dominant figure in philosophy from the 1930s onward" (in Creath, earlier, p. 463). However, in spite of Camap's stature and influence, the philosophical program of "verificationism" collapsed resoundingly in the 1950s (Galison, P., *Sci. Context*, 2, 197, 1988; Rouse, J., *Stud. Hist. Philos. Sci.*, 22, 141, 1991. It was officially pronounced dead in the Encyclopedia of Philosophy in 1967 (Popper, K. R., *Unended Quest: An Intellectual Autobiography*, Collins, Glasgow, 1976, 87). There now appears to be nothing in the philosophy of science that is as uniformly rejected as the possibility of a logically verifiable method for the natural sciences. The reason is clear: natural systems are never closed.

27. For example, *Webster's Seventh New Collegiate Dictionary*, Merriam, Springfield, MA, 1963, gives the following definition of validation: "to make legally valid, to grant official sanction to, to confirm the validity of (for example an election)." Random House similarly cites elections, passports, and documents (*Random House Dictionary of the English Language*, Random House, New York, 1973).
28. For example, a widely used and extensively debugged package such as MODFLOW (McDonald, M. G., and Harbaugh, A. W., *U.S. Geol. Sur. Tech., Water Resour. Invest.*, book 6, 1988, chap. A1, p. 1) or WATEQ (Truesdell, A. H., and Jones, B. J., *Nat. Tech. Inf. Serv.*, PB2-20464, 1973, 1) might be valid, but when applied to any particular natural situation would no longer necessarily be valid. C. F. Tsang has argued that models should be validated with respect to a specific process, a particular site, or a given range of applicability. Unfortunately, even with such a degree of specificity, the elements of the model (the conceptualization, the site-specific empirical input parameters, the estimated temperature range) are still undetermined. Furthermore, he notes that establishing "the range of application" of a model cannot be done independently of the desired performance criteria. "There is the possibility that a performance criterion could be defined in such a way that the quantity of interest can never be predicted with sufficient accuracy because of intrinsic uncertainties in the data…. Thus, one has to modify the performance criterion to something more plausible yet still acceptable for the problem at hand" (Tsang, C. F., in Reference 14, p. 87). However, this conclusion begs the question, Who decides what is plausible and what is acceptable?
29. "Environmental Assessment: Yucca Mountain Site, Nevada Research and Developmental Area, Nevada," Vol. 2 of U.S. Department of Energy DOE/RW-0073, Office of Civilian Radioactive Waste Management, Washington, D.C., 1986. This definition conflates the generic numerical simulation code with the site-specific model. A site-specific model might accurately represent a physical system, but there is no way to demonstrate that it does. A code is simply a template until the parameters of the system are put in, and therefore could not, even in principle, accurately represent a physical system.
30. "Radioactive waste management glossary." IAEA-TECDOC-264, International Atomic Energy Agency, Vienna, 1982. A recent summary of European work in this area in the context of radioactive waste management is given by Bogorinski, P. et al., *Radiochim. Acta*, 44/45, 367, 1988.
31. In defining model "validation," these workers use the descriptor "adequate" rather than "good," presumably because they recognize the difficulty of defining what constitutes a "good" representation. They propose that a model need only be "adequate" for a "given purpose," in this case compliance with federal regulations. However, this definition begs the question of whether the regulations are adequate. Furthermore, because these workers recognize that models cannot be validated but refuse to relinquish the term "validation," they end up with an almost incomprehensible statement of their goals: "(M)odels can never be validated, therefore validation is a process of building confidence in models and not providing "validated models" (Davis, P. A. et al., in Reference 1, p. 8.)
32. For example, in the guidelines of the U.S. Nuclear Regulatory Commission Radioactive Waste Management Program, NUREG-0865, U.S. Nuclear Regulatory Commission, Washington, D.C., 1990, "verification" of a code is described as "the provision of an assurance that a code correctly performs the operations it specifies. A common method of verification is the comparison of a code's results with solutions obtained analytically." However, a certain confusion in the literature over terminology is made evident by comparison of Anderson and Woessner[4] with Wang and Anderson.[3] Previously, Anderson had referred to this process as validation, and more recently and more misleadingly, as verification.
33. Admittedly, computer programmers engage routinely in what they call program verification. However, the use of the term "verification" to describe this activity has led to extremely contentious debate (see Fetzer, J. H., 1988, in Reference 12 and letters in response in *Commun. ACM32*, 1989.) One striking feature of "verification" in computer science is that it appears to be motivated, at least in part, by the same pressure as in the earth science community: a demand for assurance of the safety and reliability of computer programs that protect public safety, in this case those controlling missile guidance systems (Fetzer, J. H., in Reference 12, p. 376.) For an interesting historical paper on the problem of establishing certainty in the manufacture of weapons systems, see Burgos, G., *Soc. Stud. Sci.*, 23, 265, 1993.
34. A good example of van Fraassen's concept is the view expressed by de Marsily, G., Combes, P., and Goblet, P., *Adv. Water Resour.*, 15, 367, 1992, who claim that they "do not want certainty, [but] will be satisfied with engineering confidence. [W]e are only [trying] to do our level best." This is a commendably honest approach but one that invites a very different public reaction than claims about verified models.

35. Using postaudits of validated models, Konikow and co-workers have shown that even models that produce a good history match of past data often do terribly when extended into the future (Konikow, L. F., and Bredehoeft, J. D., *Water Resour. Res.*, 10, 546, 1974; Konikow, L. F., *Groundwater*, 24, 173, 1986, and Person, M., *Water Resour. Res.*, 21, 1611, 1985; Konikow, L. F., and Swain, L. A., in *28th Int. Geol. Cong. Selected Papers on Hydrogeol.*, Hiese, V. H., Ed., Hanover, West Germany, 1990, 433. Typically, this occurs either because the conceptualization of the system built into the numerical model was incorrect or because modelers failed to anticipate significant changes that subsequently occurred in the system (for example, changes in climatic driving forces). Postaudit studies by these and other workers have been reviewed by Anderson, M. P., and Woessner, W. W., *Adv. Water Resour.*, 15, 167, 1992. Of five studies reviewed, not one model accurately predicted the future. In several cases, models were calibrated on the basis of short-term duration data sets that inadequately described the range of natural conditions possible in the natural system. The issue of temporal variation becomes particularly important for modeling the long-term disposal of nuclear wastes. Changes in the geological conditions of the repository site, which could lead to changes in the dynamics and structure of the system, are not only possible but also are almost certain given enough time.
36. Various philosophers, including A. J. Ayer, W. V. O. Quine, I. Lakatos, and T. S. Kuhn have questioned whether we can in fact prove a hypothesis false. Ayer emphasized that refutations, no less than confirmations presuppose certain conditions (Ayer, A. J., 1946 Reference 13, especially p. 38.) Quine, Lakatos, and Kuhn emphasized the holistic nature of hypotheses and the flexible options for modifications to "save the phenomena."[19,21] However, none of these moves really undermines Popper's argument that it is still possible in principle to prove a theory false, but not possible even in principle to prove a theory true (Popper[19]).
37. Note that this is just one view. Many philosophers have disputed the hypothetico-deductive model.
38. The notion of diversity in confirmation helps to explain why it is important to test a model in a wide variety of circumstances at the modeled site, despite apparent arguments to the contrary. For example, Davis and co-workers[1] have argued that testing the performance of a model in areas not relevant to regulatory compliance is a waste of resources and can lead to the needless rejection of models that are adequate to the task at hand. While this may sometimes be the case, confirmation theory suggests that successful testing of a model in a variety of domains provides important support for the conceptualization embodied in the model. Failed tests help to establish the limits of model adequacy, and may cast legitimate doubt on the model conceptualization of the physical or chemical processes involved.
39. In his classic account of the principle of verification, A. J. Ayer, 1946, in Reference 13, opened the door to undermining his own position by recognizing that empirical statements could never be proved certain but only probable. He called this condition "weak verification," an obvious oxymoron. In hindsight it is easy to see that weak verification is probabilistic confirmation (Ayers, A. J., 1946, in Reference 13, pp. 99 and 135). Popper preferred the term "corroboration" to emphasize that all confirmation is inherently weak (Popper, 1959, in Reference 19). For a recent perspective on probabilistic confirmation, see Franklin, A., and Howson, C., *Stud. Hist. Philos. Sci.*, 19, 419, 1988; and Howson, C., and Urbach, P., Scientific Reasoning: *The Bayesian Approach*, Open Court, La Salle, IL, 1989.
40. Camap therefore argued that all inductive logic was a logic of probability (Camap, R., in *The Problem of Inductive Logic*, Lakatos, I., Ed., North Holland, Amsterdam, 1968, 258).
41. An example is the evidence of faunal homologies in Africa and South America, before the acceptance of plate tectonic theory. These data, which were used as an early argument in favor of continental drift, were considered to be equally well explained by the hypothesis of land bridges (Oreskes, N., *Hist. Stud. Phys. Sci.*, 18, 311, 1986).
42. This is an obvious example of Ptolemaic astronomy, which was extremely well confirmed for centuries and then overturned completely by the Copernican revolution. See Kuhn, T. S., *The Copernican Revolution*, Harvard University Press, Cambridge, MA, 1957. Indeed, every scientific revolution involves the overturning of well-confirmed theory. See Cohen, I. B., *Revolution in Science*, Belknap Press, Cambridge, MA, 1985.
43. Konikow and Bredehoeft,[2] on the basis of their extensive experience with both scientists and government officials, emphasize that the language of verified and validated models is typically interpreted to mean that the models under discussion are, in essence, true. It is also clear that this is the intent of many authors who claim to base results on validated models.

44. We have never seen a paper in which the authors wrote, "the empirical data invalidate this model."
45. Another example is found in the environmental assessment overview for Yucca Mountain, Reference 29, p. 4. The task of site selection, as defined in this report, consisted of "evaluat[ing] the potentially acceptable sites against the disqualifying conditions...." The authors concluded that the Yucca Mountain site was "not disqualified." That is, the null hypothesis is that the site is safe; the burden of proof is on those who would argue otherwise.
46. Cartwright, N., *How the Laws of Physics Lie*, Clarendon Press, Oxford, 1983, 153.
47. This article was prepared for a session on hydrological and geochemical modeling in honor of David Crerar at the American Geophysical Union, May 1993. We thank the organizers, A. Maest and D. K. Nordstrom, for inviting us to prepare this article; J. Bredehoeft for stimulating our thinking on the topic; J. H. Fetzer, L. Konikow, M. Mitchell, K. Nordstrom, L. Sonder, C. Drake, and two reviewers for helpful comments on the manuscript; and our research assistant, D. Kaiser. We dedicate this paper in appreciation of the work of David Crerar.

6 The Value of Natural Ecosystems and Natural Resources

Only after the last tree has been cut down,
Only after the last river has been poisoned,
Only after the last fish has been caught,
Only then will you find that money cannot be eaten.

Old Cree Indian Prophecy

6.1 SETTING THE VALUES

There is an unfortunate assumption among many of the nation's political and economic leaders that when something lacks a price, it must not be worth anything. A plant, for example, that provides neither food nor building material, but is critical to the stable functioning of an ecosystem, has value that is not represented in our current market systems.[21] This troubling mindset is reinforced by what has been the primary scorecard of the nation's economic well-being — the gross domestic product (GDP). Originally introduced during World War II as the gross national product (GNP) to measure wartime production capacity, the GNP (now GDP) was subsequently pressed into a role for which it was not intended, that of barometer for the nation's economic well-being. The GDP is merely a gross tally of products and services bought and sold with no distinction made between transactions that add to our well-being and those that distract from it. In fact, the GDP ignores everything that happens outside the realm of monetary exchange, regardless of its importance to our health or welfare. Instead of separating costs from benefits and productive activities from destructive ones, the GDP assumes that every monetary transaction adds to our well-being. Thus, the breakdown of social structure, the destruction of ecological infrastructure, and the depletion of our natural resources are, in a perverse way, considered an economic gain. Crime, for instance, has added billions to the GDP due to the need to buy more security measures, increase police on the street, repair property damages, and spend money on the care of crime victims. Likewise, divorce also has added billions of dollars to GDP calculations thanks to lawyer fees and the costs associated with setting up second households. When Hurricane Andrew devastated South Florida, the GDP recorded it as a $15 billion economic boon. It is clear that those who calculate the GDP continue to assume that there are no costs associated with the loss of natural resources or the disruption of the beneficial services provided by our ecological infrastructure; this unrealistic assumption until recently has been challenged only occasionally. The general lack of understanding and attention to the critical role of such resources and services is perhaps somewhat understandable, because these resources and ecosystems were in place hundreds of millions of years before the advent of humanity. Because they are so fundamental and operate on such a large scale, they have been easy to take for granted.

Let us briefly take a look at some of those ecosystem services to refresh our memories as to how important they really are and how strongly they influence our day-to-day lives. You need to look no further for the first than what is lying beneath your feet, Earth's soil. When was the last time you gave it even the slightest thought? Perhaps you might have, if only briefly, when you last applied fertilizer to your lawn. The moment you did, the soil began its work. Tiny soil particles, primarily bits of clay and humus loaded with negative electrical charges, began to latch onto some of those nutrients you had applied, holding them near the root zone so that the roots of your grassed lawn could eventually take them up. Were it not for these soil particles, all that fertilizer would have quickly leached away. Similarly, if you had applied grass seed, the soil would have kept it moist and warm, thereby encouraging its germination. Later, it would have provided support for the new sprouts and eventually an anchor for the mature plants. Enormous populations of bacteria, fungi, algae, protozoa, crustaceans, mites, termites, millipedes, and earthworms live within the soil matrix. These organisms help decompose all sorts of materials, including wood, leaves, dead animals, and plants; in doing so they release valuable nutrients previously locked up in cell tissue thus allowing them to once again be used as building blocks by other plants and animals. This helps perpetuate the continuous cycle of life and death on which this planet depends. If this decomposition did not take place, nutrients would be absorbed by living organisms, but no substances would be released when they died. In a short while Earth's surface would be littered with perfectly preserved corpses and eventually life on the planet would cease entirely. Another critical service supplied by our ecological infrastructure is the pollination of flowering plants. One third of our food is derived from plants that require pollination by wild pollinators. Over 100,000 different animal species including bats, bees, beetles, birds, butterflies, and flies provide pollination services free of charge.[22] In the United States alone the agricultural value of these native pollinators is estimated to be in the billions of dollars a year. Unfortunately, thanks to our indifference, the number of natural pollinators is dropping rather precipitously and you do not find any concern for their welfare listed anywhere in the calculations of the nation's GDP.

As we approach what economist, Herman Daly, calls the "full world" and what ecologists in their jargon call the "maximum carrying capacity" (humanwise, that is) in many parts of Earth, a heretofore, unlikely alliance is beginning to gel among economists and ecologists concerned about our day-to-day ignorance of ecological goods and services. Costanza[23] and a historic consortium of ecological and economic experts undertook the task of calculating a monetary value for the world's ecosystem services — an undertaking of no small effort or importance (Tables 6.1 and 6.2). Accordingly, the reactions to that undertaking were global, with the majority of experts from both disciplines generally supportive of this effort to some extent. If nothing else, the work of Costanza et al. provides us with some insight as to the breadth of those ecosystem functions and services — a powerful piece of knowledge by itself. An open online forum on the Costanza et al. work and on ecovaluation generally sponsored by the International Society for Ecological Economics (ISEE) in 1997 stimulated considerable commentary. Some of the responses* follow so that the reader can get some flavor for the fervor currently surrounding this topic.

> Over the years I've given many talks to resource managers and the public about conservation and the value of biodiversity. Wherever possible, I've told audiences "these species do something for you," and I will use the numbers generated by Costanza et al. to reinforce that in a dramatic way. But my experience with audiences leads me to strongly believe that what people are really responding to when told about their interdependences with nature isn't an appeal to their pocketbooks. Rather, it is the affirmation of what they know in their hearts, but has been beaten down in this culture: we are part of something wild, intricate, mysterious, powerful, beautiful, and much larger than we are. It will feed and shelter us into the future if we honor it with respect. This, in my opinion, is the primary message to be delivered.**

> Putting price tags on the living systems that provide life support for all of us is, I think, a good intention gone too far. Yes we need new ways to get out of the growth vs. conservation logjam. Yes we need new ways to agree on comparative values of projects and policies. Yes it is very difficult to see or agree that soil, forests, oceans, estuaries feed and care for us and that we cannot replace, engineer, manufacture, invent, design, buy or sell them. But hard as this is, given our drive to see ourselves so free from and powerful over Nature, I think it is a better way to go. Rather than go the easy route, join the market system and hope that the invisible hand will do an adequate job by cold, impersonal calculation, I think

* All comments reprinted here are with permission from ISEE and authors where confirmed.

** This is from Dave Perry, Professor, Oregon State University, with permission.

TABLE 6.1
Ecosystem Services and Functions as Described in Table 6.2

Number	Ecosystem Service	Ecosystem Function
1	Gas regulation	Regulation of atmospheric chemical composition
2	Climate regulation	Regulation of global temperature, precipitation, and other biologically mediated climatic processes at global or local levels
3	Disturbance regulation	Capacitance, damping, and integrity of ecosystem response to environmental fluctuations
4	Water regulation	Regulation of hydrological flows
5	Water supply	Storage and retention of water
6	Erosion control and sediment retention	Retention of soil within an ecosystem
7	Soil formation	Soil formation processes
8	Nutrient cycling	Storage, internal cycling, processing, and acquisition of nutrients
9	Waste treatment	Recovery of mobile nutrients and removal or breakdown of excess or xenic nutrients and compounds
10	Pollination	Movement of floral gametes
11	Biological control	Trophic–dynamic regulations of populations
12	Refugia	Habitat for resident and transient populations
13	Food production	That portion of gross primary production extractable as food
14	Raw materials	That portion of gross primary production extractable as raw materials
15	Genetic resources	Sources of unique biological materials and products
16	Recreation	Providing opportunities for recreational activities
17	Cultural	Providing opportunities for noncommercial uses

Source: Adapted from Costanza, R. et al., The value of the world's ecosystem services and natural capital, *Nature (London)*, 387, 253, May 1997. Copyrighted by *Nature (London)*, 1997, Macmillan Magazines Ltd. With permission.

we need to grapple with how to debate and compare inner human values such as quality, health, responsibility and integrity. We also need to understand how these apply in the context of ecology (the study of home); as the world shrinks, it is seen to have finite resources and capacities, and is still our one and only home.

Science can help — if ecologists know that ecosystems and the life support they provide are irreplaceable and unexchangeable — very truly priceless — we need to have the courage to stand up and say it. If we can measure or estimate the amount of life support needed to sustainably and reliably support 6 or 10 or 12 billion people, we could publish that. Once known, that much biospheric heart and lung need to be set aside, preserved, kept healthy and not put up for sale at any price.

In the dance between the partners of ecology and economics, and also that of ecology and health sciences, ecology must lead. This may not have been necessary 20 or 50 years ago, but it certainly is now…."*

Costanza et al. summarily dismiss arguments against what they have attempted to do, with the statement, "although ecosystem valuation is certainly difficult and fraught with uncertainties, one choice we do not have is whether or not to do it." This strikes me as thoroughly disingenuous. Of course we have the choice. Indeed, not only do we have the choice of whether or not to do it, we also have the choice (if we decide to do it) of what value system to use. It is in no way preordained that it should be the present economic system which, … has brought us to the brink of the precipice on which we now stand. Indeed, there is every reason to suppose that the system that got us ***in***to the mess is, by its very nature, incapable of getting us ***out***. If work like that of Costanza et. al. helps to persuade our political decision-makers otherwise, then it will have done humanity a grave disservice.**

* Author is unconfirmed.

** Author is unconfirmed.

TABLE 6.2
Average Global Values of Annual Ecosystem Services

		1	2	3	4	5	6	7	8	9	10	11	12	13	14	15	16	17	
Biome	Area (ha × 10^6)	Gas Regulation	Climate Regulation	Disturbance Regulation	Water Regulation	Water Supply	Erosion Control	Soil Formation	Nutrient Cycling	Waste Treatment	Pollination	Biological Control	Habitat Refugia	Food Production	Raw Materials	Genetic Resources	Recreation	Cultural	Total $ Value ($ ha^{-1} yr^{-1})
Open ocean	33,200	38							118			5		15				76	252
Estuaries	180			567					21,100			78	131	521	25		381	29	22,832
Forest	4,855		141	2	2	3	96	10	361	87		2		43	138	16	66	2	969
Grass/ rangelands		7			3		29	1		87	25	23		67			2		244
Wetlands	330	133		4,539	15	3,800				4,177			304	256	106		574	881	14,785
Swamps/ Floodplains	165	265		7,240	30	7,600				1,659			439	47	49		491	1,761	19,581
Lakes/rivers	200				5,445	2,117				665				41			230		8,498

Note: Values are in 1994 US$ ha^{-1} $year^{-1}$.

Source: Adapted from Costanza, R. et al., The value of the world's ecosystem services and natural capital, *Nature (London)*, 387, 253, May 1997. Copyrighted by *Nature (London)*, 1997, Macmillan Magazines Ltd. With permission.

While such global values are initially helpful, one of the respondents to the Costanza article correctly opined that, "...we need to break down the valuation into smaller components so that local decision makers (taxing jurisdictions, etc.) can incorporate them and disseminate the ideas widely so that their inclusion is accepted as realistic."*

6.2 LOCALIZING VALUES

While Costanza and colleagues readily admit that their estimates are "crude first approximations," this does not diminish their utility to local land planners: first, because these valuations strengthen and encourage greater recognition of ecological infrastructure and its functions, heretofore, afforded inadequate assessment and evaluation during site plan deliberation; and second, because such calculations assign at least some monetary value, crude as it may be, to ecological functions, where none existed previously. This author, however, is not ready to fully endorse their inclusion into the computation of cost–benefit analyses for individual land development projects, until and unless such calculations prove more capable of accurately accounting for the long-term, nearly infinite, common good benefits provided by our natural ecological infrastructure. Historically, the ecological infrastructure has not fared well at the hands of such cost–benefit analyses. Additionally, there is the question as to whether the Costanza et al. values, when applied on a per hectare or per acre basis should be used to exact compensation for the loss of a public benefit. Such exactions would tend to further legitimize the current economic practices that condone and encourage the selling and destruction of our ecological infrastructure. This is not to imply, however, that such an idea should be discarded wholesale. Instead, there may be occasions when that may be the only logical alternative, but it is a choice that for the present should be used as the choice of last resort. Primacy always should be given to the protection and preservation of intact ecosystems.

Municipal planners also may find the categories of biomes to which Costanza et al. have assigned values too generalized for practical application at the site plan level. The absence of values in some of the categories may further thwart the effective utilization of the work by Costanza et al. for municipal land planning purposes. More localized (e)valuations are now possible utilizing new methodologies with a spatial applicability as small as 5 acres. Annual aquifer recharge potential, for example, whose equivalent might be reflected in part, under the Water Supply category in Table 6.2, can now be calculated with some reasonable accuracy, for virtually any sized parcel of land, thanks to a new methodology developed by the New Jersey Geological Survey within the New Jersey Department of Environmental Protection[36] (see Appendix D). One of the major advantages of this method over the Costanza et al. data is that more detailed, site-specific characteristics — such as soil type, geology, land use, and land cover — are incorporated into the assessment.

If, using New Jersey's methodology, we were able to compute for a particular portion of a proposed development site an annual recharge rate of 10 in./year, we could begin to assign some economic value to that amount of recharge. In this case we would take that 10-in. value and multiply it times its associated acreage. This would give us acre inches. Knowing how many gallons there are in an acre inch, we can then estimate the total number of gallons of water that are expected to be recharged annually on that particular portion of the site.

Example calculations — 10 in./ year calculated recharge × 15 acres = 150 acre in. × 27,156 gal/acre in. = 4,073,400 gal recharged/ year.

By multiplying this quantity of water times an average finished water cost of $4.45**/1000 gal, we can calculate the value of the water produced at virtually no cost, by the 15-acre portion of this particular site. That value is approximately $18,126/year. Local land planners would now have multiple, ecologically related values (i.e., total annual quantity of potential recharge and the street value of that quantity of treated and stored water) to weigh into their deliberations. Armed

* Christina C. Forbes, senior attorney, affiliation not identified.

** This value is for finished potable water delivered to a household tap. Values can range from $1.55 to $5.37/1000 gal.

with this type of site-specific data the municipal planning board would have a substantial basis to request an applicant to reconfigure his or her site plan design so that most of the proposed development would be directed away from that area of the site having the highest recharge capability. In the event this cannot be done, the municipality would then have the further option of levying an assessment for the loss of a public service supplied by the existing ecological infrastructure. The question then becomes how much of a levy should actually be imposed. Should it be based on a single year's loss of service, a 20-year loss, or even a 60-year generational loss? By using the values derived in the prior example, this could mean a potential assessment of $18,126 (1-year loss), $362,520 (20-year loss), or $1,087,560 (60-year loss).

Had this detailed information been prepared previously by the municipality and incorporated into its master plan, as we are strongly urging in this book, the prospective developer would have been put on notice of this ecological value long before he had put a dime into the project, thereby sparing himself the loss of time and money spent on redesign and potential costs of value loss assessments. In this particular example, we were fortunate that methodologies were available to calculate an ecological *and* monetary value. That is not always the case, as we see in the later chapters of this book, where we purposely rely more on ecological values alone. This strategy and preference is in keeping with the major objective of this book, which is to assure that the protection and preservation of intact ecological infrastructure become the primary considerations in all local land use planning activities.

7 Private Property Rights and Public Trust Resources

Municipalities who desire to craft a local master plan, for which the major premise is the protection and preservation of the community's ecological infrastructure, can expect some significant and vocal opposition from organized, vested interest groups that have historically included lawyers, realtors, developers, and less frequently farmers. The cause for their concern is, of course, the issues of private property rights and regulatory takings. It is important, however, for community planners not to let such opposition deter them from their obligation to safeguard the public health, safety, and welfare. Some very imaginative, personal interpretations of the constitutional protection of private property rights are sure to be profusely argued by these vested interests to sway (perhaps *intimidate* would be a better word) municipal planners from such a bold and courageous course. The most common argument is that *any* restriction on a person's right to do whatever he wants with his property is a *taking* that requires the municipality to pay compensation. Let me assure you that such an argument is not in comport with the state of the law for the last 200 years. The true test, and one that this author does support with certain limitations, is that if a regulatory agency takes *all reasonable use* of a person's property for the public good, the owner should be compensated for the value of the property taken. However, it should not be forgotten that the courts also have repeatedly held — in decisions going back hundreds of years — that no person has a right to make a profit from the sale of their land and no person has the right to change the essential character of his or her property to use it. If you bought a swamp, you must have wanted a swamp.[24]

I do not mean to imply that municipalities always need to be on the defensive — quite the contrary. There is more than sufficient scientific evidence (as we see in later chapters) and moral and philosophical argument to sustain a considerable offense. Not the least is the argument that the development value and tempting profits that attach to open land are less a result of the owner's skill and sacrifice than the result of the municipality's assignment of value primarily through artificial and arbitrary parcelization called zoning. What landowner can, in good conscience, argue that he or she has had a direct hand in creating the community's ecological infrastructure, such as its geology, soils, and waterways? Given this truism, it would seem that fairness is not violated if the community denies an owner some of the *fabricated* values he or she did nothing to create. Farmers, whose family may have owned and operated a farm for generations, might be an exception. They, at least, have contributed to the upkeep of the land. Certainly, their interests would be treated preferentially over those of late arriving land speculators, developers, and attorneys who have only recently purchased a parcel of land or have simply purchased an option on a piece of property, conditioned on receiving site plan approval. Furthermore, it seems just as logical to conversely argue that the public should be compensated, if the landowner degrades, destroys, or seriously alters some or all the community's ecological infrastructure or its processes that are yielding common good benefits. You recall that we touched briefly on this very issue in Chapter 6, where we derive some value for products and services rendered by certain components of our ecological infrastructure. As I stated there, I am still unwilling to recommend that such calculated values be used as a buyoff to justify further fragmentation and degradation of our ecological resources. Instead, such values would be best used, at least initially, to dramatize the true worth of the community's natural resources to community residents, especially during the public participation phase of the community's preparation of an ecologically based municipal master plan (MMP).

Vested interests also can be expected to offer excerpts from recent, and some not so recent, common law cases concerning takings and individual property rights to induce (scare?) municipalities into retreating from any aggressive campaign to subordinate current zoning ordinances to more inclusive master plans. In this regard, therefore, it is worthwhile to examine some of the case law and some of the more significant decisions that have been issued by the United States Supreme Court pertaining to the issues of takings, private property rights, and constitutional protections.

The proposal of environmentally-related land use regulations often inspires immediate concern in the public mind over the issue of confiscation of private property without just compensation, as provided by the Fifth Amendment (for federal regulations) and the Fourteenth Amendment (for state and local regulations).

These potential regulatory takings are quite apart from the traditional takings that government can exercise through the power of eminent domain, where property may be physically seized and taken for a public use, such as a roadway, as long as just compensation is paid to the owner. The intent of the Fifth and Fourteenth Amendments was to bar the government from forcing individuals to bear public burdens that should rightly be shared by the community at large, certainly a proper temperance of the police powers of governmental authorities. The evolving expansion of the government's police power role through regulatory acts instead of traditional condemnation and physical seizure has not set well with some landowners, who see this expanded regulatory role and environmental regulations, in particular, as a significant interference with the bundle of rights they acquired with the purchase of their properties. While these property owners may chafe at such expanded environmental regulation, it is clear that the judiciary accepts some diminution in private property rights without concurrent compensation, where there is a demonstrated danger to the public health, safety, or welfare. In *Keystone Bituminous Coal Association vs. DeBenedictis*, 480 US 470, 107 S. Ct. 124, 94 L. Ed. 2d 472 (1987), the court reaffirmed, "...long ago it was recognized that all property in this country is held under the implied obligation that the owner's use of it shall not be injurious to the community."

The issue of regulatory takings was likewise adjudicated very early on by the American judiciary, with a landmark decision emanating from the federal Supreme Court's decision in *Pennsylvania Coal vs. Mahon*, 260 US 393, 43 S. Ct. 158, 67 L. Ed. 322 (1922). In this case the court set down, for the first time, some benchmark language against which a regulatory taking might be determined. The court declared, "The general rule at least is, that while property may be regulated to a certain extent, if regulation goes too far, it will be recognized as a taking...."

Significantly, 4 years later, in another federal Supreme Court decision, *City of Euclid vs. Ambler Realty Co.*, 272 US 365, 47 S. Ct. 114, 71 L. Ed. 3303 (1926), local zoning was held not to be a taking per se.

Two subsequent cases, also decided by the federal Supreme Court some 50 years later, laid further groundwork to determine when a regulatory taking has occurred. The first case, *Penn Central Transportation Co. vs. City of New York*, 438 US 104, 123–124, 98 S. Ct. 2646, 57 L. Ed. 2d 631 (1978) established a three-part formula that is still used by the courts to evaluate the worthiness of a taking claim.

The formula includes consideration of:

1. The character of the governmental action
2. Its actual economic impacts
3. The extent to which the governmental action interferes with investment-backed expectations

In regard to this last factor, it is important to remember that an owner's investment-backed expectations begin at the time he or she takes ownership of the property. If you are buying a swamp or wetland, you must have wanted property with such natural features or you would have purchased another less environmentally constrained parcel.

A lesser known case, *Agins vs. City of Tiburon*, 447 US 255, 100 S. Ct. 2138, 65 L. Ed. 2d 106 (1980), established a two-part formula for evaluating regulatory takings. That formula considers whether a land use regulation:

1. Eliminates economically viable use of private property
2. Does not substantially advance a legitimate government interest

Since the Penn Central and Agins cases, three additional cases decided by the federal Supreme Court have been the cause of some trepidation, albeit probably undeserved, among those who support valid environmental regulation. The first, *Nolan vs. California Coastal Commission*, 483 US 825, 107 S. Ct. 3141, L. Ed. 2d (1987), resulted from a condition inserted into a building permit by the California Coastal Commission to the plaintiffs that required the property owners to grant an easement to the state to connect public beaches to the north and south of the Nolan property. This commission's rationale for requiring such an easement, was to assure the public visual access to the ocean; however, the proofs presented by the state did not link this objective specifically to the granting of the easement. The court, therefore, found there was no direct nexus between the granting of the easement and the commission's desire to maintain visual access to the ocean. Such an exaction would have been upheld if it had been shown there was a substantial causal relationship between the exaction imposed and the burdens and needs produced as a result of the new development being permitted.

The second case, *Lucas vs. South Carolina Coastal Commission*, 112 S. Ct. 2886 (1992), drew even more attention than the Nolan case. In this case, plaintiff David Lucas had, in 1984, purchased two vacant oceanfront lots on a barrier island near Charleston for $975,000. At the time, neither state or local law prohibited the construction of homes that Lucas planned to build. Subsequent to Lucas' purchase, in 1988, the South Carolina legislature passed the Beachfront Management Act that was designed to limit construction within the beach–dune system in designated critical beach areas. Under the new regulations, Lucas was limited to constructing only a small deck or walkway on his properties. Lucas subsequently sued the South Carolina Coastal Council, asserting that the use restrictions of the new law and regulations denied him economically viable use of his lots, and therefore, was a regulatory taking of private property requiring that he justly be compensated. The lower court found in Lucas' favor and awarded him $1.2 million. The South Carolina Coastal Council appealed to the South Carolina Supreme Court who ruled that the use restrictions were not a regulatory taking because the Beach Front Management Act was established to prevent serious public harm. Lucas continued his appeal to the U.S. Supreme Court, which eventually held that the state supreme court had erred in its interpretation of prior case law on nuisance regulations declaring "...where regulation denies all economically beneficial or productive use of land or where regulation requires physical invasion of property, the owner may be compensated without case-specific examination of the public interest behind the regulation." The U.S. Supreme Court did not, however, define the property unit (because property ownership is imbued with a bundle of rights) against which a taking is measured or the extent of interference necessary for a taking to occur. It instead accepted the finding of the South Carolina Court of Common Pleas that Lucas' property had been deprived of all value. It left unclear whether the extinction of all value of property is synonymous with the extinction of all development value, if other uses of the property in its natural state remain.

On remand, the South Carolina Supreme Court ruled the Beach Front Management Act regulations did constitute a temporary taking that required compensation. Instead of going through further trial to establish damages, South Carolina settled with Lucas. Despite this apparent adverse ruling, there have been subsequent court decisions based on Lucas that have had some beneficial interpretations for municipalities (see especially *Preseault vs. United States*, 27 Fed. Cl. Ct. 69 (1992)), which held, "An owner's expectancy for the property is measured at the time the owner takes title to the property."

The third case, *Dolan vs. City of Tigard*, 114 US, S. Ct. 2309 (1994), involved a challenge by property owner Florence Dolan to certain conditions imposed within the language of a building permit issued to her to expand her hardware store operations. The existing hardware store, approximately 9700 square ft in size, and situated on a 1.67-acre parcel, was to be razed and a new 17,600-square ft store constructed in its place. The site plan was approved with two conditions. Dolan was to dedicate part of her property, about 7000 square ft, for a public greenway and a 15-ft-wide, paved public bike and pedestrian walkway along an adjoining floodplain area. The city's justification for these dedications was to promote flood control, especially because the paving of a sizable portion of the plaintiff's property would facilitate flooding. The city further justified the bicycle pathway as a way to facilitate nonmotorized transportation to the nearby downtown area, on the presumption that vehicular traffic in the area already was congested and the increased traffic from the Dolan site would aggravate the situation. The pathway was a means to mitigate congestion. Dolan contended that these permit conditions nullified her right to exclude others from her property and that she should be paid just compensation. Dolan appealed to Oregon's lower courts, where she lost and then to the Oregon Supreme Court where she again lost. At both levels, Dolan acknowledged the city's authority to exact some form of dedication as a condition for the permit, but challenged the city's showings justifying these particular exactions. She ultimately appealed to the U.S. Supreme Court, which made the following findings. With respect to the nexus between these public purposes and the means proposed to achieve them, the court concluded there was a nexus between the prevention of flooding and the limitation of development within the specified area of the floodplain. The court further noted that there was a nexus between the permit conditions for a bicycle path and the public purpose of mitigating traffic congestion, because the expressed purpose for the plaintiff's expansion was to attract more customers and hence more traffic. The city's exactions, however, foundered on the adequacy of those nexuses. The court reviewed state case law for some guiding principles to determine the adequacy of such nexuses. It declined to adopt, as too stringent, the *specific and uniquely attributable* test utilized in some state courts. This test requires that the government demonstrates the action is directly proportional to the specifically created need. The court also rejected the *reasonable relationship* test utilized by some other states, which requires a balancing between the dedication of property and the impact of the proposed development. The court instead chose to create its own standard, which it called the *rough proportionality* standard. In doing so, it proclaimed that although, "...no precise mathematical calculation is required...the city must make some sort of individualized determination that the required dedication is related, both in nature and extent to the impact of the proposed development."

Under this standard, the court found insufficient justification for requiring Dolan to dedicate a portion of her property as a greenway. The court also concluded that the city had failed to quantify its findings that the public bike path would offset the increased traffic expected to be generated.

Clearly, as municipalities begin to prepare ecologically based MMPs, there are legal precedents and constitutional law concerning private property rights and takings that need to be considered. None of these issues, however, seem insurmountable. In fact, in retrospect, it appears that each of the cases cited here might have been resolved without the need for judicial intervention, had the principles, practices, and strong scientific foundation accompanying an ecologically based MMP been crafted, adopted, and effected at the time in each of these jurisdictions.

As review, some of the more salient legal points to come out of this current case law that municipalities ought to pay particular attention to follow:

1. Municipalities must show that the means are roughly proportional to the end and must quantify such.
2. The factual showing of an unsafe condition necessary to support a state or municipality's exercise of its police powers must be more than a facial showing that one property owner's development desires are inconsistent with the public interest.
3. An environmental regulation of real property is not a regulatory taking per se if it does no more than curb what could have been controlled by the state or municipal courts under common law nuisance rules. To meet this test, however, a municipality must amass ample evidence to show that, in fact, such a nuisance is threatened and that, therefore, such statutes or ordinances are required to abate this widespread nuisance.[25]

8 Getting Ready

Before we begin assembling our ecologically based municipal master plan (MMP), it is important that we briefly review and reinforce some of what we have learned in previous chapters and at the same time lay out some additional suggestions, criteria, and ideas that may enhance the process.

First, as we indicated in Chapter 5, a tremendous amount of scientific research and monitoring data has been collected and stockpiled. Despite this, however, we still have not completely monitored, characterized, and defined every single component, function, and interaction of our ecological infrastructure; therefore, gaps in our knowledge do exist and some scientific uncertainty is expected. This does not mean that we should be stymied in our decision-making processes or in the compilation of a satisfactory and scientifically defensible MMP. It is our responsibility to glean from this surfeit of scientific data the information that can support, with some recognized but acceptable uncertainty, the objectives of an ecologically based MMP. Where individual pieces of scientific documentation are less than robust, we should use what the scientific community now refers to as a *weight or preponderance of the evidence approach*, utilizing multiple metrics to make our assessments. Note that this is a much different proof than *beyond a reasonable doubt*. We always need to keep in mind, however, that every piece of scientific data, even if generally regarded as irrefutable, can be legally challenged. This should not be an excuse for temerity, especially if we are trying to bring the municipal land use planning process into the 21st century.

Second, we should, wherever possible, rely more on field-collected data instead of data generated by theoretical models (see Oreskes et al. Guest Essay, Chapter 5), Section 5.2. In this regard we strongly recommend the use of what are being referred to as *indicators*, the number of which has grown considerably in the last decade. Indicators, as we shall shortly see help us to characterize the condition and vulnerability of a community's ecological infrastructure. Indicators can be classified into three general categories; *stressor, environmental* or *exposure*, and *response*. Stressor indicators are, as their name implies, those that place some stress on community ecological infrastructure. An example would be the pounds of various pollutants discharged into community waterways or airsheds by industrial facilities, landfills, commercial facilities, roadways, or Superfund sites. An environmental or exposure indicator is usually one of an abiotic nature, in that it either measures the accumulation of pollutants or demonstrates potential risk, damage, or interference with normal ecological functions. A response indicator is usually of a biotic nature perhaps best described as follows. Once a discharge of pollutants has occurred (stressor), and it has accumulated in the environment (environmental or exposure), then those organisms occupying niches in that affected portion of the environment show some type of response, usually but not necessarily of a negative nature. Unfortunately, we have an abundance of negative response indicators including a multitude of fish species whose flesh has accumulated so many pollutants that they are dangerous to eat, waterfowl whose eggs do not hatch or whose offspring are born deformed, and populations of cold water fish so stressed by changes in waterway conditions and chemistry that their survival as a species is seriously in question.

Third, thanks to new national initiatives to revive watershed or river basin based planning to promote a much broader approach to environmental protection, the water resources of the community, particularly the surface waterways and their concomitant watersheds (drainage areas), have major importance in the ecologically based MMP. Other than air, no other resource impinges so directly on the day-to-day well-being of every living creature of the biosphere including man. We can live 30 days without food, but only 3 days without water. Furthermore, the branching network of surface waterways is generally the common denominator of the local community; this network binds in one way or another all other components of a community's resources, both natural and otherwise, thus abetting the much sought after goal of *holistic* management. That watershed boundaries are much easier to define and recognize than the sometimes ephemeral boundaries of other ecological components makes management by watersheds even more appealing.

Fourth, we must not forget that the major premise of our ecologically based MMP is *preventative planning*, instead of mitigation and remediation through best management practices (BMPs) or other means after the fact. Aside from the fact that BMPs are often not fully successful, they also require substantial amounts of external energy to keep them maintained and operational, thereby causing higher costs to the municipality and its citizens in the long run.

Fifth, we need to be sure to utilize, to the maximum extent possible, a community-wide public participation process that goes beyond even that normally prescribed in MMP review and adoption ordinances; we need to recodify them if necessary to solicit resident and other stakeholder input so that any final MMP is a bona fide consensus of the entire community. It is important to solicit input from every interested party as early in the process as possible, allowing them to work directly with any hired experts so that some feeling can be gained for what is and is not important to the community as a whole. Working committees of residents, elected officials of the community, and other stakeholders should be set up to assist in the gathering of information — including as much of the scientific data as possible — the analyzing of it, and the setting of specific objectives for the MMP. Initially, diverse opinions, perhaps with polarization and friction between various groups, are to be expected; however, as the process proceeds, each interest group is forced to look beyond its own short-term needs to the needs of the community in the long term. Certainly, individual vocabularies are to be expanded to include, what have been to many, rather foreign terms such as sustainability, transgenerational planning, and ecological footprints. Finally, we need to be sure to also include in the public participation process, representatives from neighboring municipalities who might share common ecological infrastructure. It is within the realm of possibility that a joint ecologically based MMP could be the result of such a collaboration.

9 Getting Started

9.1 DEVELOPMENT OF THE COMMUNITY-WIDE PARCEL BASE MAP

The first essential ingredient of our ecologically based municipal master plan (MMP) is a community-wide property parcel map (Figure 9.1). This map is the foundation document with which we eventually overlay and evaluate subsequent layers of environmental data. Historically, the municipal tax map, which is a plotting of the property boundary lines of all real estate parcels within the municipality, has been used most frequently for this purpose, although perhaps with a lesser degree of accuracy than we might like or require for an ecologically based MMP. Tax maps are chosen because of their standardized scale and because of some statutory requirements for minimum standards of accuracy, and most importantly because of the inherent importance of property parcel boundaries in land use planning.

Generally, except for the smallest communities, such maps are composed of a series of 24 × 36 in. sheets at a scale ranging from 1 in. = 100 ft to 1 in. = 400 ft. Our preference would be to combine all these separate sheets into a singular map; however, due to their relatively large scale, such compositing could produce a map of unwieldy proportions, even in an electronic format. Our alternative then is to combine the sheets at a much smaller scale, for example, 1 in. = 800 ft or 1 in. = 1000 ft. This is not to say that you cannot utilize the individual tax sheets at their present scale. Combining them into a singular map, however, makes it more convenient to review, retrieve, and evaluate data.

We have mentioned previously that our requirements for accuracy would most likely be a bit more rigorous than those that have been historically required, for example, for such purposes as the support of zoning ordinances. Consequently, the plottings of the deeds for each individual parcel must usually be at a higher degree of precision and therefore must require the application of coordinate geometry and resolution of any discrepancies found between individual property lines. In addition, street centerlines and, where practical the centerlines of surface waterways (if only at stream crossings), need to be accurately located and plotted. This is now being done by many local, county, and state governments through the use of global positioning units (GPUs), which are generally capable of submeter accuracy and with some extra effort, centimeter accuracy.

Once the deed plottings and the stream and street centerlines have all been rectified, current black and white (and color also) aerial photography of the community, at a scale common to the parcel base map, should be obtained. Such photography should be in the form of digital orthophotos, which enable the accurate interpretation and interpolation of vertical elevations, topographic contours, and landscape features. Digital orthophotos are simply digital images of aerial photographs in which displacements caused by the camera and the terrain have been removed. Such digital images are produced by scanning aerial photographs that meet National Aerial Photography Program Standards, with a precise, high-resolution scanner. The digital images are then rectified to an orthophotographic projection by processing each image pixel (picture element) through photogrammetric equations using ground control points that can be identified from photographs, camera calibration and orientation parameters, and digital elevation models (DEMs). Once finished, the orthophoto is a spatially accurate image with planimetric features represented in their true geographic position. Many county, state, and federal agencies already may possess such photography and might provide it to you for no charge or for a nominal fee. The U.S. Geological Survey (USGS) is an excellent source of digital orthophotos at a scale of 1:12,000 (3.75 minute quadrangles) or

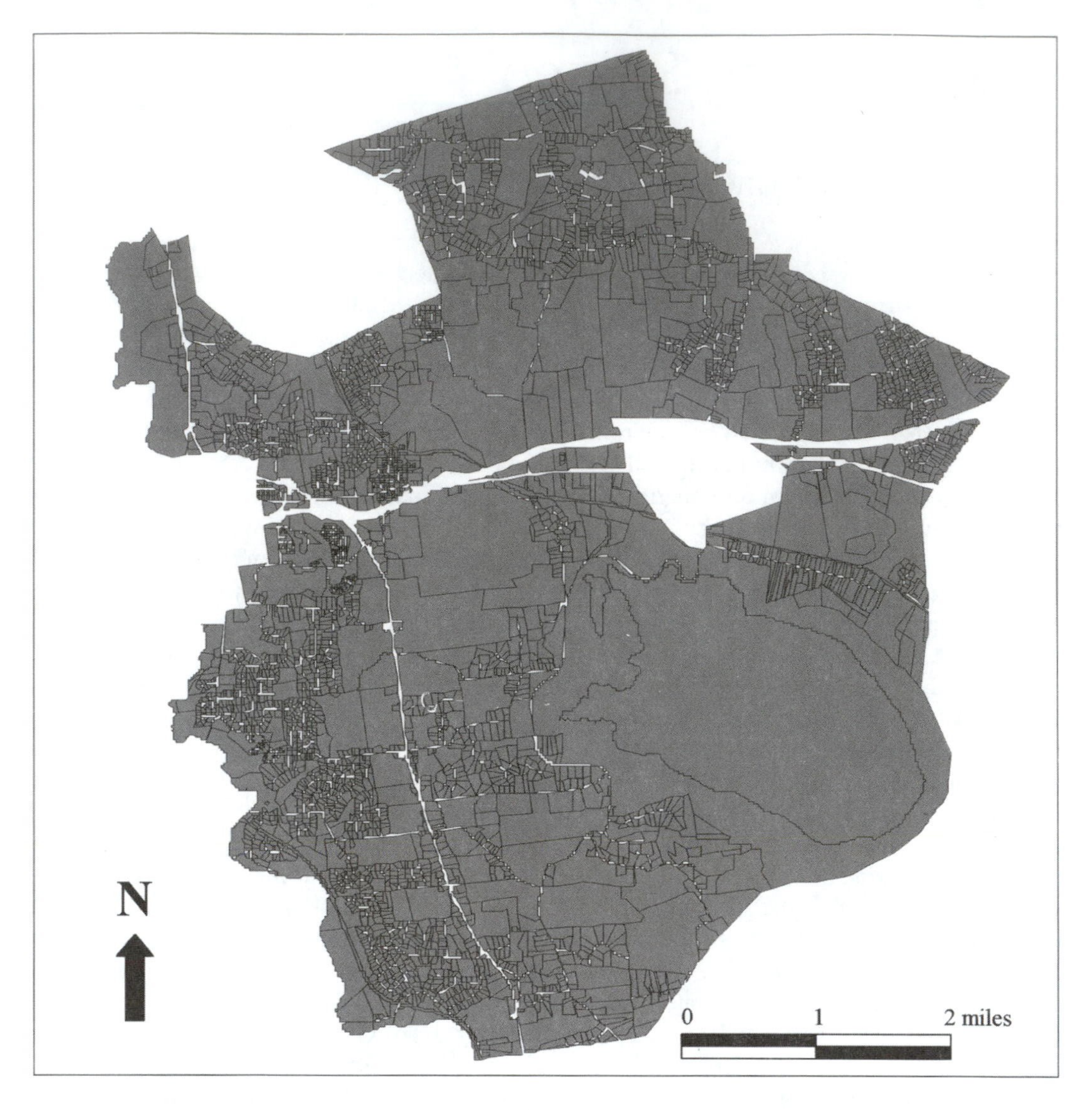

FIGURE 9.1 Typical township property parcel map. This map was developed using Hunterdon County, New Jersey, Geographic Information System digital data, but the secondary product has not been verified by Hunterdon County and is not county-authorized. (Reprinted with permission of the Hunterdon County GIS.)

1:24,000 scale (7.5 minute quadrangles). In any event, once obtained, the rectified community parcel map should be overlain (electronically or manually as a mylar) on the digitized aerial photography to further validate the accuracy of street and stream centerline locations. This process is known as conflation.* Once you have verified the joint accuracy of the two maps, the community parcel map should receive final digitization.**

* Conflation is the process of creating and updating a new master set of data from the best spatial and attribute qualities of two or more data sets. The objective of conflation is to combine or fit two digital geometric data sets together into one, reconciling the feature geometry of one to the other. This may involve rubber sheeting of lines and points and the matching of line segments and shapes.

** An excellent treatise on GIS parcel mapping can be found in the *Digital Parcel Mapping Handbook*, 1998, prepared by the New Jersey State Mapping Advisory Committee.

9.2 INCLUDE SOME HISTORICAL PERSPECTIVE OF THE COMMUNITY

Much as the passage of time clouds the memories of many of the prior events in our personal lives, so to do the prior events in a community's life become obscured, if not completely forgotten, with the passage of time. As residents come and go and local planning boards members come and go, fewer and fewer residents remain to recall what the character and conditions of the community once were. For instance, did earlier inhabitants actually swim in local waterways that are now so polluted that swimming is banned? If so, where were those "swimming holes"? Were native brook trout once the dominant species of fish in some of community waterways? Where were those streams located? What the dominant land uses were in the community, decades earlier, also would be of special interest in light of current research that indicates land uses, particularly, agricultural land use, as much as 40 years prior, still could be negatively influencing the current health of the community's aquatic ecosystems.[26] Those same prior agricultural uses also could pose a risk to the occupants of the community's newest suburban developments that might have been built on such previous farmland, due to the presence of high levels of insecticides, herbicides, and fungicides in soils previously under cultivation. Questions such as these barely rise to the surface in those communities currently undergoing rapid sprawl-type development. Contemporary residents are usually more focused on finding ways to get around the severe traffic congestion on local roadways that is constantly impeding their 40-mi commutes to work. They may even believe that swimming always has been in the backyard pool and never in the backyard stream. They may further believe that fish life is best observed and preserved in a 20-gal tank.

Communities need to create a place to permanently preserve some institutional memory of what a community's character once was, if only to serve as a permanent reminder of what the community's prior land planning efforts have wrought. The perfect repository for such historical data is within the pages of the modernized, ecologically based MMP. The contents need not be lengthy or detailed, and can be as simple as successive black and white aerial photography of the community at 10- or 20-year intervals (Figure 9.2) and some population growth data (Figure 9.3) derived from the U.S. Census Bureau databases. This information can be further supplemented with narrative descriptions of the prior extent and quality of a community's natural resources, retrieved from archived county and state databases, local historical groups and societies, colleges and universities, nonprofit environmental groups, or even newspaper archives. In addition, anecdotal information and perhaps old photographs can be obtained from older residents who may have stayed long enough to witness firsthand the decline in environmental quality.

9.3 INVENTORY AND IMPORTANCE OF COMMUNITY RESOURCES

The local municipality that keeps an up-to-date inventory of its ecological infrastructure and its associated components, specifically for planning purposes, is more rare than common, although many municipalities can be credited with at least having made an initial inventory of such resources at one time or another. While MMPs and zoning ordinances have been reexamined and updated on a regular basis (usually every 6 years), most of these initial ecological or natural resource inventories have not been kept similarly current. As a result many are decades old and need some serious reexamination. Furthermore, the scope of these inventories often was limited to only a few categories of resources, primarily soils, steep slopes, open space, wetlands, and vegetation; and the details on their condition or health was notably absent — not quite the holistic, state of the environment report envisioned to become the foundation of the new MMP. Note also that we are including the community's *man-made* infrastructure in this same inventory. This is due to the close cause and effect relationship that exists between these two types of infrastructure.

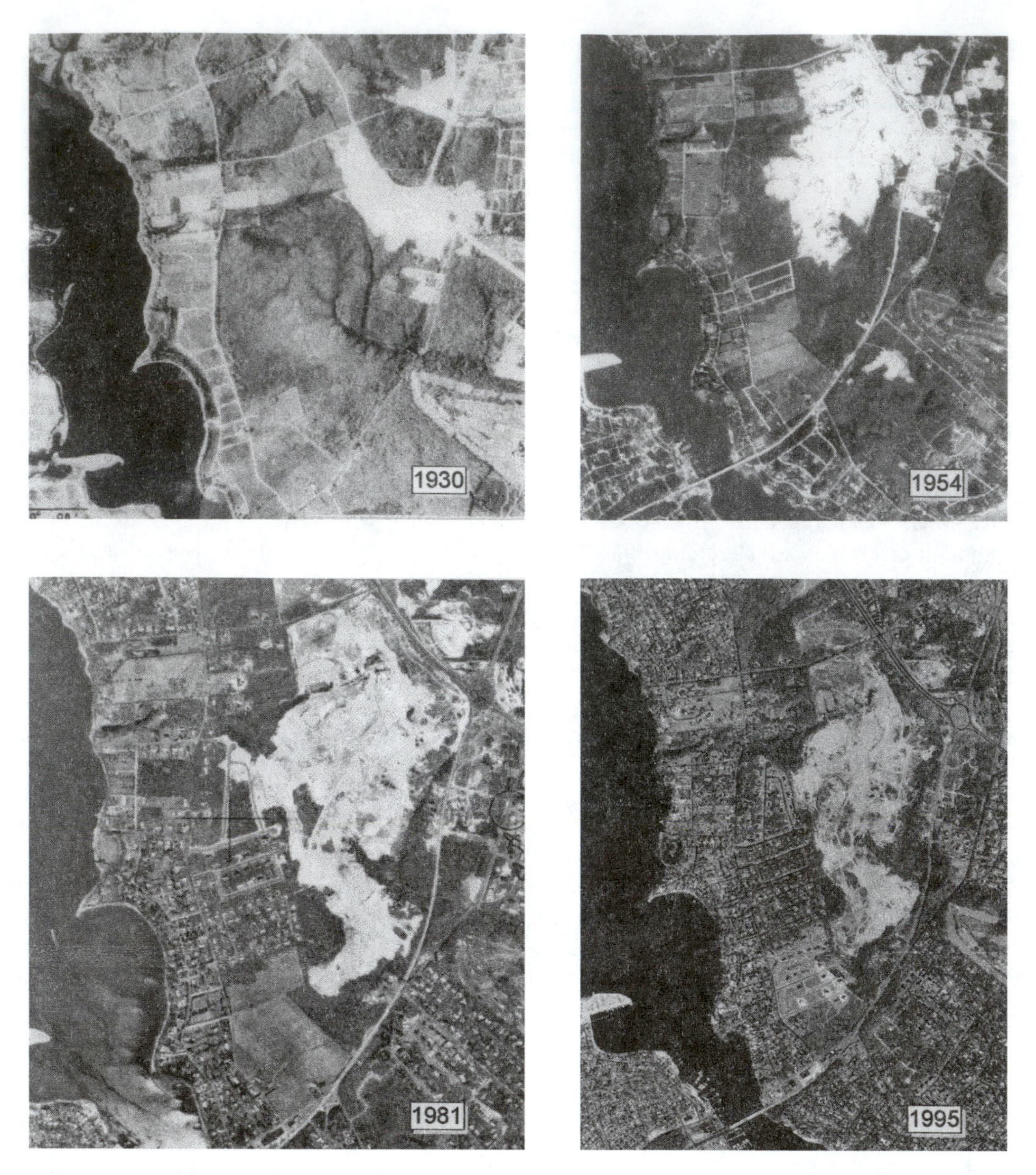

FIGURE 9.2 Where has all the open space gone? The pictures tell the story. (Courtesy of the New Jersey Department of Environmental Protection, Bureau of Tidelands).

Because of space limitations here, we can discuss only some of the components of these two types of infrastructure. As you go on to prepare your own MMP, your selection of components may vary from those we have selected here. This is due to the fact that your selection process is dependent, in large part, on the intensity and amount of land development already in place in your community. As a metropolitan center (i.e., inner city or downtown), for example, your ability to practice preventative planning, a major premise of this book, may be severely constrained (although rehabilitation is certainly not out of the question). A more suburban or rural community, on the other hand, may have considerably more opportunity to protect a larger proportion of its natural resources from the impact of further development.

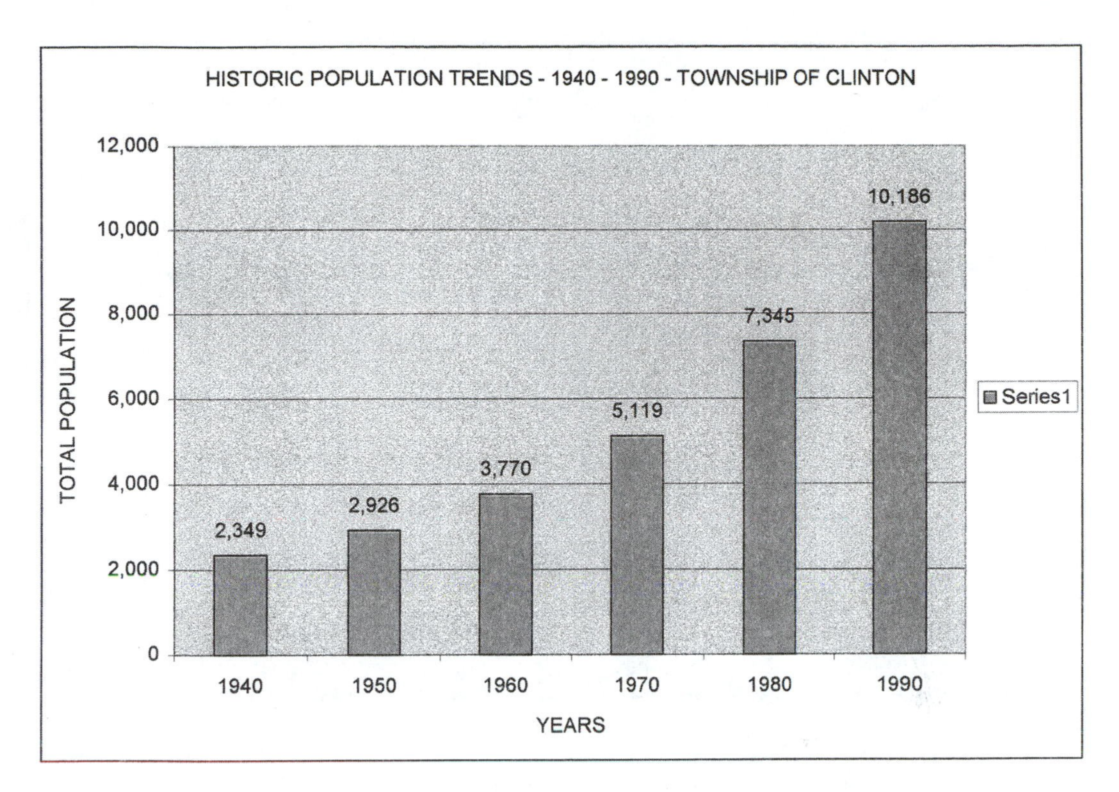

FIGURE 9.3 A growing population means growing demands on a community's finite ecological infrastructure.

9.3.1 Ecological Infrastructure and Associated Components

9.3.1.1 Water Resources

9.3.1.1.1 Introduction

Every day 4.2 trillion gallons of water (in the form of precipitation) falls on the United States. This precipitation is part of a complex process of recirculation that has existed for millions of years. The molecules of water that fall to Earth's surface today, are very likely the same ones, or a portion at least, of the ones that fell to Earth millions of years ago. This recycling process has been termed the *hydrological cycle* and is illustrated in Figure 9.4. There is no replacement of the world's water supply from beyond Earth's atmosphere, despite perceptions to the contrary.

Water covers 71% of Earth's surface and this apparent superabundance has fostered an illusion of inexhaustibility. Of the total world supply of water, however, 97% is in the oceans and is generally unutilized due to its salinity. The remaining 3% — the freshwater, nonsaline portion — is stored temporarily in the polar ice and glaciers, lakes, rivers, soil, groundwater, hydrated earth minerals, plants, animals, and of course the atmosphere. Three quarters of this freshwater is locked up as ice and likewise unavailable. The remaining one-quarter fraction, which is largely groundwater, has been the major source of water for human needs. Thus, despite the appearance of a copious supply of water, human life is precariously linked to a very tiny fraction of Earth's total water supply.

9.3.1.1.2 Surface Waterways

The role of surface waterways in the hydrological cycle is obvious to most everyone. They are downgradient interceptors, collectors, and conveyors of fallen precipitation that, on impact, has somehow escaped infiltration or absorption into the ground and instead, propelled by gravity, flows

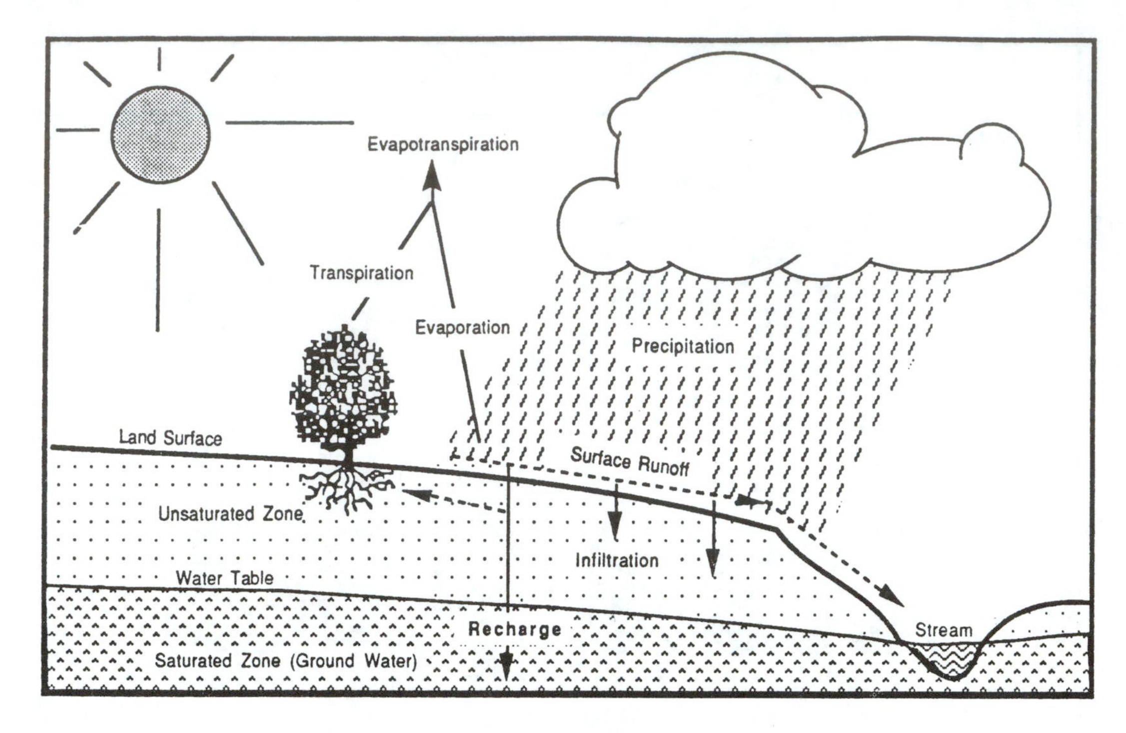

FIGURE 9.4 Hydrological cycle. (From New Jersey Department of Environmental Protection, A Method for Evaluating Ground Water Recharge in New Jersey, New Jersey. Geological Survey, 1993.)

downhill over its surface. As this fallen precipitation (more commonly referred to as runoff) makes its way downhill, it may pluck and transport materials lying at or near the ground's surface, whether that surface is bare, covered with asphalt, or heavily vegetated. It also may interact chemically with some of those same materials or with other constituents in the top layer of soil, all of which can alter the runoff's original purity (both negatively and positively). This intimate contact between the runoff and the land surface makes the receiving waterways an excellent ecological barometer of the general condition (health) of the contributing watershed. Any degradation, alteration, or disruption of the watershed's physical or ecological character directly affects the quality of the water running off the land and ultimately collected in these open channels of conveyance (Figures 9.5 and 9.6). The same applies to the organisms who take up residence in these watercourses. Their diversity and abundance (as we shall shortly see) also is affected directly by the quality (and sometimes quantity) of the water emptying into these arterial networks. Because most local communities are, in reality, nothing more than an aggregation of contributing watersheds (drainage areas) and their accompanying watercourses, they are a crucial data layer and hence one of the first ecological properties to be included in our inventory.

This data layer has special significance because of the current and growing national interest in reviving a more holistic, *watershed-based* management of the nation's water resources. Those old enough to recall the provisions of the 1972 federal Clean Water Act and its 1977 amendments, remember that a watershed or river basin management scheme was strongly encouraged by such sections as 201, 208, 209, and 303(e) of that act. Unfortunately, while much work was accomplished in this regard in the decade that followed, the concept of watershed-based management eventually languished in deference to very parochial legislative initiatives such as the Toxic Substances Control Act (TSCA), the Resource Conservation and Recovery Act (RCRA), and the Comprehensive Environmental Response Compensation and Liability Act (CERCLA) also known as the Superfund Act. In hindsight, we can see clearly now the advantages of management by watersheds, not only for the well-being of the nation's water resources but also for the protection and preservation of other natural resources.

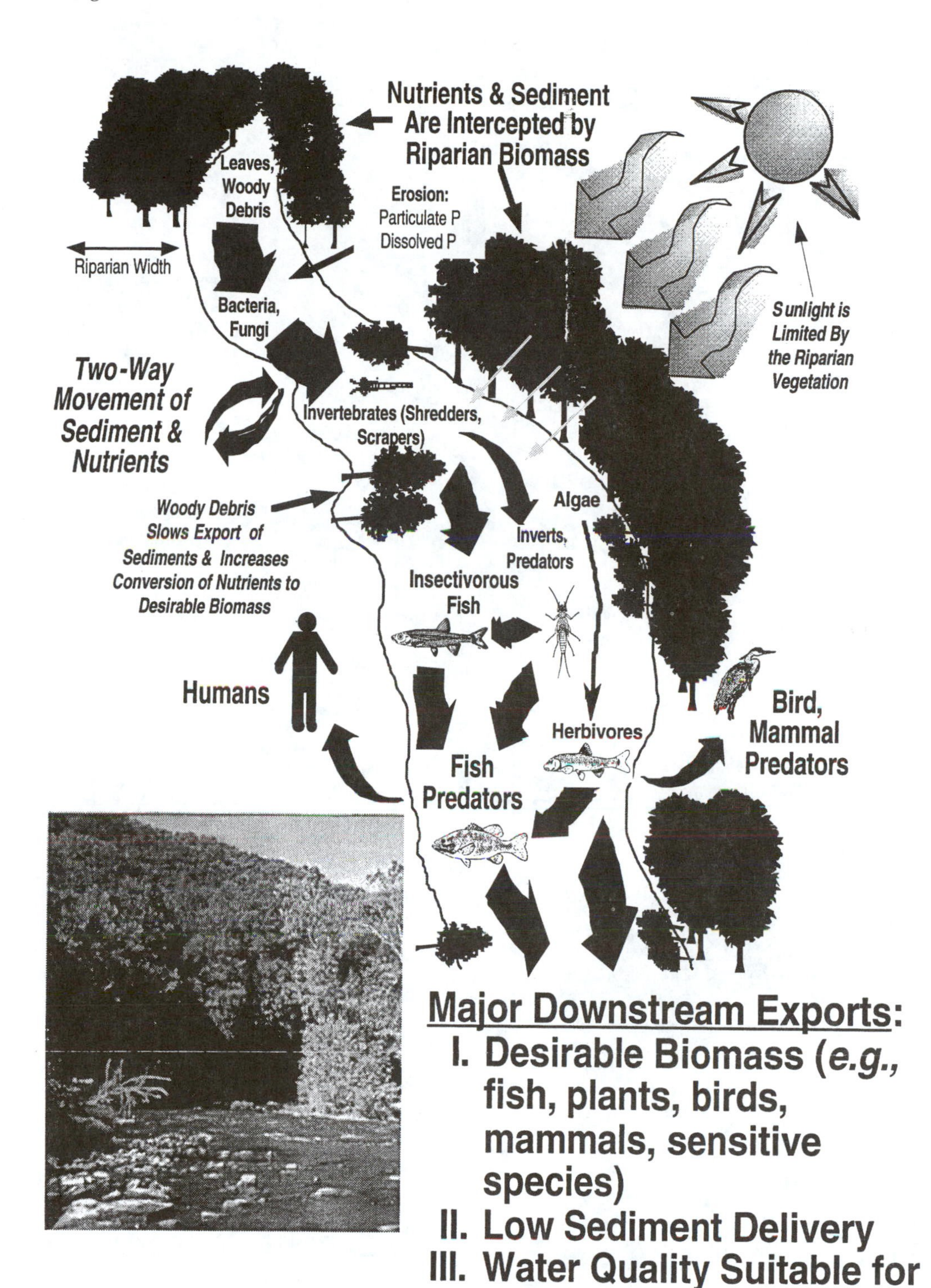

Good Stream Habitat

OhioEPA

FIGURE 9.5 (From Ohio Environmental Protection Agency. With permission.)

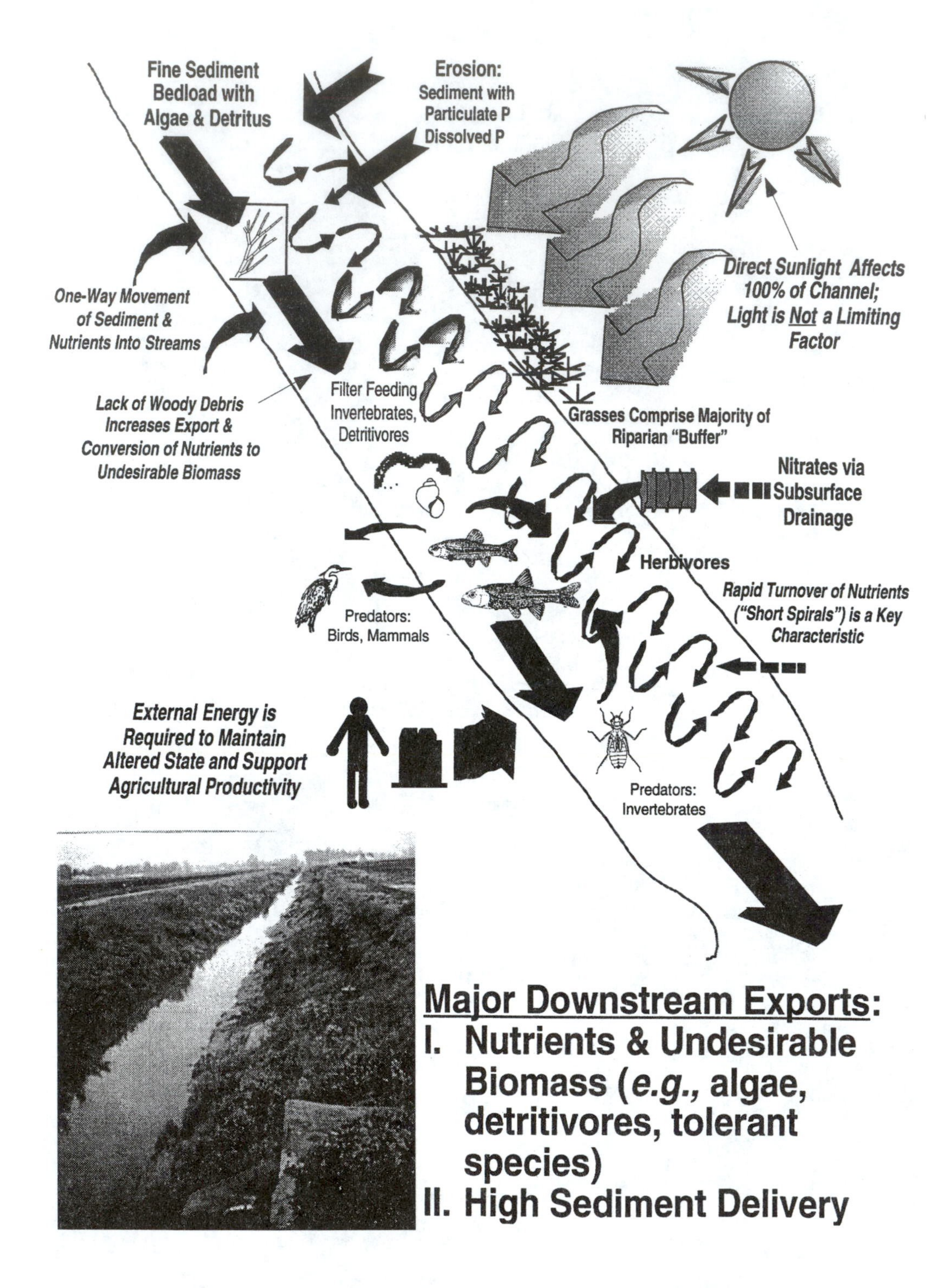

FIGURE 9.6 (From Ohio Environmental Protection Agency. With permission.)

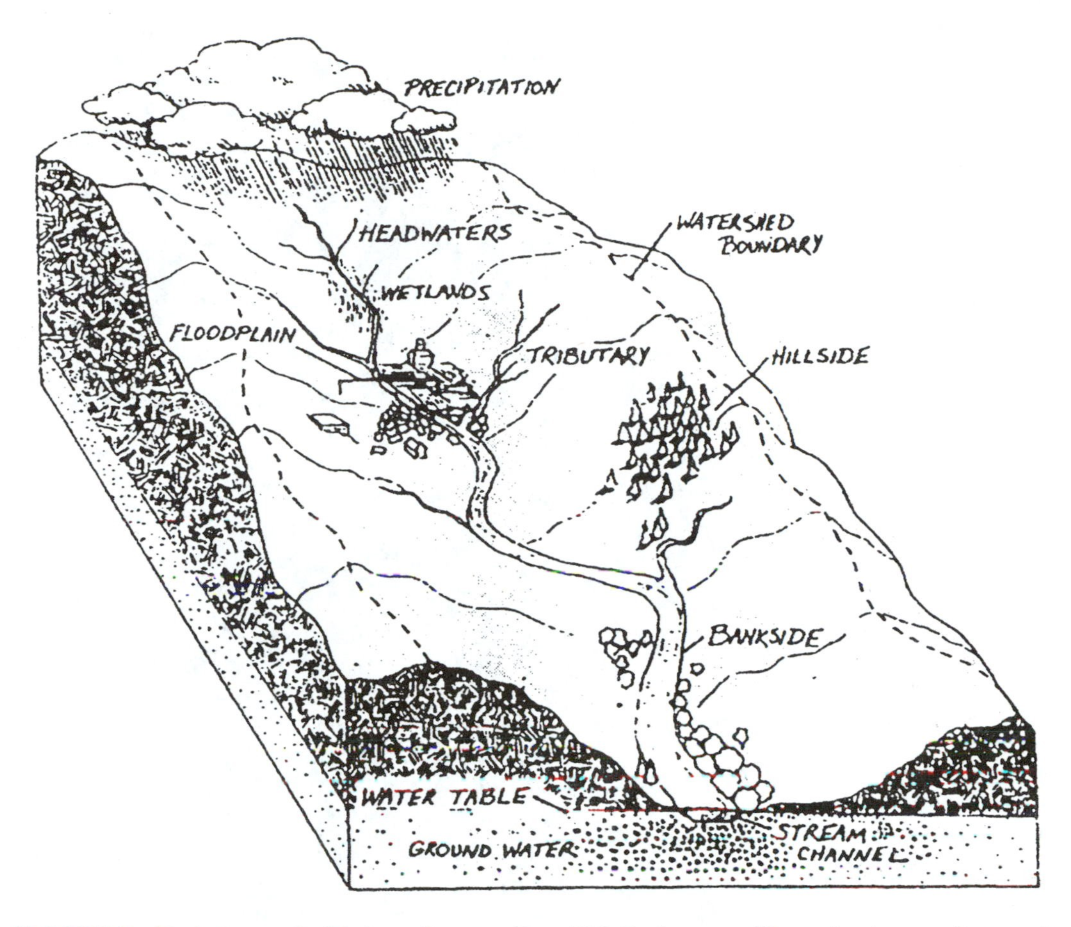

FIGURE 9.7 Typical watershed drainage features. (From U.S. Environmental Protection Agency, Community Based Environmental Protection: A Resource Book for Protecting Ecosystems and Communities, EPA 230-B-96-003, 1996.)

The rather easily discernible physical boundaries (Figure 9.7) of a surface waterway's watershed or drainage area make proposals for such watershed-based management initiatives even more appealing to those having difficulty in distinguishing the boundaries of other, more complex, ecological units and processes.

Thanks to the efforts of some federal agencies, the USGS and the U.S. Department of Agriculture (USDA) Natural Resource Conservation Service (NRCS), in particular, most of the larger watershed boundaries already have been delineated (Figure 9.8). A classification system was developed by the USGS in 1972, that divided the United States and the Caribbean into 21 major regions, 222 subregions, 352 accounting units, and 2150 cataloging units for the purpose of delineating river basins with drainage areas greater than 700 square mi. Each cataloging unit was assigned a unique eight-digit *hydrological unit code* (HUC). For example, a cataloging unit with an assigned eight-digit HUC number of 02040201 would be broken down as follows:

02 = region

0204 = subregion

020402 = accounting unit

02040201 = calculating unit

FIGURE 9.8 Watershed boundaries for New Jersey as defined by hydrological unit codes. (From U.S. Geological Survey.)

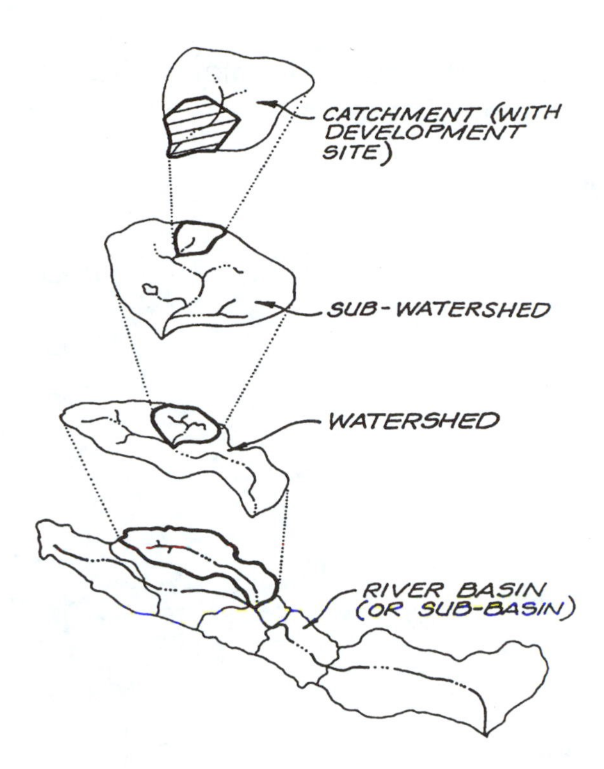

FIGURE 9.9 Nested watershed management units. (From Center for Watershed Protection, *Rapid Watershed Planning Handbook — A Comprehensive Guide for Managing Urbanizing Watersheds*, Ellicott City, MD, October, 1998. With permission.)

Since that time these eight-digit delineations have been used extensively by state, federal, and local agencies for the purpose of water resource management. In 1992, however, some states determined that there was a need for the delineation of even smaller sized watersheds. In that year the NRCS developed guidelines for extending the 8-digit classification system to an 11-digit and a 14-digit system, called *watersheds* and *subwatersheds*, respectively. The USGS subsequently set about delineating these smaller watersheds. An assigned 14-digit HUC number such as 02040201010351 would now be broken down as follows:

02 = region

0204 = subregion

020402 = accounting unit

02040201 = calculating unit

02040201010 = watershed

02040201010351 = subwatershed

By definition, the size of an HUC 14 subwatershed is generally between 12.14 and 161.87 square km (3,000 to 40,000 acres or 4.69 to 62.50 square mi), although some smaller sized watersheds with an HUC 14 designation do occur. For our purposes the HUC 14-sized watershed is, in most instances, an acceptable management size for our MMP. Occasionally, however, a smaller hydrological unit called a *catchment* may be more efficacious (Figure 9.9). A typical catchment size would range from 0.05 square mi (32 acres) to 0.50 square mi (320 acres).[19] Many states are

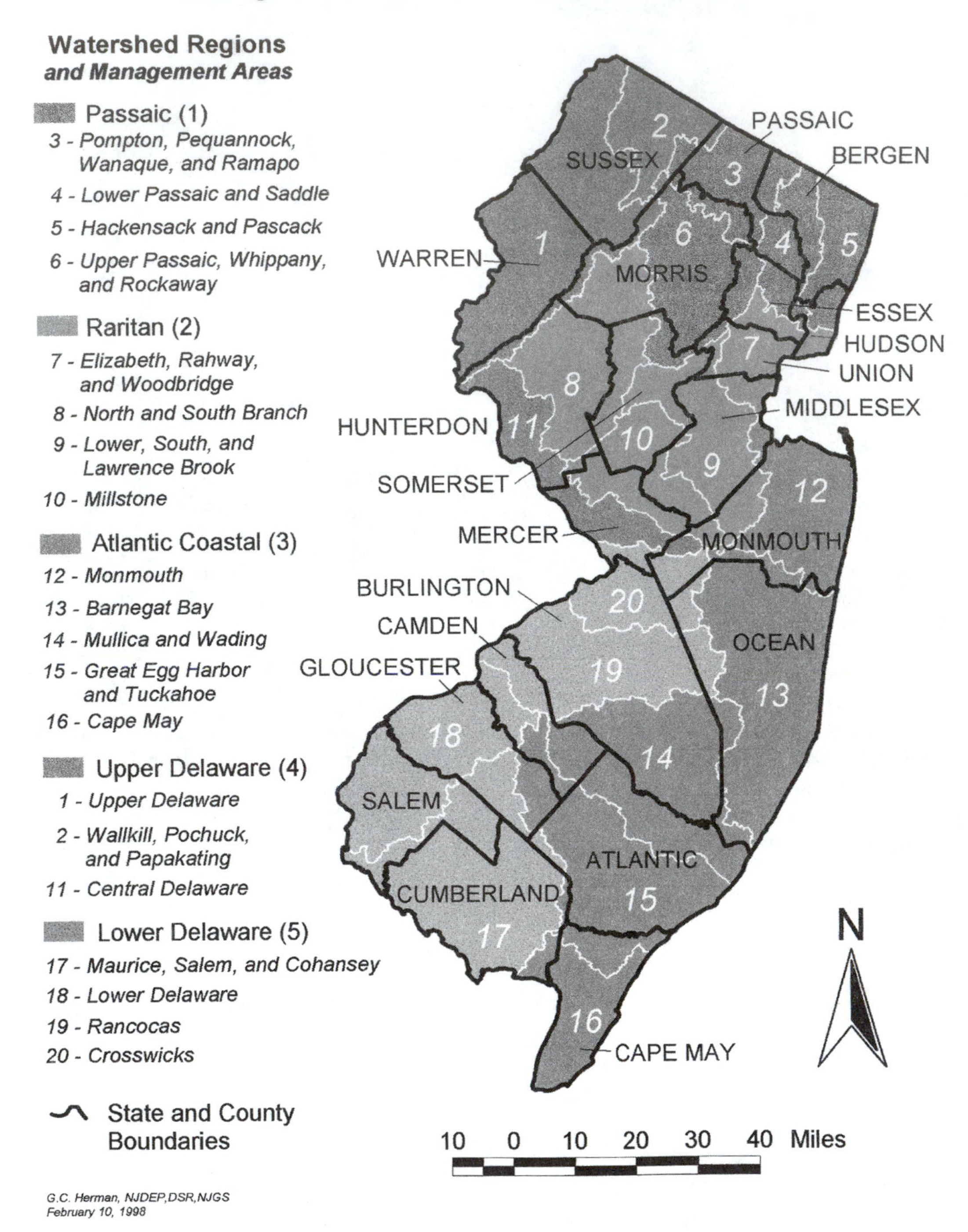

FIGURE 9.10 New Jersey Department of Environmental Watershed Regions and Watershed Management Areas with New Jersey county boundaries. (From New Jersey Department of Environmental Protection.)

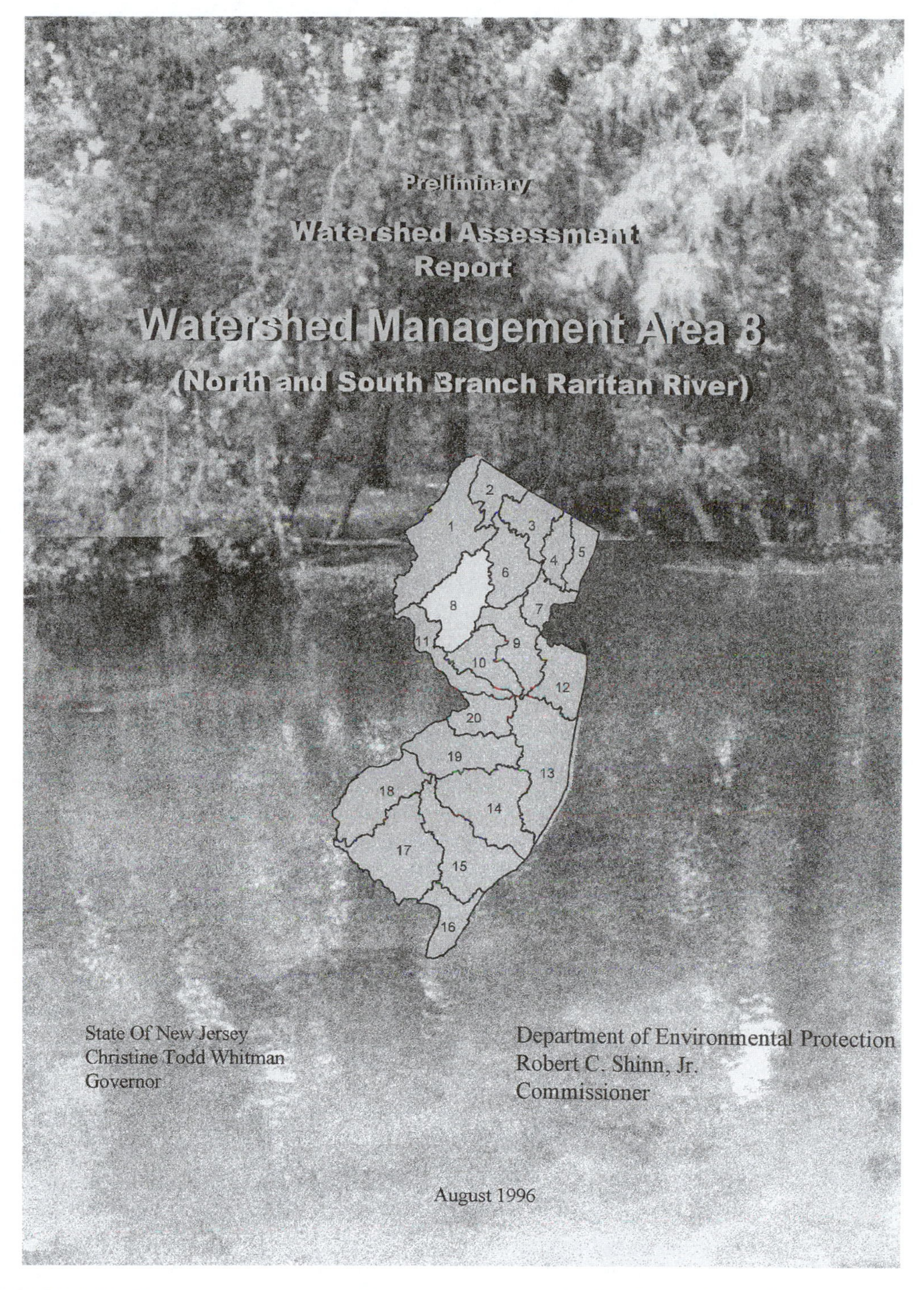

FIGURE 9.11 Typical preliminary watershed assessment report for New Jersey. (From New Jersey Department of Environmental Protection, Bureau of Freshwater and Biological Mont.)

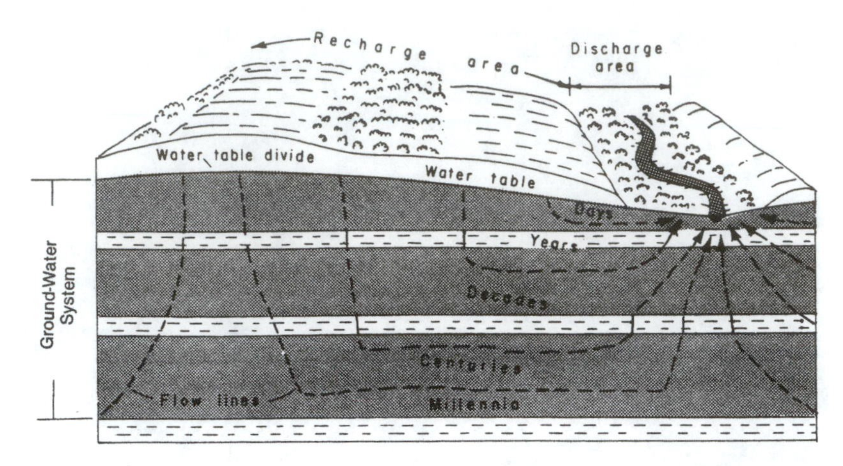

FIGURE 9.12 The groundwater system. (From U.S. Environmental Protection Agency, Protection of Public Water Supplies from Ground Water Contamination, EPA/625/4-85/016, September 1985.)

now actively pursuing watershed-based development strategies. New Jersey, for example, which is the nation's most densely populated state, is aggressively pursuing a statewide watershed-based management program, having adopted legislation in 1996 and 1997 that requires the appropriation of a small percentage of the state's corporate business tax, on an annual basis, to support watershed-based management. Right now that amounts to about $5 million per year. This state with 526 municipal and 21 county governments, has divided its 7800 square mi into 20 watershed management areas (WMAs), which are aggregations of its HUC 11 and HUC 14 watersheds (Figure 9.10*). A key component of New Jersey's program is the production of an individual watershed management area *characterization and assessment report* (Figure 9.11) followed ultimately by a holistic management plan for each WMA. While the protection of the state water resources is the primary focus of its watershed management strategy, New Jersey also is evaluating other natural resources at the same time and intends to prepare similar management plans for these other resources.

A frequently asked question in preparing this data layer is, how far up the watershed should we go in delineating each stream channel? The obvious answer is, as far as there is a discernible channel. However, I contend it should go beyond, and, let me explain why. While we have initially emphasized the role of surface waterbodies as major collectors and transporters of surface runoff, it also should be recognized that water in these same streams is not all runoff all the time. When it does not rain, water in the stream channel comes from another major source and that source is *groundwater* (barring of course, man-made discharges from industrial or domestic sewage treatment plants). This is the precipitation that did infiltrate the ground surface following impact (Figure 9.12); then, pulled downward by gravity, the precipitation made its way down through underlying soil layers and rocks to eventually flow back out to the ground surface at a much lower elevation, appearing in the form of a spring or a seep (Figure 9.13) Often these springs and seeps supply the initial water to a stream channel. This is not to imply, however, that groundwater enters a stream channel only at these points of discharge. To the contrary, groundwater often can be found exfiltrating directly into the bottom or bank of the existing stream channel far downstream of a spring or seep. These springs and seeps, however, are most important, because in addition to providing the initial water to a surface watercourse, they also provide other valuable ecological benefits, including wintertime foraging habitat for a variety of birds and mammals such as the wild turkey and the white-tailed deer. These areas also provide habitat for many species of amphibians. However, their role as a source of relatively clean water is most important. Yet these areas often are regarded with nearly total disdain during land development. They are seen more as a nuisance than as the

* All color figures appear after p. 82

(a)

(b)

FIGURE 9.13 Springs and seeps such as the one shown here are vital components of aquatic ecosystems and need to be protected from intrusion, especially from man-made discharges of storm water runoff.

asset they really are. As a result, they are all too frequently covered over, routed through mazes of underground, concrete storm drains, and mixed with contaminated storm water runoff — all of which destroy their original purity. This premature degradation might seem innocuous to some, but to users downstream it is a matter of critical importance. Sewage treatment plants, for instance, rely on the purity of upstream water to further treat and dilute any residual pollutants being discharged in their treated wastewater. With stream water arriving in an already degraded condition, these treatment plants may be forced to upgrade to a higher, more costly treatment level or in the worst case may be prevented from expanding to accommodate further development.

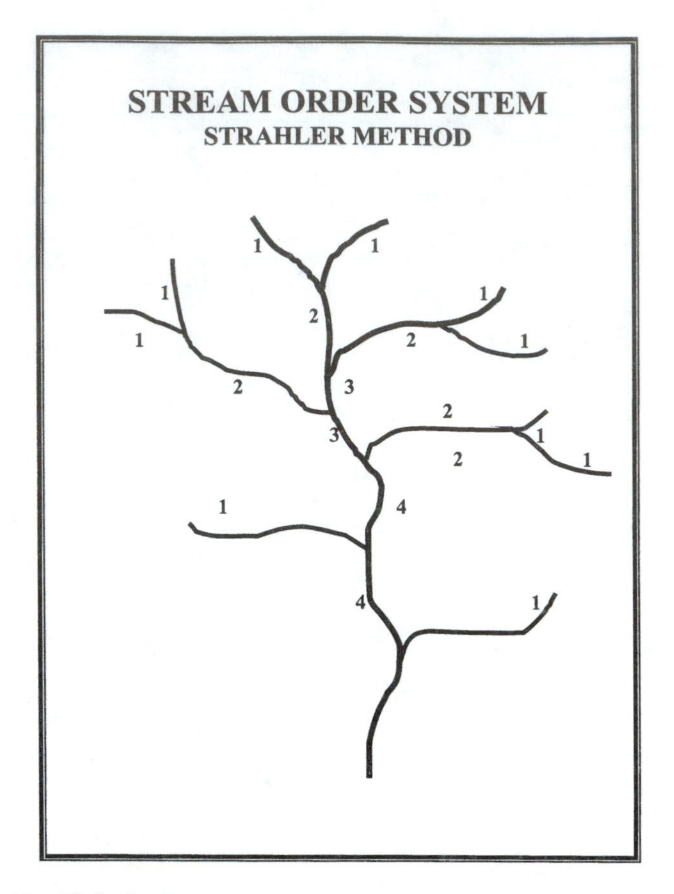

FIGURE 9.14 Strahler Method

After this somewhat long-winded discourse, I think my intentions are clear: springs and seeps should be included as part of any stream channel delineation despite the fact that such areas may already be included in areas designated and regulated as wetlands.

Before we move on, it is important that we talk briefly about *stream hierarchy.* The branching patterns of streams can sometimes be confusing, when one is trying to determine which channel is the *main channel* and which is a *tributary channel*. For this reason, a number of schemes have been devised that break the network of streams into a classification hierarchy. In Figure 9.14, for example, we see one popular method developed by Strahler, which employs the common nomenclature of *stream order* and identifies all initial sections of stream channels (i.e., those in the headwater or uppermost part of the watershed with no contributing tributaries) as *first order* (identified by a 1). From there we see *second-order* streams (those with a single tributary) and eventually as we wend our way downstream, we see *third-* and *fourth*-order stream segments. This stream hierarchy is important in our health assessment and long-term ecological planning strategies that follow. First-order streams with a mean drainage area of about 1 square mi (640 acres) are of particular importance because they are usually the most vulnerable to human intrusion, and represent the largest number of streams nationwide, some 1,570,000 out of an estimated 2,023,400. For comparison there are only 350,000 second-order, 80,000 third-order, 18,000 fourth-order, and 5,200 fifth-order streams nationwide.[19] Interestingly, there is a single tenth-order stream that has a mean drainage area of 1,250,000 square mi.

9.3.1.1.3 Lakes and Ponds

Lakes and ponds often are regarded as merely ancillary components of the surface water system, which, thanks to the way many of them were created, has some validity. In the northeastern United

FIGURE 9.15 Weed-choked and full of sediment — an all too common scene at many of the nation's lakes and ponds.

States, for example, many lakes and reservoirs were created simply by damming up a river's main channel, thereby creating an instream impoundment. These impoundments were sources of power for various mills as well as cooling water for manufacturing, for drinking, for diluting industrial wastes; and finally, for ice to fill turn-of-the century kitchen ice boxes. A secondary use was recreation, which today, in a curious reversal, has become, in many instances, the number one use. Unfortunately, these impoundments also became major areas of deposition for various debris and waste (mostly from upstream land development) that made its way into the upper reaches of the stream channel and then was carried downstream by the flowing water. Washed away soil (erosion) was a particular problem, so much so that one of the largest problems facing lakes, reservoirs, and ponds today is the accumulation of depth-reducing, contaminant-laden sediment.

Unlike our highly prized and utilized coastal waters, these inland bodies of water have been given little or no attention in the way of funding to rehabilitate and restore them to their original capacity and attractiveness (the Great Lakes are an exception). Many have become an embarrassing centerpiece of the community — filled with debris and silt, choked with weeds (Figure 9.15), and covered with unsightly and smelly blooms of blue green algae. Lakes exhibiting such characteristics are described as *eutrophic*. While such eutrophication is usually a long-term process, that process has been accelerated considerably, thanks to man's cultural pollution (Figure 9.16). Often, the best we can say about them is that they are truly a tribute to the way in which we carried out our prior land use planning activities.

These impoundments, contrary to popular beliefs often have ecological characteristics much different than their parent waterways. This is due in part to their greater depth, and the considerable decrease in the velocity of the water passing through them (the latter characteristic being the major reason they collect so much extraneous debris). There are, of course, lakes and ponds not necessarily created by impounding a stream. These are generally formed as a result of some prehistoric geologic event that created a natural dam or depression. Some have been enhanced by a man-made dam and some may be fed solely by springs. These geologically or naturally occurring lakes and ponds may be much deeper than their instream cousins.

In conclusion, the eutrophication status of lakes and ponds is an excellent barometer of what is happening in the watershed and is, therefore, an important part of our natural resource inventory.

9.3.1.1.4 Groundwater

You may recall our earlier discussions of springs and seeps, which we described as the product of precipitation that had fallen onto the ground surface and instead of running off, had infiltrated into the underlying soil. Our earlier, much simplified description of this downward progression of water needs some further elaboration. As we previously explained, the water is propelled vertically

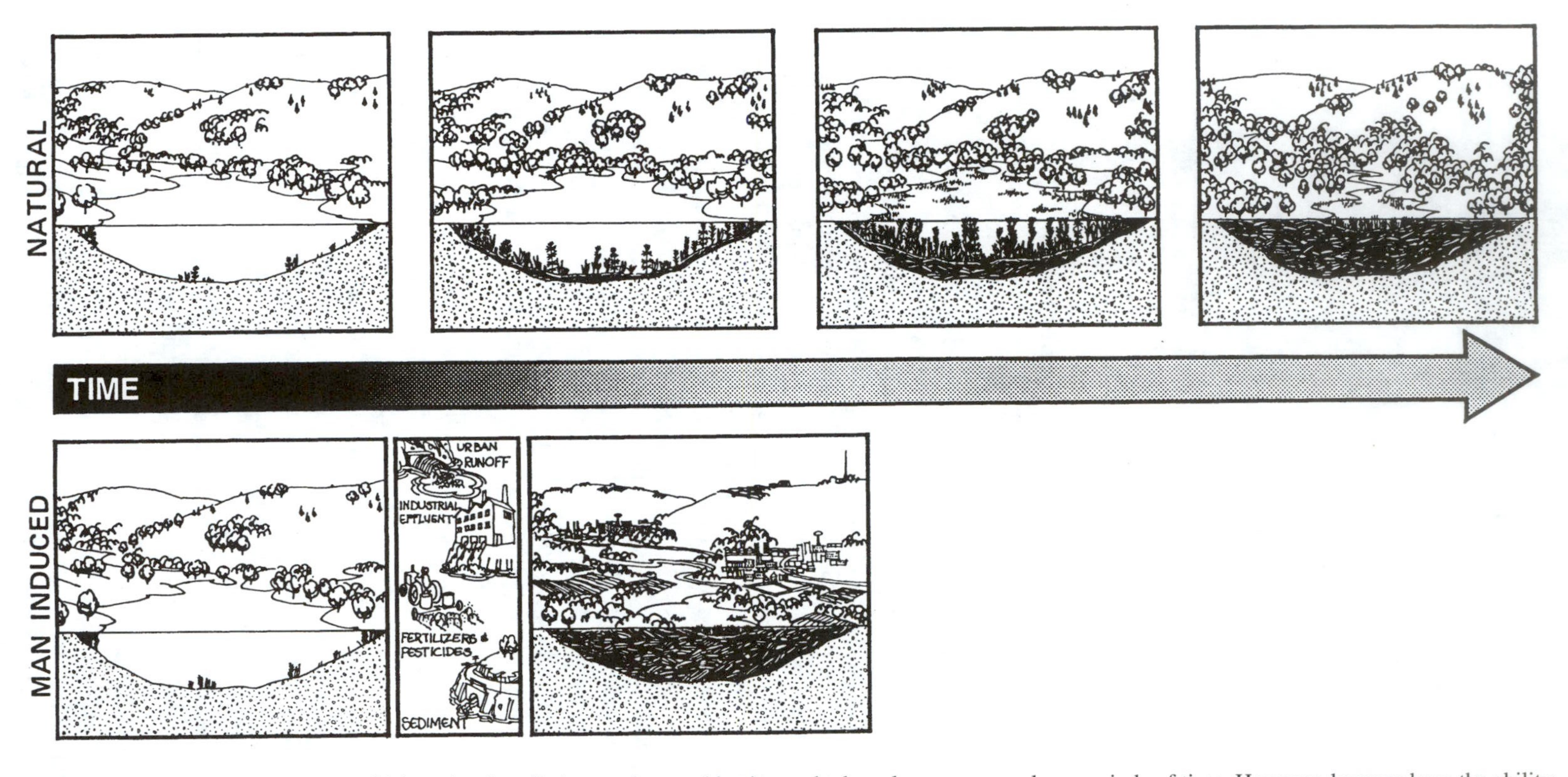

FIGURE 9.16 The normal process of lake aging is called natural eutrophication and takes place over very long periods of time. However, humans have the ability to accelerate that process considerably through cultural pollution, which is often the result of poor land use planning. (From Born, S. M., and Yanggen, D., *Understanding Lakes and Lake Problems*, Upper Great Lakes Regional Commission, University of Wisconsin, Madison, 1972. With permission.)

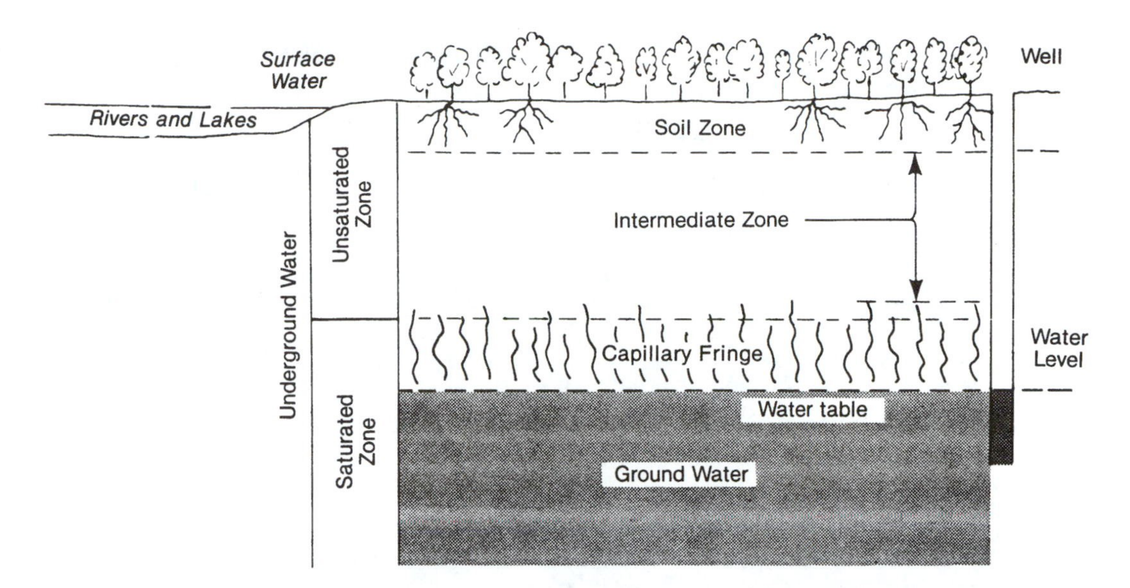

FIGURE 9.17 Typical cross section of the groundwater system showing various zones. (From U.S. Environmental Protection Agency, Protection of Public Water Supplies from Ground Water Contamination, EPA/625/4-85/016, September 1985.)

downward by gravity and passes through the soil column, whose depth may range from a few inches to hundreds of feet. Once through the soil the water may flow into fractures, cracks, crevices, and cavities in any underlying bedrock formations, or in their absence, into deep formations of sand, gravel, or other porous soils. This process is called *recharge*. The water continues its downward movement until it reaches a depth where the rock is no longer fractured (i.e., lacks further storage space) or reaches a soil strata whose density prevents or retards further downward penetration. Once prevented from moving downward, the water begins to accumulate, filling the voids and spaces between the soil particles or filling the fractures and cracks in the rock. The height to which this accumulated water rises depends largely on how much precipitation percolates into the ground. That portion of the soil or rock where the spaces, cracks, or fractures are completely filled with water is known as the *saturated zone*. This saturated zone is often referred as an *aquifer*. The top of this saturated zone is called the *groundwater table*. Figure 9.17 illustrates a typical cross section of the groundwater system.

The top of the groundwater table, much like the ground surface above it, is sloped, and generally parallels ground surface contours, though not always. Once prevented from moving any farther downward vertically, the water shifts to a flow in a more horizontal direction, often exhibiting its presence with springs and seeps on the sides of slopes. The speed or velocity at which this water moves through the ground is substantially slower (except perhaps for cavernous rock areas) than that of surface streams and can range from several feet per day to several feet per year. In fact, in some areas the movement is so slow that it is not unusual for many residents to find themselves making lemonade with 100-year-old water. Because groundwater is often the only reasonable source of potable water for many municipalities, residents need to concern themselves with the following issues. First, where in the municipality are the areas of greatest infiltration (recharge) and should they be placed off limits to development to protect this capability? Second, are there areas of soil in the municipality that are excessively porous and, therefore, provide immediate entryways for pollutants to contaminate the underlying aquifer? Third, what is the areal extent of the aquifer, what is the quality of the water (Figure 9.18), who else is using it for a water supply, and does the aquifer have the capacity to sustain the current and future withdrawal rates?

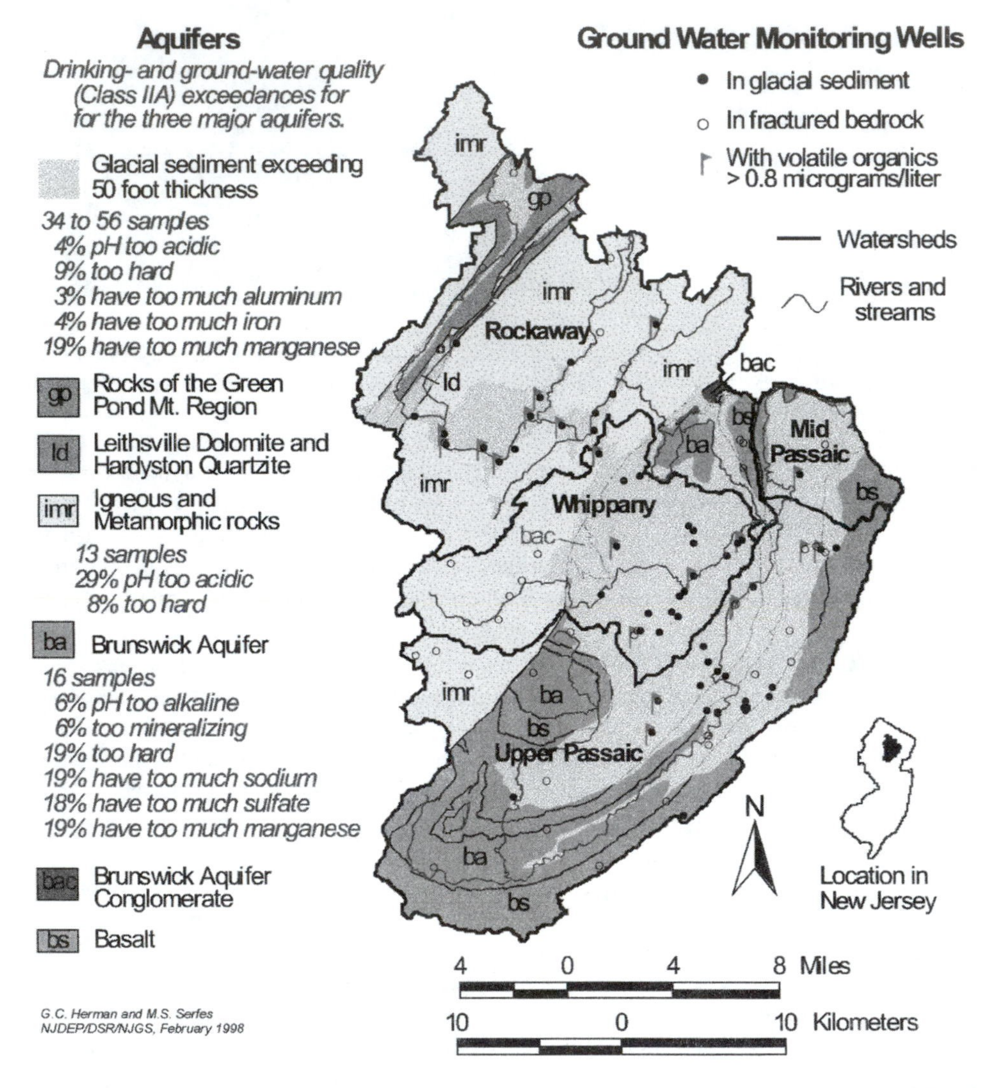

FIGURE 9.18 A community should be compiling information about its groundwater quality well before it begins to approve individual projects that may use that groundwater. (From New Jersey Department of Environmental Protection, New Jersey Geological Survey.)

9.3.1.2 Wetlands

Wetlands is the collective term used to describe marshes, swamps, bogs, and similar areas found in flat vegetated areas, in depressions in the landscape, and between dry land and water along the edges of streams, rivers, lakes, and coastlines. Wetlands are among the most biologically productive natural ecosystems in the world and are often compared with tropical rain forests and coral reefs in the diversity of species they support. Their ecological benefits are numerous. Wetlands are vital to the survival of various animals and plants. The U.S. Fish and Wildlife Service estimates that up to 43% of the threatened and endangered species rely directly or indirectly on wetlands for survival. Wetlands furnish a wealth of natural products, including fish, timber, wild rice, and furs. Wetlands are also important spawning and nursery areas for commercially important fish. In 1991, the dockside value of fish landed in the United States was $3.3 billion, which served as the basis of a $26.8 billion fishery processing and sales industry,[27] which in turn employs hundreds of thousands of people. An estimated 71% of this value is derived from fish species that depend directly or indirectly on coastal wetlands during their life cycles. Wetlands often function as natural tubs or sponges, storing floodwater and slowly releasing it. This combined action, storage and slow release can lower flood heights and reduce the erosive potential of water, thereby helping to control increases in the rate and volume of runoff from urban areas. Wetlands also improve water quality, including that of drinking water, by intercepting surface runoff and removing or retaining its nutrients, processing organic wastes, and trapping loads of sediment before it reaches open water. Finally, wetlands provide opportunities for popular activities such as hiking, fishing, boating, hunting, and photography of wildlife.

Over the past few years the subject of wetlands protection has become a common and controversial topic, well covered in the media. At the heart of this controversy is the competition for the use of these areas. Some see wetlands only as areas to be converted to dry land so that they can be used for development or for agriculture. Others see them as vital ecosystems that if eradicated or impaired seriously jeopardize our quality of life. The responsibility for oversight on these unique areas lies largely in the hands of state and federal regulators (see Section 404 of the federal Clean Water Act). At the federal level the U.S. Army Corps of Engineers and the U.S. Environmental Protection Agency (USEPA) have jurisdiction. In fiscal year 1994, over 48,000 applications were made to the Army Corps for a Section 404 permit. Of these, only 358 (0.7%) of the permits were denied. In addition to these approved permits, there are nearly 50,000 other activities permitted in wetlands that do not require notification to the corps at all. While jurisdiction over wetlands lies principally in the hands of federal and state regulators, municipalities should not abrogate all responsibility to these valuable resources. Absentee regulators, while successful in delineating the areal extent of wetlands, may not be as successful in relating the significance of those same wetlands to the community at large. This is the job of the municipal land planner and the appropriate venue is in the MMP.

9.3.1.3 Wildlife

In spite of the fact that 141 million Americans participate in wildlife-related activities, the nation's wildlife resources continue to be assigned an extremely low, nearly nonexistent value in the land planning process, especially at the local level. The number of projects denied nationally by local planning boards due to adverse impacts on wildlife or its habitat could probably be counted on two hands or less. Yet municipal ordinances calling for an assessment of the impact on wildlife as a result of man's "land improvements" are abundant. Why then does wildlife continue to receive such feeble recognition in the land planning process? The answer is complex. One of the major reasons, however, is that many land planners (and many Americans as well) are unfamiliar with the life cycles or habitat requirements of wildlife in their immediate geographic area. As a consequence, they must rely heavily on documentation provided by "experts" as required by their ordinances.

Unfortunately, many of these required assessments, usually in the form of environmental impact statements (EISs) leave much to be desired. They are not only often poorly written but also often lack both substance and accuracy. The following excerpts from six EISs, submitted as part of some land subdivision projects, give you some idea of the quality of information being provided to municipal land planners:

Example 1 — "The existing wildlife on this site seems to be of a visiting nature, rather than that of actual inhabitance. Small animals such as rabbits or birds such as crows would visit the site to satisfy their eating habits only. Because of the barrenness of the land, inhabitance is unlikely."

Example 2 — "An area now serving as habitat for many wild animals and birds will no longer be available for this purpose. It is expected that many animals now dwelling at the site will seek new homes."

Example 3 — "The species threatened most seriously are those that are most conspicuous and those that are sensitive to changes in land use. Hawks and some owls may be harassed, but will be affected most by the conversion of cultivated and abandoned fields to home sites, roads, lawns, and parks with a resultant reduction in rodent population."

Example 4 — "The proposed construction on this site will force most of the wildlife to leave the immediate area, although the proposed use should allow most of the wildlife to live adjacent to this site in the proposed County Park."

Example 5 — "There will be a substantial reduction in wildlife visiting the area."

Example 6 — "No known wildlife systems will be destroyed other than rodents."

Your attention is directed, in particular, to Examples 2 and 4, which indicate that wildlife is to somehow find new habitat in adjoining land areas — a convenient assumption, but one that is more fiction than fact. Wildlife moves when forced too, but whether it can survive that displacement is dependent on two variables — *availability* and *suitability* of the new habitat. Rarely are these two items adequately addressed in contemporary environmental impact statements. In some parts of America, our wildlife has simply run out of places to move; yet this *displacement myth* persists and few land planners have risen to challenge its validity.

America was not particularly kind to its wildlife resources as a developing nation; however, I do not mean to imply that the nation was totally indifferent to them. Some of the nation's earliest laws pertaining to wildlife even predate the establishment of an American nation. In 1667, for example, Connecticut prohibited the export of game across its borders; in 1738, Virginia banned the harvest of doe deer. One of the most significant laws, however, was passed by Rhode Island in 1846 when that state established the first seasonal regulations to protect waterfowl from spring shooting. The concept of daily bag limits was established by Iowa in 1878, and by 1900, 12 additional states had set up similar regulations. That same year Congress passed the Lacey Act, which outlawed market hunting; and by 1910, 33 states were obtaining revenue for wildlife restoration and protection from the sale of hunting licenses.

The preponderance of hunting-related regulations among these early American wildlife laws reflects the dominant uses Americans have traditionally made of their wildlife resources, namely, sustenance and sport hunting. Today these two uses are under intense criticism, especially from a number of groups that have formed within the past decade and a half and that strongly oppose the hunting or killing of any wildlife. Obviously this rigid opposition to harvesting wildlife is an irritation to America's 50 million plus hunters, fishermen, and trappers who for over 50 years have contributed enormous sums of money (some $3 billion since the 1920s) to assist in the restoration and management of American wildlife. While these contributions have admittedly been self-serving, all wildlife species — game and nongame alike — have been beneficiaries.

The contention by some of these antihunting groups that legalized sport hunting (Figure 9.19) and fishing are wildlife's greatest threat is an irksome allegation, even to environmental planners. *I can assure you that legitimate sport hunting and fishing will not result in the demise of America's wildlife.* Unfortunately, such a groundless accusation not only lessens the chances for a much needed collaboration of all groups concerned with the welfare of the nation's wildlife but also

FIGURE 9.19 Sport hunting has been and will continue to be a legitimate method for controlling and managing various populations of wildlife. (Courtesy New Jersey Department of Environmental Protection, Division of Fish, Game, and Wildlife.)

FIGURE 9.20 This suburban-dwelling raccoon has lost much of its fear of humans and boldly shares dinner with this trio of domestic cats. These mammals may harbor the deadly rabies virus and therefore pose a serious threat to these cats and their owner.

(worst of all) diverts the focus of the nation away from the real threat to wildlife survival — *habitat loss*.

Man and wildlife have been and continue to be competitors for the same habitat — a contest in which wildlife has rarely, if ever, been the victor. When woodlands and trees are cut and removed, some existing and potential wildlife nesting sites and food sources are likewise removed. The filling of wetlands and the placement of concrete and asphalt pavement and buildings results in similar losses. Even agriculture contributes to these losses when brushy hedgerows between fields that serve as corridors for the passage of wildlife and nesting and resting areas are plowed under.

Not all of America's wildlife suffered as a result of habitat loss or change. Mammals such as the white-tailed deer (*Odocoileus virginianus*), the raccoon (*Procyon lotor*), and the western coyote (*Canis latrans*), have adapted remarkably well to life in American suburbs and cities. Such adaptations, however, can pose special dangers to humans because familiarity encourages the illusion that such animals have been partially domesticated.

A large population of urban- and suburban-dwelling raccoons now survives quite well in all parts of the United States, thanks to an abundance of garbage cans, bird feeders, human handouts (Figure 9.20), and thousands of miles of underground highways called storm drains. In California, the western coyote has adapted so well to the urban lifestyle that it can be found boldly walking city streets in broad daylight. In this instance, however, the coyotes are supplementing their normal urbanized fare with pet cats and dogs — a trait that has alienated a once tolerant human population.

FIGURE 9.21 A few hours after this picture was taken, this young deer finally succumbed to the ravages of starvation. Some well-meaning citizen groups had pressured refuge managers to ignore the warning signs that the ecosystem in this particular case was under great stress due to an overpopulation of these large herbivores. (Courtesy of Doug Roscoe, New Jersey Division of Fish, Game, and Wildlife.)

Out of control white-tailed deer populations in some sections of the Northeast are acting as hosts and carriers for the notorious deer tick carrier of the devastating Lyme disease.*

Americans are fortunate in that a significant number of wildlife refuges have already been set aside across the nation. These actions, however, have fostered the additional myth that wildlife can be stockpiled forever in such areas. *They cannot.* Concentrating wildlife into such confined habitats (literally oases in some areas of dense human development) can be lethal to wildlife. When these areas exceed their carrying capacity, death can result from starvation (Figure 9.21), stress, or disease (Figure 9.22). Wildlife needs areas in which it may move freely to breed, nurse, and hide its young; to rest undisturbed, and to feed. Despite an abundance of provisions for the consideration of wildlife in local land use regulations, few, if any, planning boards ever deny subdivision or site plan approval over concerns for the needs of wildlife. The now common practice of requiring donations of land from developers for open space preservation is no guarantee that such areas are suitable for the needs of wildlife. In many instances such parcels are often discontinuous, *isolated islands*** of habitat surrounded by human development. This characteristic *fragmentation* of the landscape is especially devastating to species whose life cycle is dependent on the presence of large, unbroken tracts of forested habitat. These species, sometimes referred to as *specialists*, may disappear

* Another out of control wildlife population, Canadian geese (*Branta canadensis*) may be more than just a nuisance to municipalities. They also may pose some serious health threats. Research has provided clear evidence that these birds can act as mechanical carriers of infectious oocysts of cryptosporidium parvum, disseminating these organisms to surface water bodies, including drinking water supplies, through their contaminated feces.[28]

** Habitat island size has become an important consideration in the allocation of habitat for the retention of wildlife.[29] Much of that is based on prior studies of oceanic islands and archipelagos by MacArthur and Wilson (MacArthur, R. H., and Wilson, E. O., *The Theory of Biogeography*, Princeton University Press, Princeton, N.J., 1967) who developed a

CHAPTER 9 FIGURE 13 Springs and seeps such as the ones shown here are vital components of aquatic ecosystems and need to be protected from intrusion, especially from manmade discharges of storm water runoff.

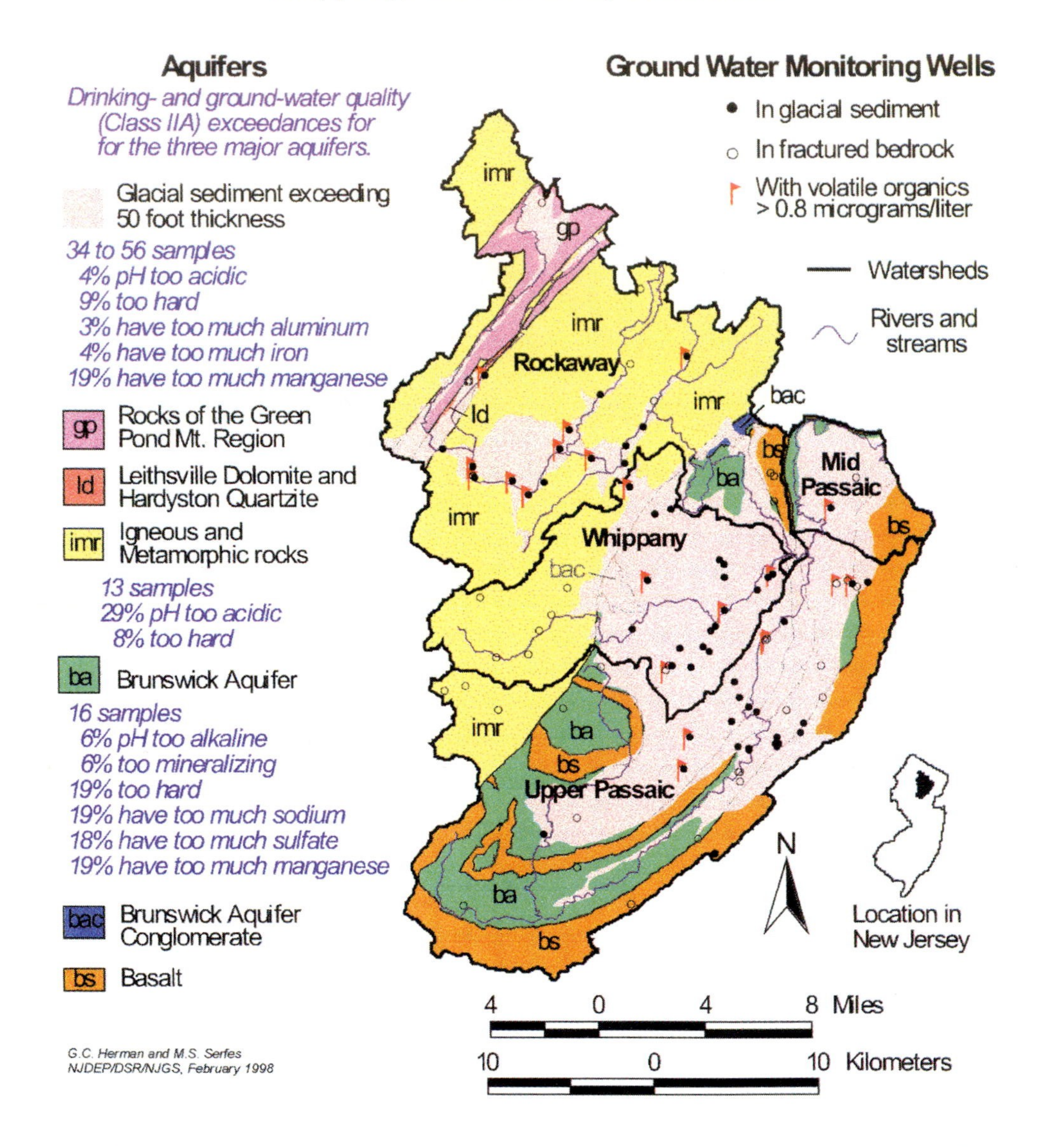

CHAPTER 9 FIGURE 18 A community should be compiling information about its ground water quality well before it begins to approve individual projects that may use that ground water. (From New Jersey Department of Environmental Protection, New Jersey Geological Survey.)

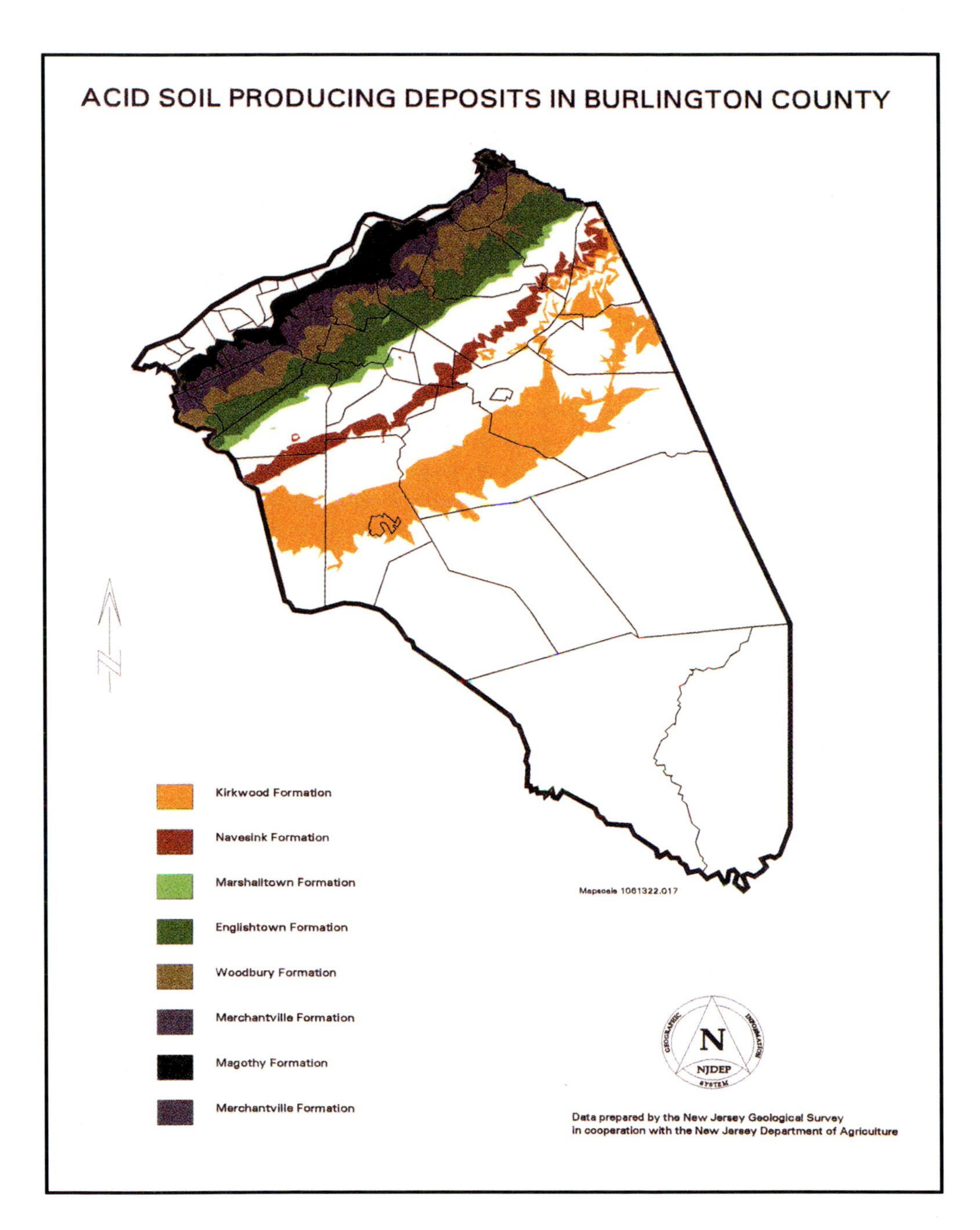

CHAPTER 9 FIGURE 28 Exposing acid-producing soils may thwart the ability of homeowners to establish lawns and other vegetation. In addition such soils may threaten nearby surface water bodies and their instream biota. (From New Jersey Department of Environmental Protection, New Jersey Geological Survey.)

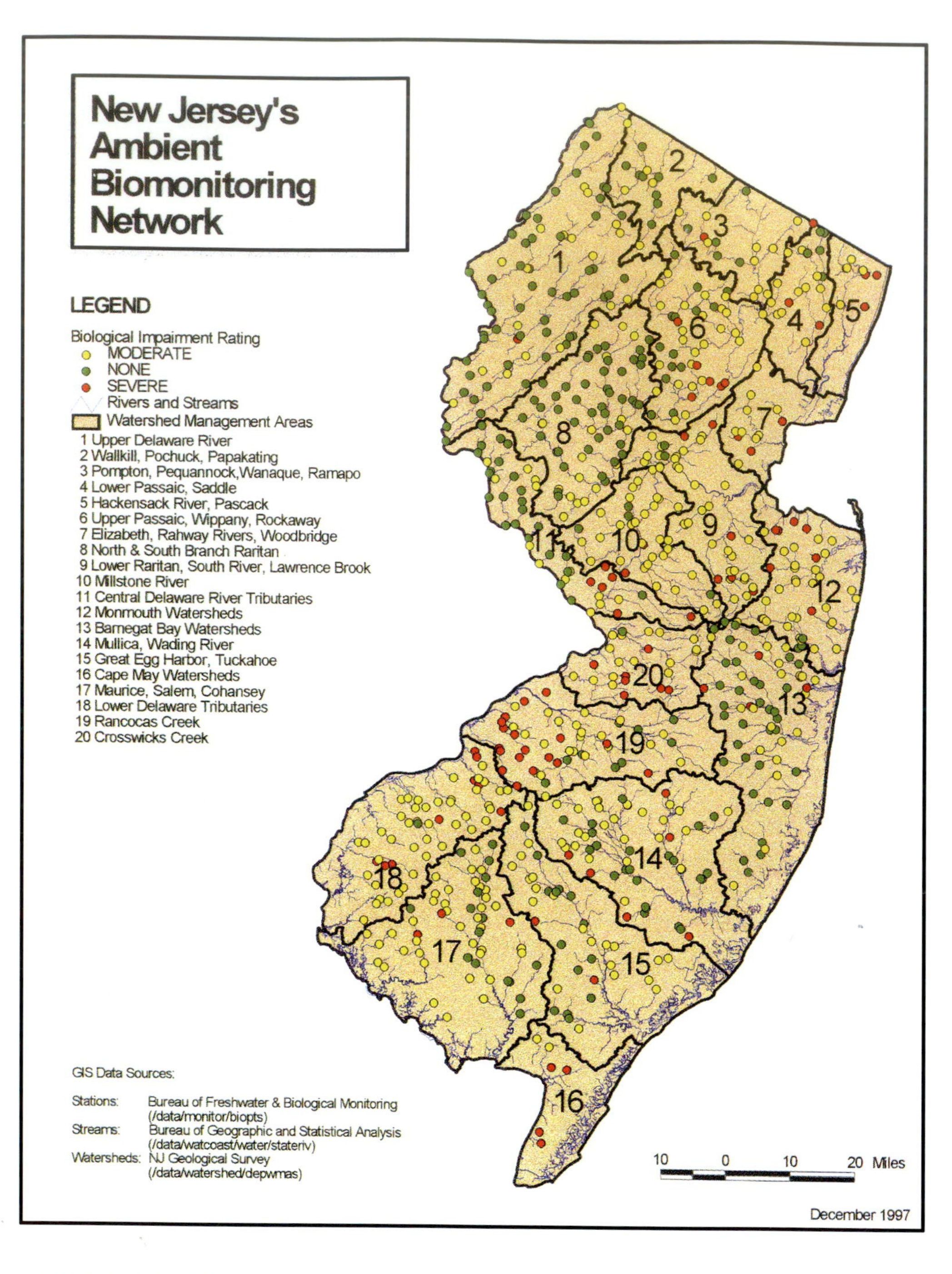

CHAPTER 10 FIGURE 3 New Jersey's Ambient Biomonitoring Network. (From New Jersey Department of Environmental Protection, Bureau of Freshwater and Biological Mont.)

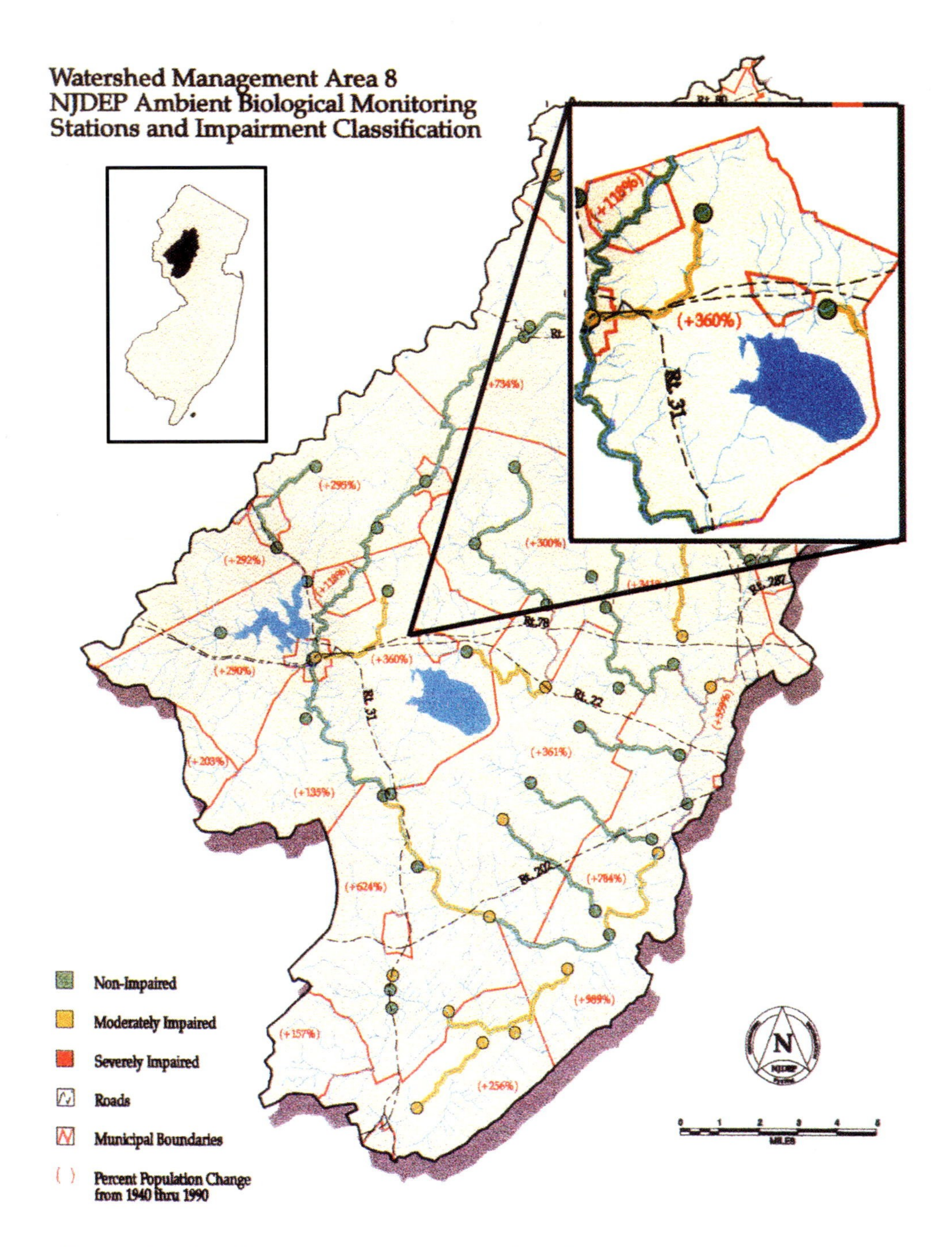

CHAPTER 10 FIGURE 4 Watershed Management Area 8's ambient biological monitoring stations and impairment classification. (From New Jersey Department of Environmental Protection, Bureau of Freshwater and Biological Mont.)

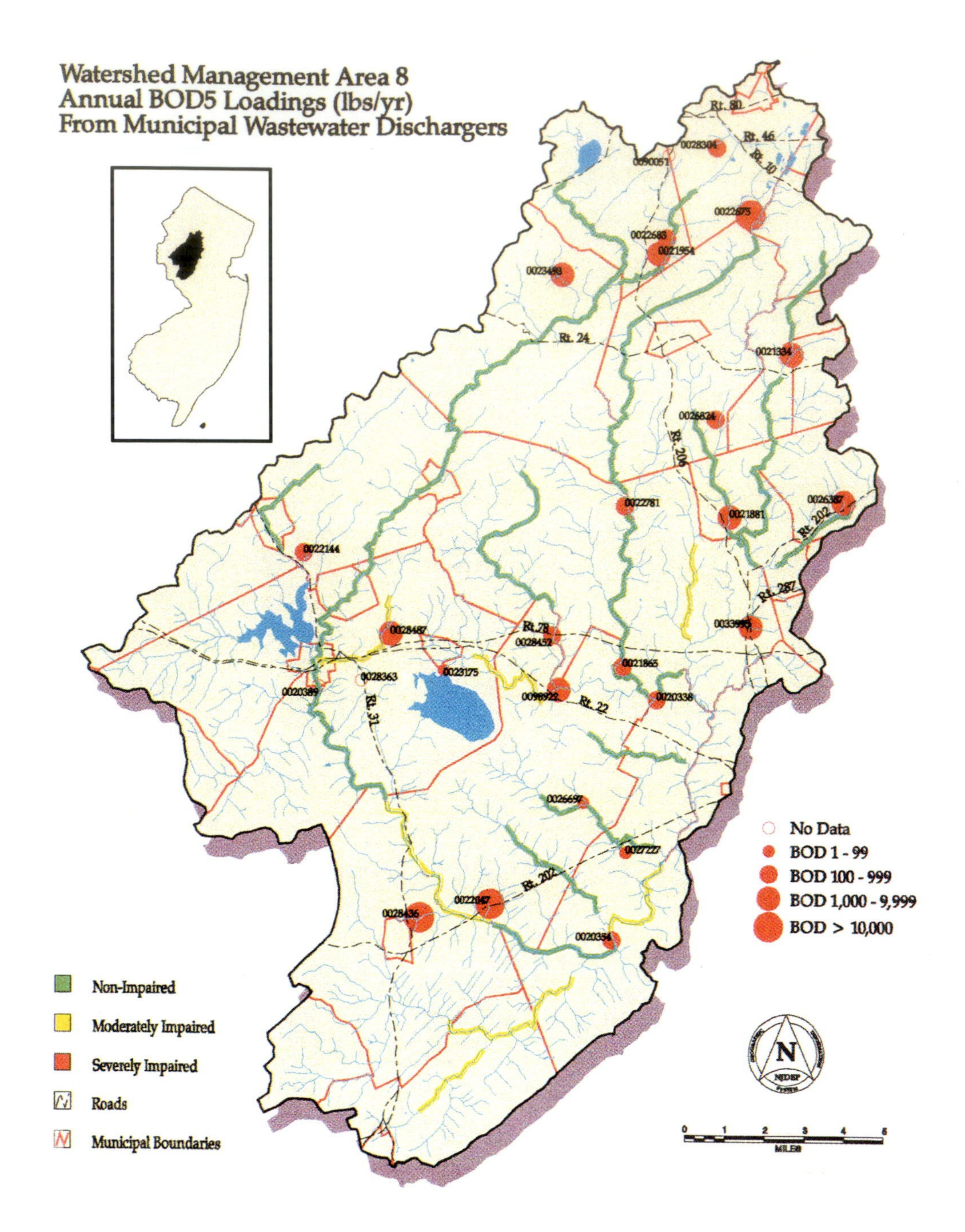

CHAPTER 10 FIGURE 5 Watershed Management Area 8's annual BOD5 loadings (lbs/yr) from municipal wastewater dischargers. (From New Jersey Department of Environmental Protection, Bureau of Freshwater and Biological Mont.)

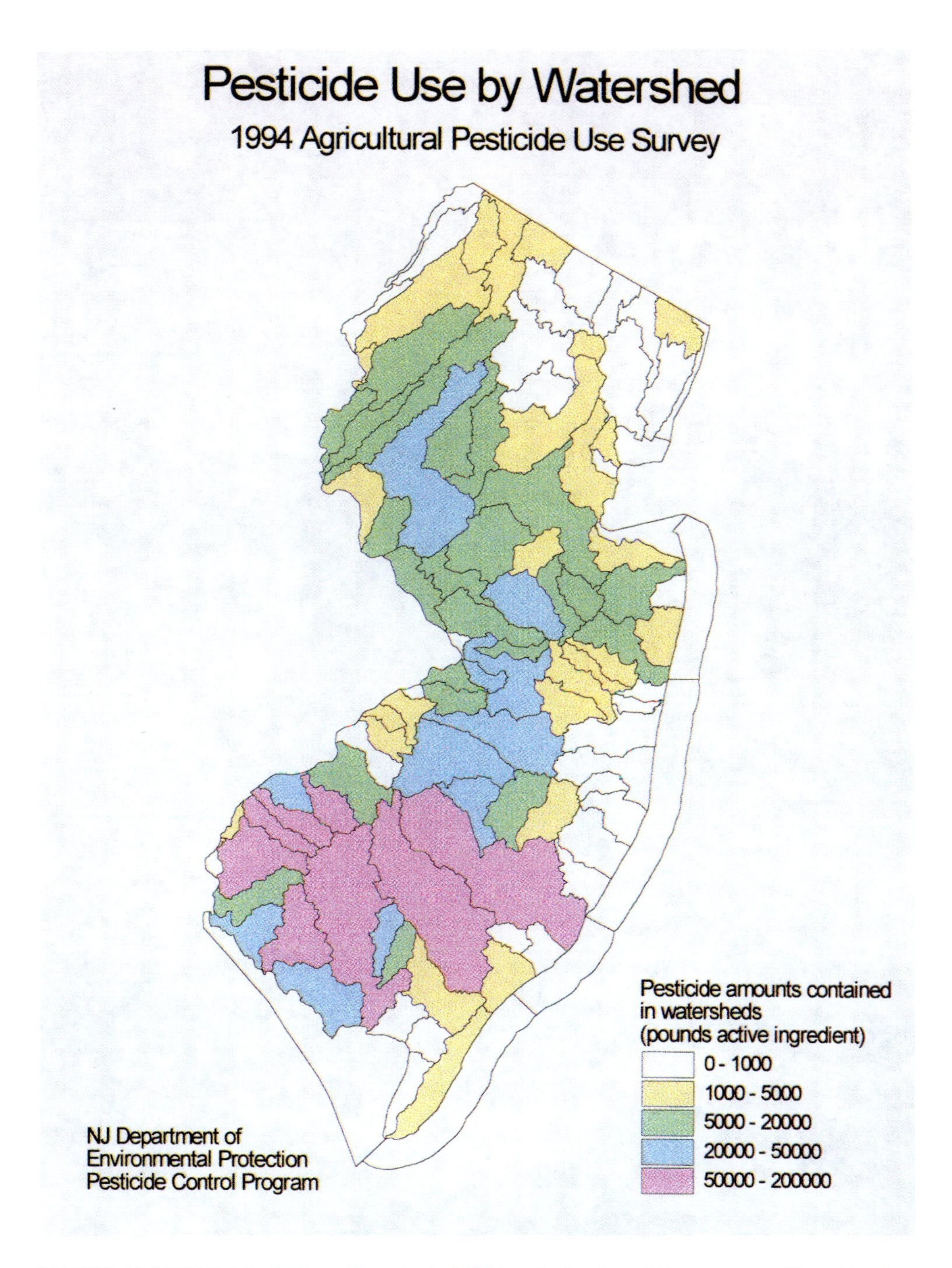

CHAPTER 10 FIGURE 6 Pesticide use by watershed. 1994 agricultural pesticide use survey. (From New Jersey Department of Environmental Protection, Pesticide Control Program.)

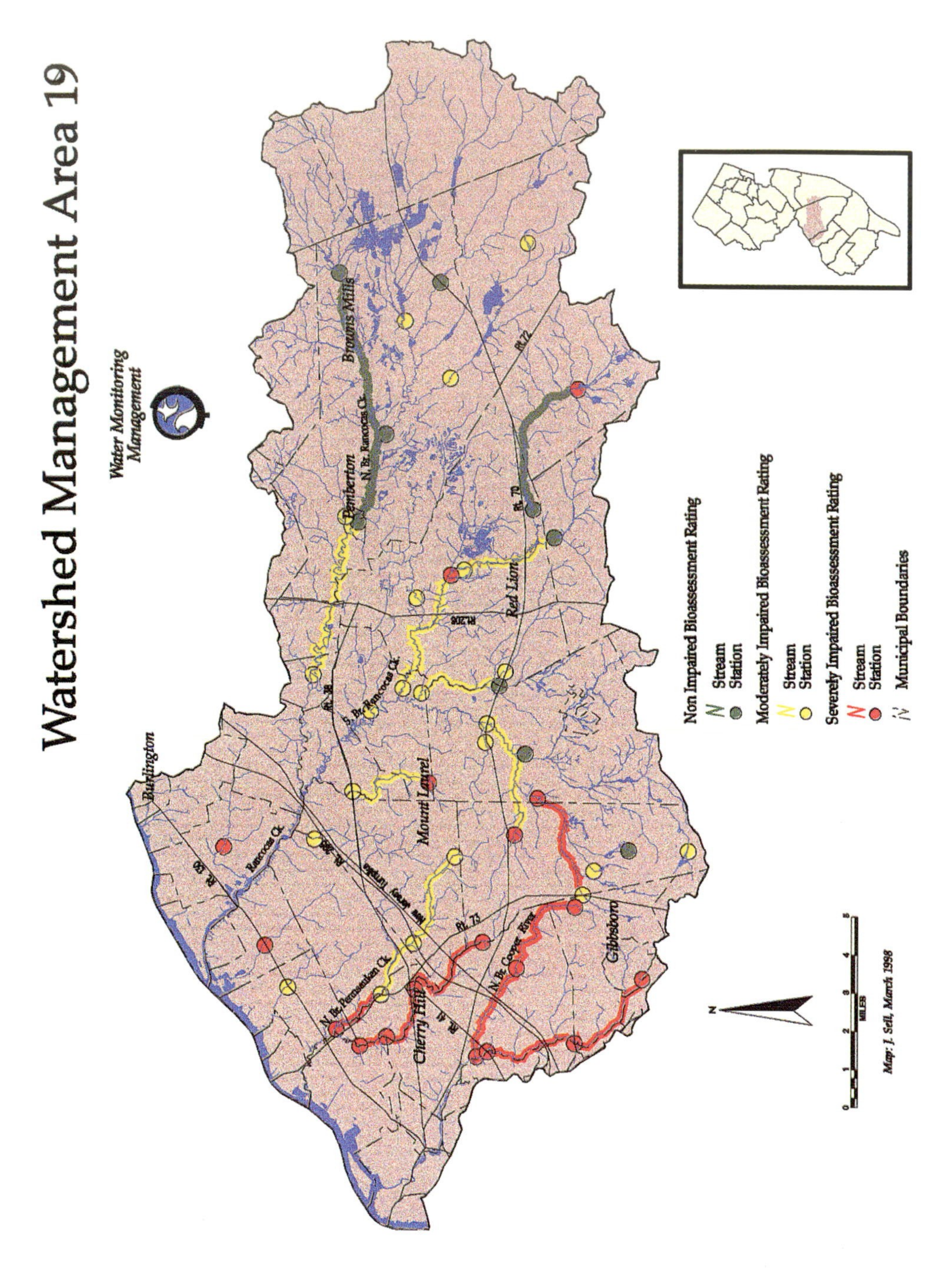

CHAPTER 10 FIGURE 12 Biological impairment in a watershed as measured by the benthic macroinvertebrates. (From Bureau of Freshwater and Biological Mont., New Jersey Department of Environmental Protection.)

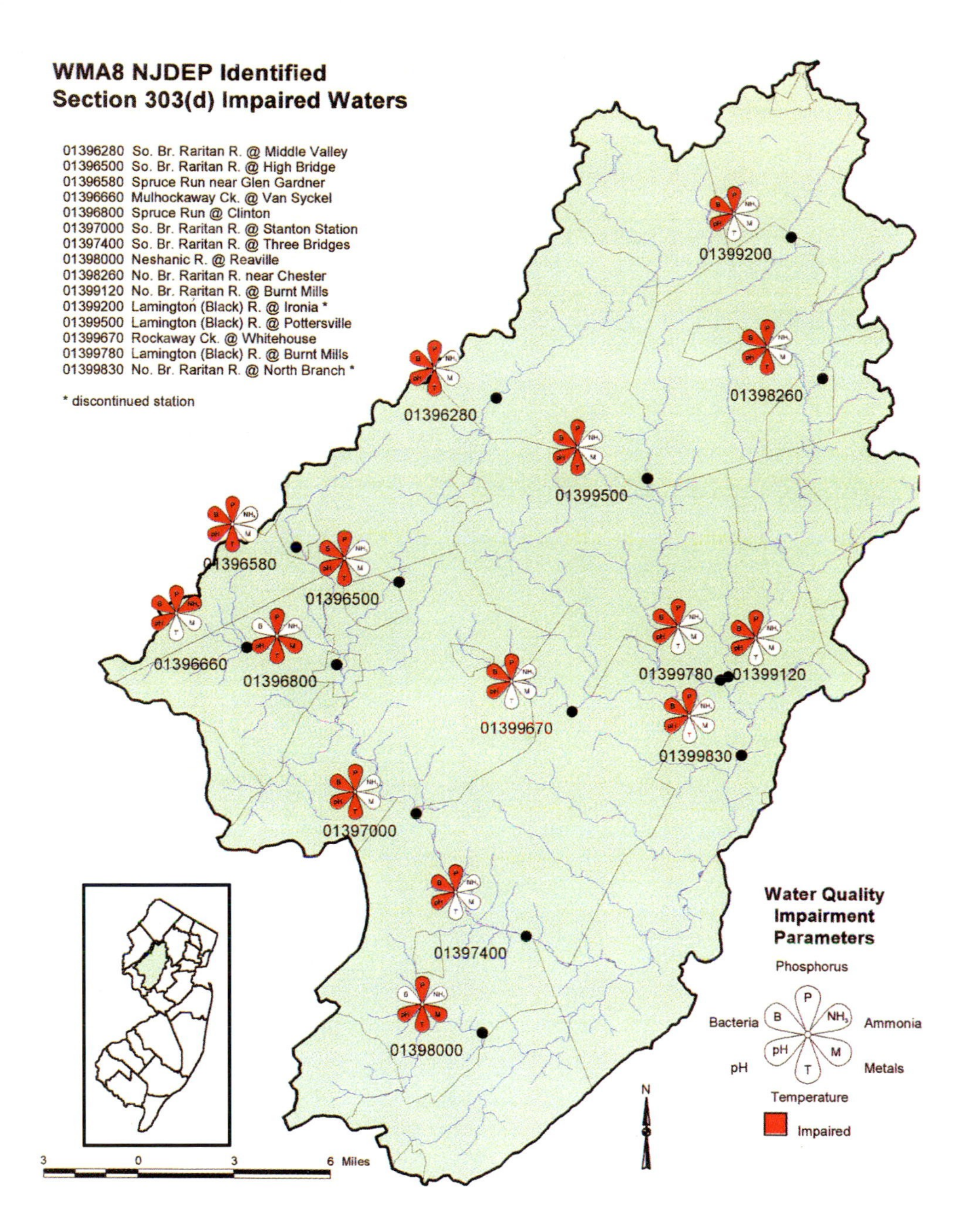

CHAPTER 10 FIGURE 22 Due to the intimate link between land use and water quality, existing water quality impairments of local streams should be clearly identified in the municipal master plan. Such data can significantly bolster community arguments for wide buffer zones or enhanced wastewater treatment. (From New Jersey Department of Environmental Protection, Bureau of Freshwater and Biological Mont.)

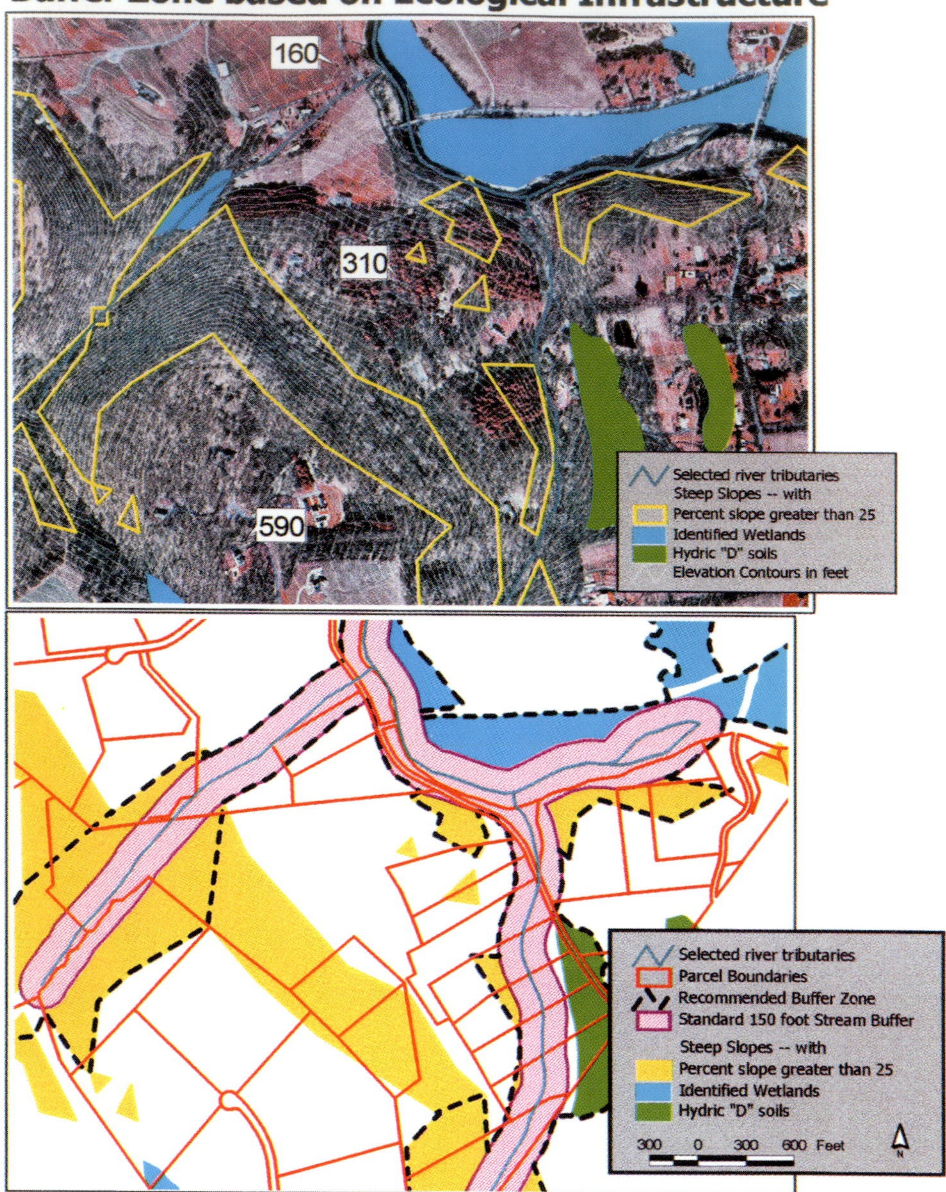

CHAPTER 10 FIGURE 30 The top figure shows the ecological data layers collected from a typical GIS databank. The bottom figure illustrates how a stream buffer boundary line is then established based on that GIS data. Note the variability in widths. (Graphics by Integrated Spatial Solutions, Moorestown, NJ.)

CHAPTER 10 FIGURE 43 Typical wellhead protection map showing the time of travel in years of ground water flowing from upgradient areas of the watershed. (From New Jersey Department of Environmental Protection, New Jersey Geological Survey.)

Groundwater Recharge Rates for a Selected Parcel

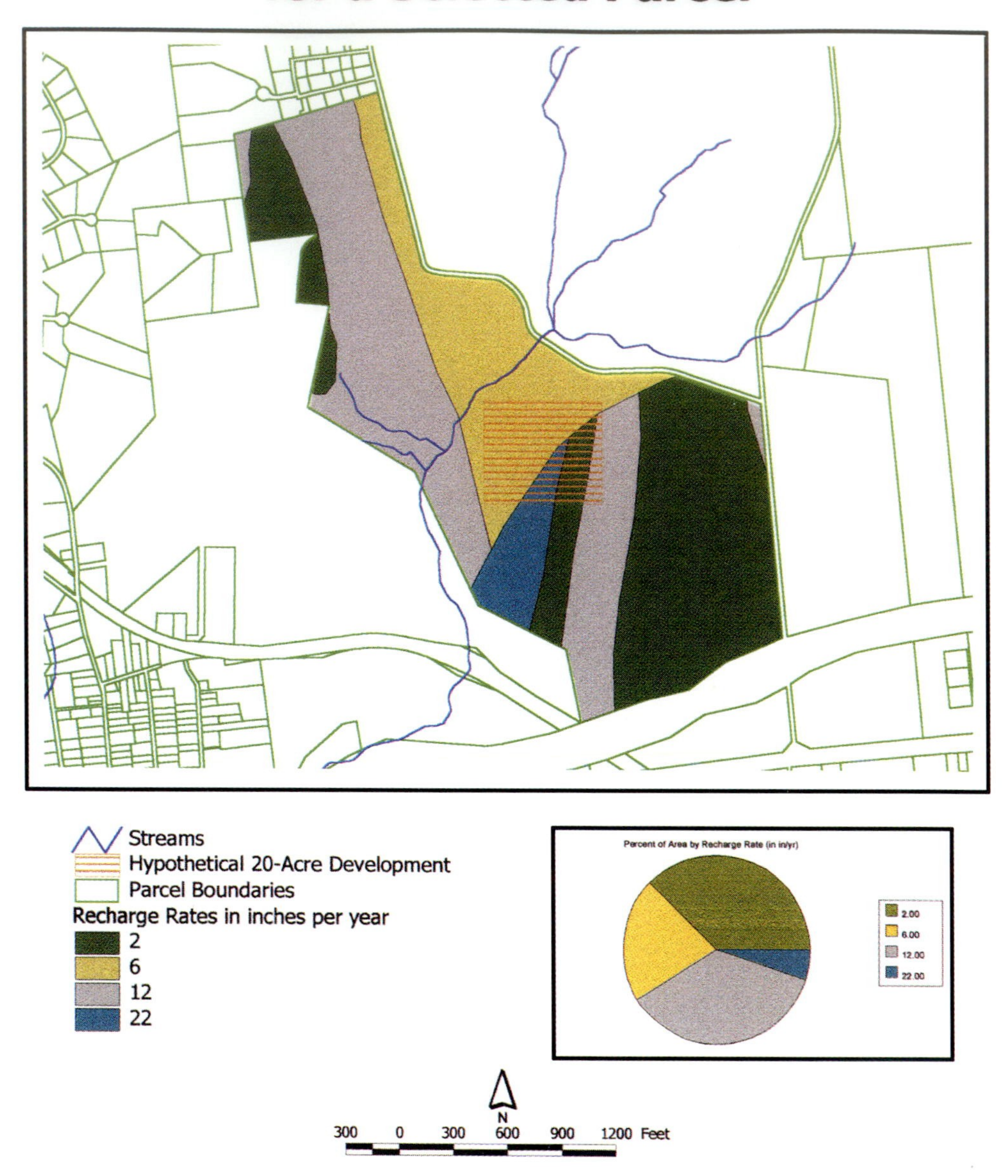

CHAPTER 10 FIGURE 44 Areas of the community with the highest ground water recharge potential need special protection and recognition in the municipal master plan so that the individual development projects can be configured so as not to impair the critical function of these particular landscape features. (Graphics by Integrated Spatial Solutions, Moorestown, NJ.)

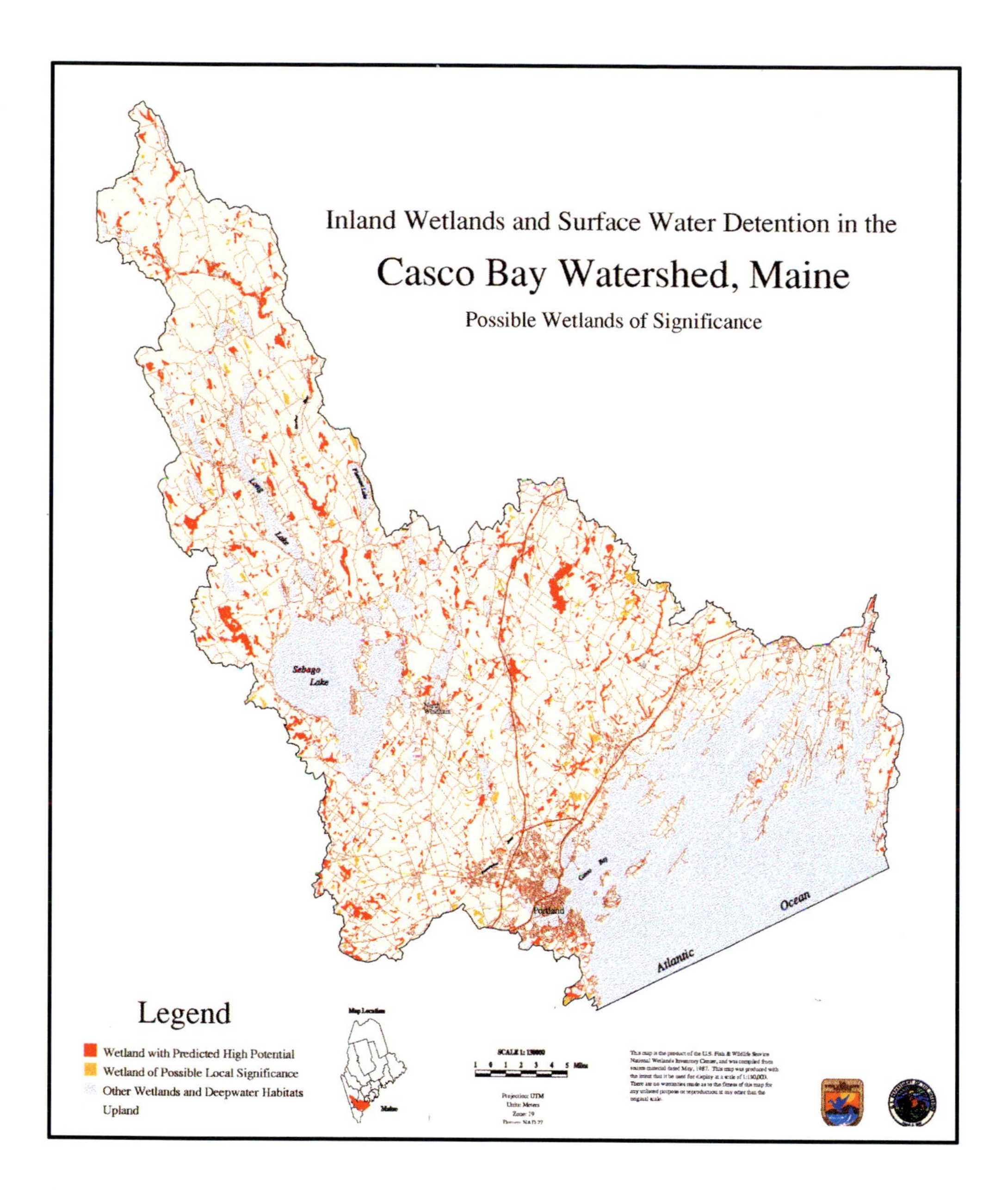

CHAPTER 10 FIGURE 49 Inland wetlands and surface water detention in the Casco Bay watershed, Maine. (From Tiner, R. Schaller, R.S., Peterson, D., Snider, K., Ruhlman, K., and Swords, J., Wetland characterization study and preliminary assessment of wetland functions for the Casco Bay watershed, southern Maine, U.S. Fish and Wildlife Service, Northeast region, National Wetland Inventory, Ecological Services, Hadley, MA, NWI Report, 1999.)

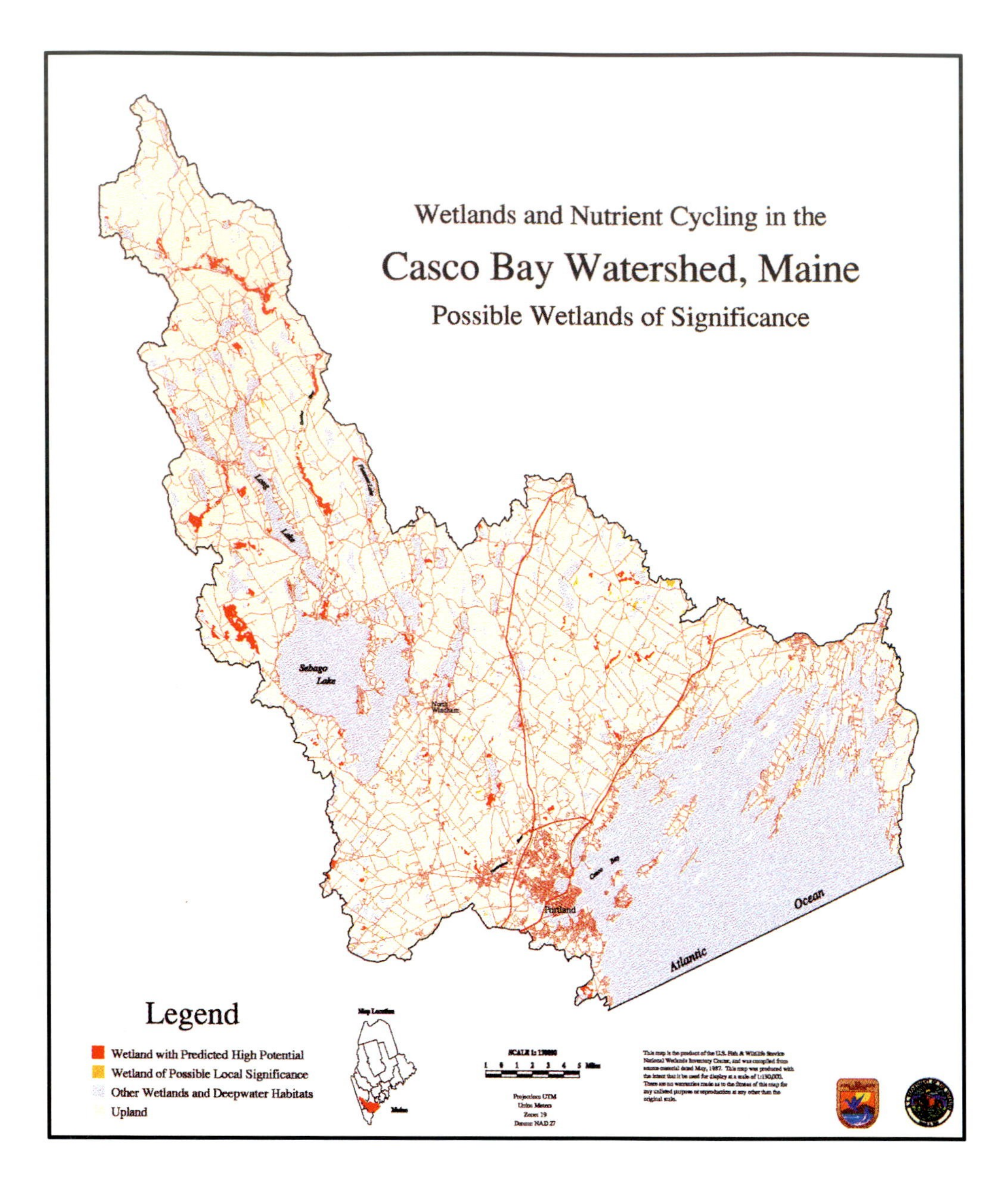

CHAPTER 10 FIGURE 50 Wetlands and nutrient cycling in the Casco Bay watershed, Maine. (From Tiner, R. Schaller, R.S., Peterson, D., Snider, K., Ruhlman, K., and Swords, J., Wetland characterization study and preliminary assessment of wetland functions for the Casco Bay watershed, southern Maine, U.S. Fish and Wildlife Service, Northeast region, National Wetland Inventory, Ecological Services, Hadley, MA, NWI Report, 1999.)

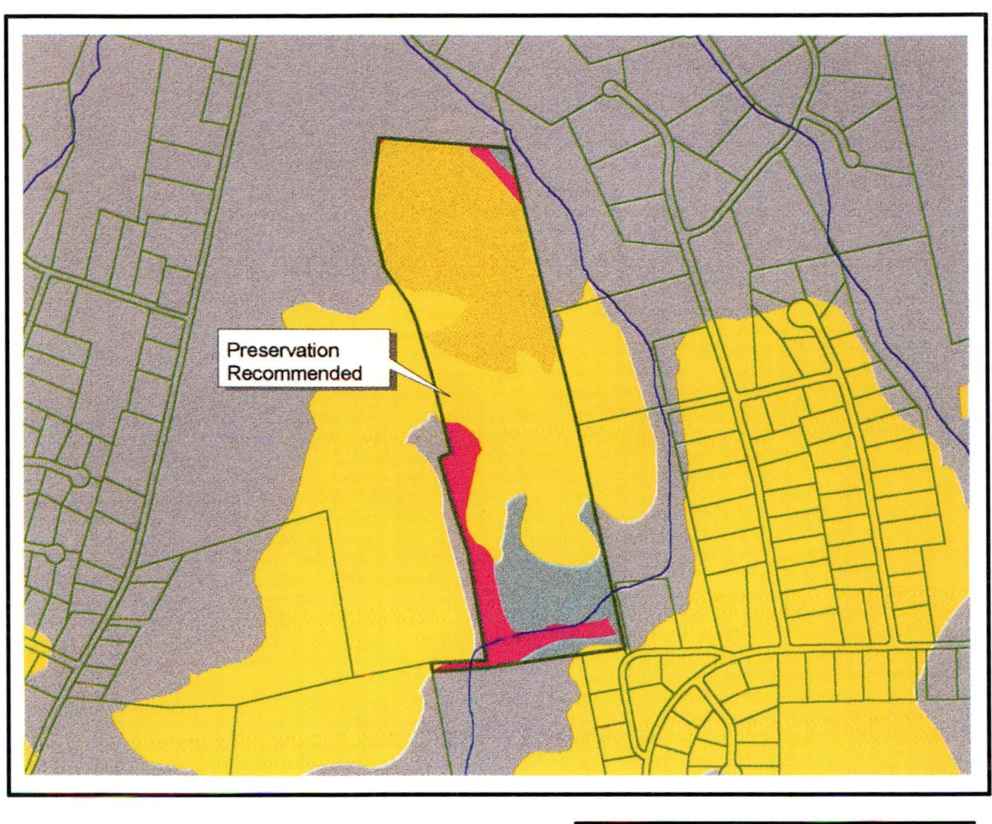

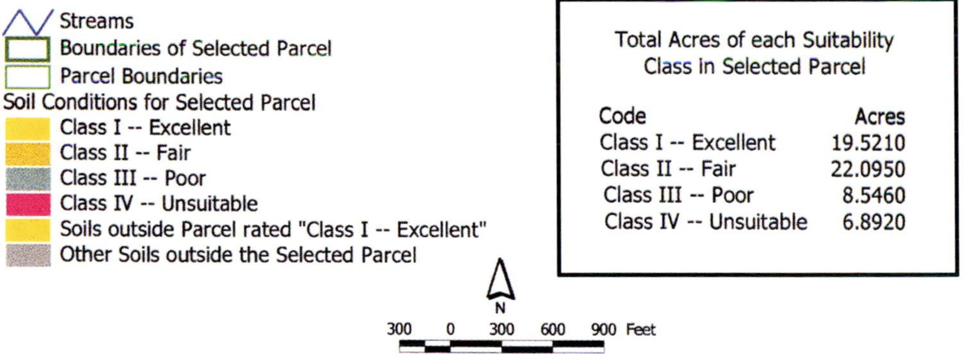

Total Acres of each Suitability Class in Selected Parcel

Code	Acres
Class I -- Excellent	19.5210
Class II -- Fair	22.0950
Class III -- Poor	8.5460
Class IV -- Unsuitable	6.8920

CHAPTER 10 FIGURE 51 Soil suitability for agricultural operations. The developer of this particular parcel has been put on notice that the community will no longer accept the routine burial of it best agricultural soils under what will probably be tons of asphalt and concrete. (Graphics by Integrated Spatial Solutions, Moorestown, NJ.)

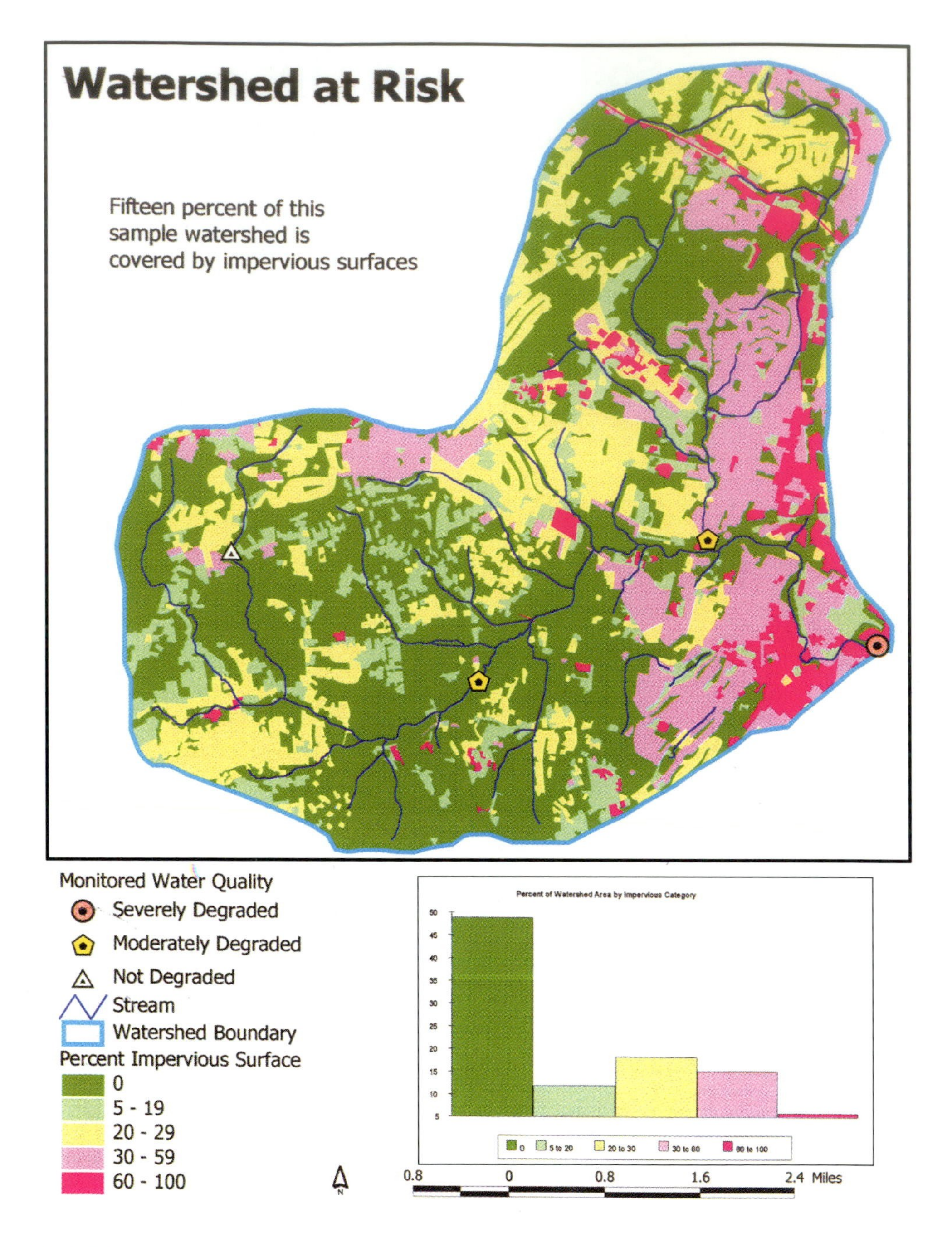

CHAPTER 10 FIGURE 52 Watershed at risk. This watershed may very well be on the brink of irreversible damage to the quality of its water resources. With this type of information now included in an ecologically based master plan, the community has the opportunity to reverse this all too common and disastrous trend. (Graphics by Integrated Spatial Solutions, Moorestown, NJ.)

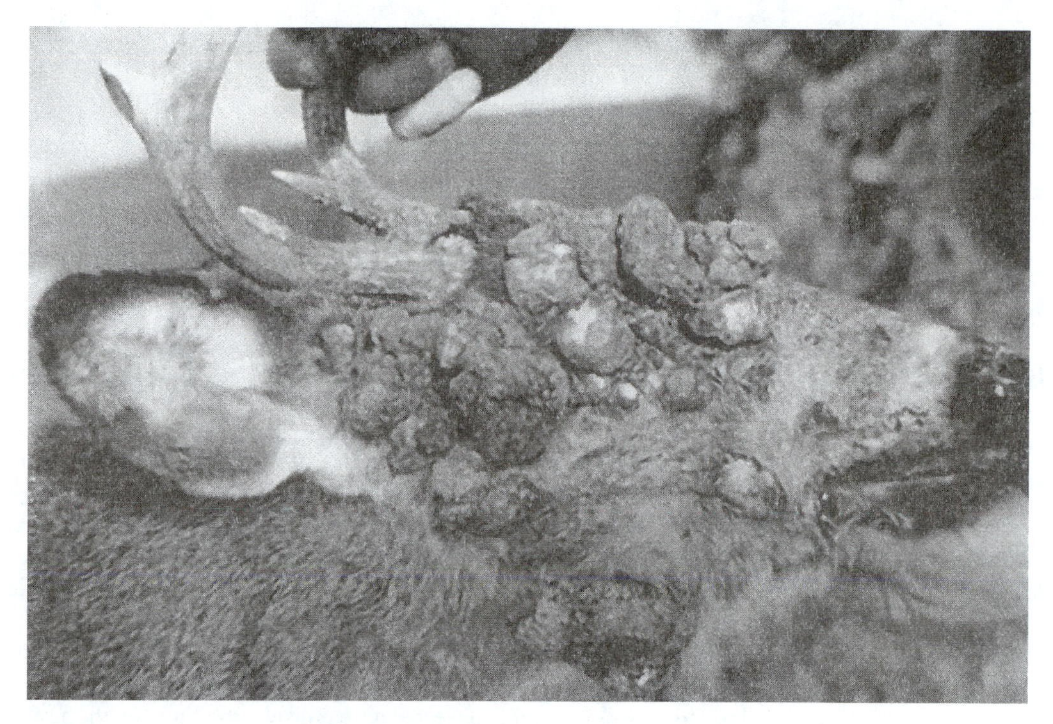

FIGURE 9.22 Uncontrolled and crowded populations of wildlife forced to share smaller and smaller habitats often exhibit debilitating diseases such as this *papilloma* shown here on this white-tailed buck deer. (Courtesy New Jersey Division of Fish, Game, and Wildlife.)

altogether, to be replaced by more adaptable *generalists species*. For example, Aldrich and Coffin[30] compared the breeding bird use of a 95-acre, mature eastern deciduous forest tract in Fairfax County, Virginia in 1942 with bird use of the same tract in 1979, after it had become a well-established residential community. Red-eyed vireos, ovenbirds, and scarlet tanagers were abundant in 1942 but not found in 1979. Other typical forest birds present in 1942 but lacking in 1979 were the Acadian flycatcher, eastern wood pewee, yellow-throated vireo, worm-eating warbler, hooded warbler, and Louisiana waterthrush. Conversely, the gray catbird, American robin, and house sparrow all absent in 1942, were numerous in 1979. Also more numerous in 1979 were blue jays, mockingbirds, starlings, cardinals, and song sparrows. While you may argue that birds, in general, have a distinct advantage over their terrestrially bound counterparts in that they can fly long distances to new habitat, this is no guarantee that they are any more successful in their escape flights to more distant habitats. Our part-time populations of neotropical birds are a classic example. Now when they fly south for their seasonal sojourn to the tropics, they are finding their formerly lush habitat similarly clear-cut, fragmented, and also scorched by land-clearing fires.

Habitat fragmentation can have consequences even more dire for the nation's many species of wildlife — potentially threatening them with extinction. Wildlife populations isolated on habitat islands and restricted in their ability to move freely throughout their former range, encounter fewer and fewer mating partners and thus are subject to a shrinking gene pool, creating what O'Brien[31] calls a "genetic bottleneck." As a result, inbreeding between adults and offspring and between siblings becomes more

theoretical model to explain the number of species inhabiting an island based on a dynamic equilibrium between immigration rates and extinction rates that are, in turn, influenced by island size and isolation among islands and the mainland. In general, they predicted that immigration rates would increase and extinction rates would decrease on larger less isolated islands, resulting in a higher number of species as compared with the number on smaller, more distant islands. This theory has since been applied to "terrestrial islands," out of which has come an important recurring observation — that habitat size or area is a major factor accounting for differences in species richness (total number of species).

FIGURE 9.23 Although pretty to look at, this piebald albino white-tailed deer may in fact be an environmental indicator signaling the presence of dangerous inbreeding.

frequent. Recessive genes, normally masked by a healthy genetic diversity, begin to dominate (Figure 9.23); the results are congenital defects (both physical and reproductive), rising infertility, and weakening of each animal's immune defense system. This latter characteristic is of particular concern, because an entire population with similar genetics could become the victim of a single fatal epidemic.* Genetic bottlenecks also may come about as a result of the efforts of some well-intentioned special interest groups. I am referring to efforts currently underway, particularly in the eastern United States, to control overpopulations of white-tailed deer; this overpopulation is largely the result of our prior failure to adequately address the needs of wildlife under our current land use planning practices.

Some groups strongly opposed to sport hunting have been zealously lobbying for the use of immunocontraception,** in lieu of sport hunting, to suppress the reproductivity rate of these prolific large mammals; this, if successfully applied, would create a series of innoculant induced genetic bottlenecks for this species throughout the eastern part of the United States. At first glance such a purported quick fix solution is admittedly very appealing and certainly not unexpected from a society whose citizens have been conditioned to believe that there is a "magic pill" solution for all of their ills, personal, ecological, and otherwise. That we would even contemplate the introduction of such innoculants into an environment (both biotic and abiotic) that is already seriously overburdened with the assimilation of a mind-boggling array of exotic man-made chemicals, including some potentially dangerous endocrine disruptors,*** is simply unconscionable. Thankfully, the impracticality of application at the spatial scale or scales needed has thus far thwarted its large-scale application. Undaunted, some of these groups are now advocating the use of abortion-inducing drugs (abortifacients), suggesting that pregnant does be treated roughly halfway through their normal gestation period, when fetuses are well formed. That time period coincides with one of the most stressful stages of the white-tailed deer annual life cycle — the postwinter season. Given the choice between having a woodlot or park full of aborted fetuses or a population innoculated with a substance potentially capable of inflicting permanent ovarian dysfunction**** on its unsuspecting

* Packer's[32] observations of an isolated lion population in the Ngorongoro Crater of Tanzania's Serengeti Plain is a particularly enlightening study in this regard.

** The principal innoculant would be porcine zonae pellucidae (PZP) glycoproteins, which on injection produce antibodies that prohibit sperm attachment and penetration of the ovum.

*** Measurable doses of human and animal medications and their metabolites have been showing up in surface water, groundwater, and even drinking water, including some natural estrogens that are commonly used in human birth control pills and hormone replacement therapies (Halling-Sorensen B., Nors Nielsen, S., Lanzky, P. F., Ingersiev, F., Holten Lutzheft, H. C., and Jergensen, S. E., Occurrence, fate, and effects of pharmaceutical substances in the environment — a review, *Chemosphere*, 36, No. 2, 357, 1998.

**** There are other data from prior PZP treatment of other species (Wood et al., 1981, rabbit [*Oryctolagus cunkulus*]; Mahi-Brown et al., 1985, dog [*Canis familiaris*]; and Dunbar et al., 1989, baboon [Papio sp.]) that show changes in ovarian morphology and function associated with immunocontraception; this may lead to long-term possibly irreversible infertility (Turner, J. W., Jr., Irwin, K. M. L., and Kirkpatrick, J. F., Remotely delivered immunocontraception in captive white-tailed deer, *J. Wildl. Manage.*, 56, No. 1, 154, 1992.

hosts, and deer harvested by one or two quick shots from a shotgun or rifle, I consider sport hunting as the most humane and ecologically sound method of control. Regretfully, there are some very specific instances where small, closely confined concentrations of these ungulates may need to be controlled by means other than sport hunting, due largely to the extreme proximity of human habitation or other human pressure. In these cases, immunocontraception, trapping and removal, and (perhaps as a last resort) injection with an immobilizing drug followed by euthanasia may be the only recourse.

While there continues to be sharp disagreement over which after-the-fact control method is most appropriate, the fact that this controversy exists at all should send all of us immediately scurrying back to the drawing board, or more appropriately, the planning board; there we should plot new strategies that put greater emphasis on preventative planning measures, so that this habitat fragmentation and subsequent isolation of wildlife populations, with all the appurtenant drawbacks of overpopulation and genetic bottlenecks, become more the exception than the rule.

9.3.1.4 Open Space

This is not a specific resource per se, but instead an aggregation of land within the community that has been set aside through deeds of dedication, fee simple purchases, or easements to remain in an undeveloped (excepting recreational uses) state. Included in this category would be state and federal parkland, forests, wildlife refuges, public hunting and fishing grounds, county and municipal parks, delineated wetlands, stream corridor easements, storm drainage easements (if appropriate), and preserved farmland on which all development rights have been sold. However, this last category is sometimes segregated from the other areas of open space because of a citizen preference to retain some agricultural uses in the community. In that event, these areas are to be treated differently in the MMP than if they were considered just open space. It is important to note here the growing national trend for all levels of government, especially local governments most recently, to set up open space and farmland acquisition funds using real estate tax levies of a few cents per $100 of assessed value. While commendable initiatives, localities need to prepare management plans for all these acquired parcels, even if the major use is only for passive recreation, such a hiking and birdwatching. Large mammal populations if present on site, such as the white-tailed deer, may need to be managed through such means as controlled hunting. Acquired farmland needs special attention, because in many instances the local support industries for agriculture have all but disappeared. The mills that used to receive and grind various grains have simply been shut down and dismantled. Suppliers of tractor parts have moved to distant locations, often in other states. Creameries and milk processors may no longer pick up milk at the farm because there are too few dairy farmers left. Given these conditions it would be foolish for a municipality to expect agriculture to be what it was when the community was primarily agrarian. If, despite this, the municipality still wants agriculture to continue on the property, it must set up a strong farm management plan that includes good soil erosion control practices such as grassed swales, buffers along streams, contour plowing, strip cropping, and crop rotation to help soil fertility. Again, wildlife populations need to be managed or crop plantings may be a futile effort.

9.3.1.5 Vegetation

I think it is fair to say that without the exploitation of the earth's vegetative resources, the human society we know today would probably be remarkably different. Perhaps, it might not exist at all.

When we were a society of nomadic hunters and gatherers, our prehistoric ancestors needed several square miles of territory for each individual to ensure an adequate supply of food. When territory was unlimited (i.e., absent competition from other humans), the system worked well. When the population of man increased, however, individual territories began to overlap, huntable food species declined, and the scattered communities of edible plants were overused. Fortunately for man, he acquired the knowledge to cultivate certain vegetation to mass-produce edible grains. He likewise learned to domesticate various wild herbivores that could be herded and maintained by the controlled grazing of vegetation. Such cultivation and domestication encouraged an economy of space and the assimilation of small scattered groups into larger enclaves, which eventually became settlements and villages. The sedentary nature of cereal grain cultivation and animal husbandry released a significant number of individuals from the process of food procurement, thereby allowing more leisure time to develop further organization and even the arts.

On the American continent, in more modern times, the vegetative resources made colonization possible. The first New England colonists simply could not have survived without American forests supplying them with logs for cabins; lumber for furniture, stockades, and fences; and fuel for heat. At that, almost one half of the American continent (some 1,065,000,000 acres) was covered by forest. Unfortunately, America's woodlands, in spite of their enormous utility to these early Americans, were often regarded as an impediment to "civilizing" the continent. As a result, these vegetative resources were severely exploited for the next three centuries. The enormous expanse of grasslands of the Great Plains met a similar fate when America's western frontier eventually pushed past the Mississippi River.

In spite of the fact that we sometimes perceive ourselves as an autonomous species, we remain inextricably bound to the earth's vegetative resources: from the cereal grain, fruits, and vegetables we consume daily to the cotton fibers in our clothing, the lumber in our homes, and the wooden poles that carry our electrical and telephone lines. Even some of our most notable medicines have been derived from vegetative sources. Curare, the powerful muscle relaxant, was first distilled from the Peruvian climbing liana vine, *Chondodendron tomentosum.* The malaria combatant, quinine, came from the bark of the *Cinchona ledgeriana* tree in Ecuador. Even more common plant species have yielded significant derivatives. For example, the common garden flower of Europe, foxglove, gave us the cardiostimulant, *digitalis.* The discovery and utilization of these naturally occurring alkaloids accelerated the development of modern medical techniques by decades, if not centuries. Even the air we breathe is dependent on the oxygen produced by the daytime respiration of plants. Scientists also are studying how "urban forests" can mitigate what is termed the *urban heat island effect* in our cities. This is the phenomenon that occurs when naturally vegetated surfaces are replaced with asphalt and concrete. The temperature of these artificial materials, when heated by the sun, can be 20 to 40 degrees higher than that of vegetated surfaces. These materials also store this heat and thus remain hot long after sunset, which in turn produces a heat dome over the city that averages 5 to 10 degrees higher than air temperatures over adjacent rural areas. Researchers have found that city parks and other urbanized areas that retain trees and grass were cooler than parking lots and areas with high concentrations of buildings. These vegetated areas were cooler because the vegetation was dissipating the solar energy by absorbing surrounding heat and using it to evaporate water from leaves, thus cooling the air. This temperature-buffering capacity of vegetation is of critical importance along stream corridors as well — a point demonstrated quite clearly by Maxted et al.[33] Maxted and colleagues examined two streams in Kent County, Delaware, one shaded (Figure 9.24) and one unshaded (Figure 9.25), and measured daily values on both streams for temperature, dissolved oxygen, and specific conductance (Figure 9.26). As you can see, the stream with the heavy forest canopy had significantly less variability in its physical and chemical conditions as compared with the unshaded stream, which not only had wide fluctuations in the parameters measured but also often had values exceeding threshold values that would be detrimental if not lethal to organisms living in the stream.

In biological terms, the human relationship with the plant kingdom largely has been predatory. Our vegetative resources, however, are renewable and thus capable of being used by successive generations of human consumers. In some parts of the world, particularly in the tropics and in what we often refer to as third world countries, man's predation on native vegetation, largely forested land, has accelerated proportionally with population increase. Unfortunately, the conservation practices that would ensure replenishment over time are rarely considered, let alone implemented.

FIGURE 9.24 Shaded stream. (Courtesy of John Maxted, Delaware Division of Water Research.)

FIGURE 9.25 Unshaded stream. (Courtesy of John Maxted, Delaware Division of Water Research.)

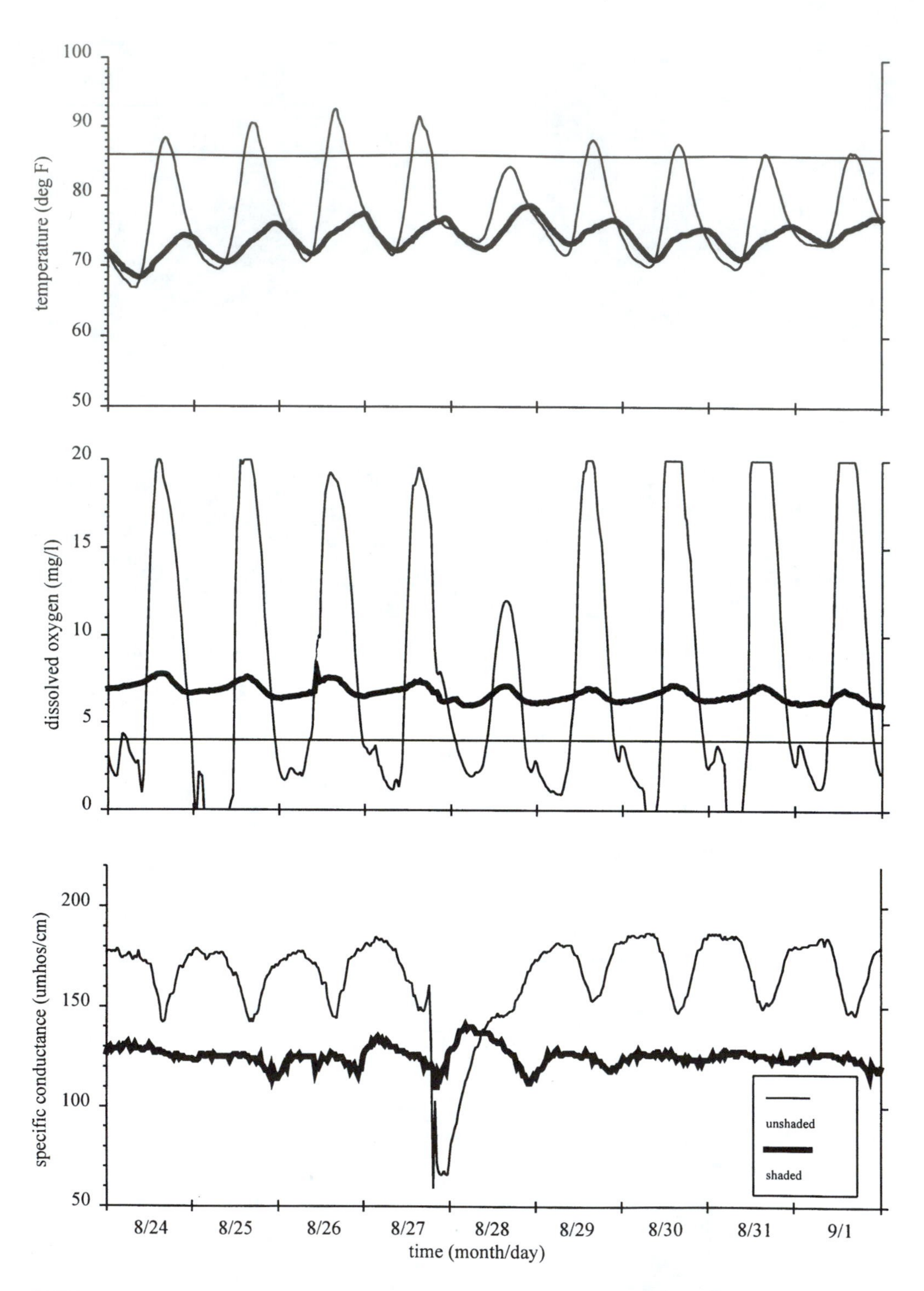

FIGURE 9.26 Temperature, dissolved oxygen, and specific conductance differences between an unshaded and shaded stream. Note the wide fluctuations in the unshaded stream. (Adapted from Maxted, J. R., Dickey, E. L., and Mitchell, G. M., The Water Quality Effects of Channelization in Coastal Plain Streams of Delaware, Division of Water Research, Watershed Asst. Section, Delaware Department of Natural Resources and Environmental Control, Dover, DE, 1995. With permission.)

We spoke previously of the benefits man has derived from the plant kingdom and of the plant-derived medicines that accelerated the treatment of human diseases by several decades or more. Many of those compounds came from tropical forests that now face wholesale destruction. We can only wonder what potential medicinal discovery will slip from our grasp completely unnoticed.

In your vegetative inventory you should not forget individual tree specimens or groups of trees, which because of their size, age rarity, or scarcity, would have some importance to the community. Residents of one small southern New Jersey town, for instance, are desperately trying to save a 300-year-old, 82-feet-high white oak; this tree has somehow managed to survive fire, wind, lightning, disease, and chainsaws, only to find itself in the path of a proposed corporate complex. Had this particular tree been specifically identified in the MMP, long before site plans were drafted, the property owners and/or developers might have been able to design their project around such a magnificent specimen or the municipality might have been able to purchase an acre around the tree to provide for its future protection. In any event, the appropriate time to raise the importance of such a unique specimen would have been during the formulation of the MMP. The community as a whole could have decided at that time if this one tree was worth saving, and if so, plot its location and establish a priority for its preservation in the MMP. As with most natural resources and certainly as we have seen in previous chapters, the cross-link between vegetation and other natural resources is common — such is the nature of ecosystems. For example, soils stripped of their vegetative cover, particularly forested cover, lose the benefits of extensive root growth that helps keep the topmost layer of soils permeable and capable of capturing and infiltrating rainfall, thereby curtailing excess runoff. Vegetation, again as a mature forest canopy, also provides nesting habitat for many threatened forest interior raptors (hawks, owls) and for other more common species such as the common crow (*Corvus brachyrhynchos*), which further utilize such mature stands of forest for communal roosts.

Other concerns might include vegetative communities not presently listed as threatened or endangered, but which, nonetheless, deserve some recognition and consideration. For example, a concern particularly dear to the heart of this author, and one that has both regional and local implications, is the plight of the American chestnut (*Castanea dentata*). This magnificent tree once dominated much of the forest canopy of the northeastern and southeastern U.S. forests. An imported blight devastated the species, beginning in the early 1920s, and now remnant populations are struggling to overcome the cruel fungal infection that allows growth to sapling or sometimes pole stage before it delivers a deadly coup de grâce. After decades of repeated diebacks, however, some seedlings and sprouts are demonstrating some increasing resistance to the point where some are capable of producing viable fruit. Scientists are hopeful that in the decades to come a blight-resistant genotype can eventually emerge and the American chestnut can regain its rightful place in the woodlands of the eastern United States. Unfortunately, before some of these scattered remnants of besieged trees have a chance to fully perfect their resistance, they may fall victim to the chainsaw or bulldozer of an unknowing land developer, farmer, or even public agency. A colleague of mine experienced such a sad loss when a grove of chestnut trees he had been observing and documenting for almost a decade fell to the hands of a contractor's chainsaw, and their former location is now occupied by a sprawling brick bi-level. My friend had assumed their isolated location, along a steep ridgeline, would protect them, but he had forgotten the resourcefulness and skill of our professional engineers who can build almost anything anywhere. If you know of similar sites where this particular species is present, I urge you to locate them by GPS and get them on the map and into the MMP.

Few of us may have the opportunity to actively involve ourselves in the global protection of our vegetative resources. We can, however, continue to make significant contributions on a more localized and national scale.

9.3.1.6 Soils

The definition of soils varies somewhat from profession to profession. For instance, geologists describe soil generally as *regolith* that is further defined as, "the noncemented rock fragments and

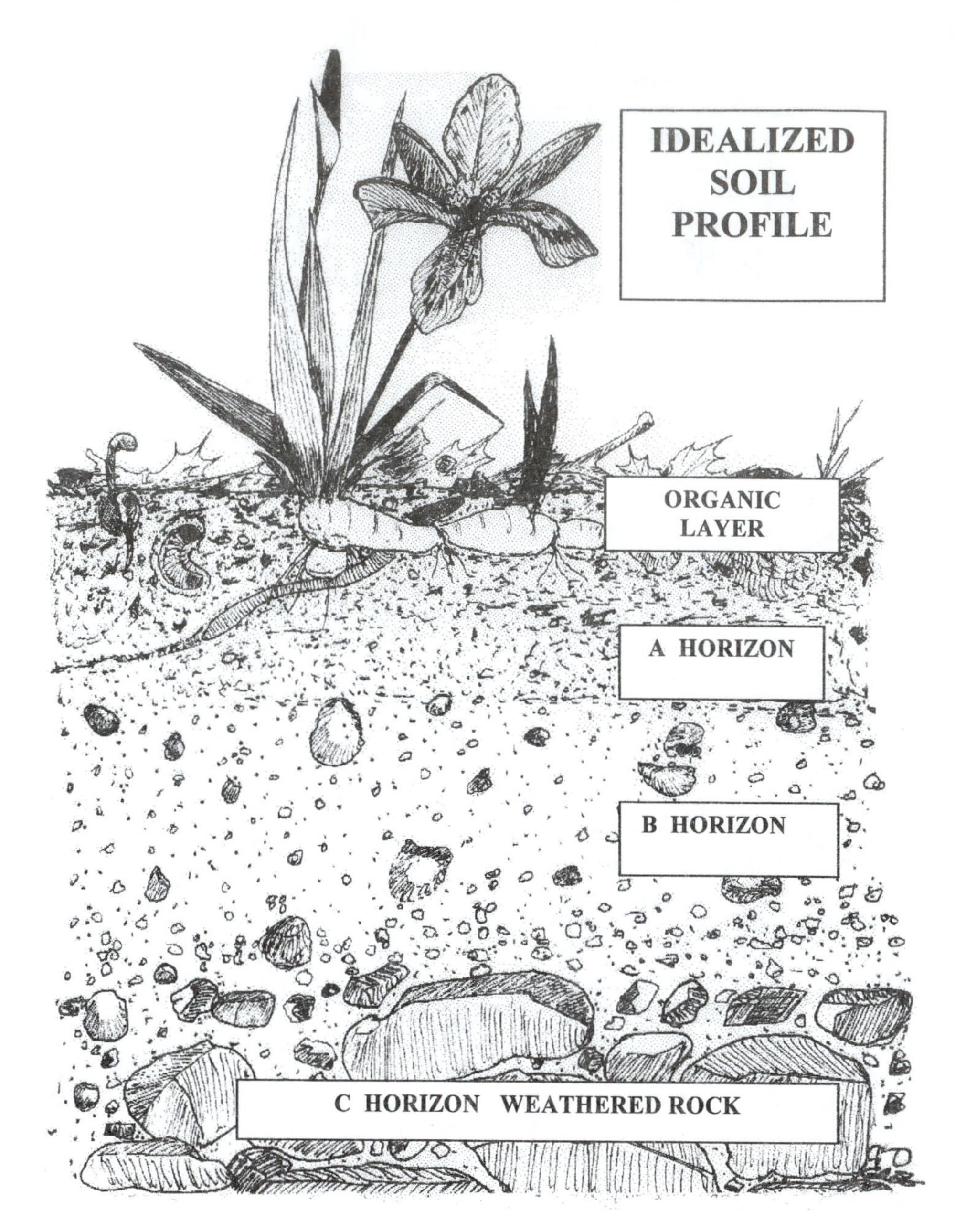

FIGURE 9.27 Our soils and the critical function(s) they perform are all too often taken for granted, much like the other components of our ecological infrastructure. (Graphics by Greg Honachefsky.)

mineral grains derived from rocks, which overlies bedrock in some places." Soil scientists, on the other hand, have defined soil as, "earth material that can be removed without blasting."

The formation of soil usually begins with the physical and chemical weathering of rock, which is related directly to the interaction between rock and the hydrological cycle. Subsequent biological, chemical, and physical processes complete the transformation of rock to a mature soil. The soil may remain in place and therefore be classified as residual soil or may be transported from its source by other natural processes and labeled a transported soil. Transported soil may undergo further changes once removed, thereby forming an entirely new soil.

Soil formation is an extremely slow process, often measured in terms of thousands and even millions of years. It is likewise a continuing process. Figure 9.27 illustrates a typical soil profile of a mature residual soil. Horizontal layers called *horizons* can be clearly identified in the profile based on their texture, color, chemical makeup, and mineral constituents. Common nomenclature for identifying these horizons are O, A, B, and C in a descending order. These general horizons can be further subdivided into narrower bands to provide more detail about the profile. The A horizon has lost part of its original substance through leaching of soluble materials by either carbonic or humic acid or through mechanical removal by percolating water. The B horizon, on the other hand, is a zone of accumulation, having gained part of the constituents that the A horizon has lost. Iron oxides and clay are common accumulations in the B horizon. The C horizon consists of the parent material — either rock or transported soil.

While parent materials strongly influence the character of soils, the influence of climate can be an even stronger influence in determining the final soil characteristics. For instance, in the northeastern part of the United States the A horizons of the mature soils have lost large fractions of their calcium and magnesium carbonates due to leaching. Consequently, soils tend to be acidic. In addition, the B horizons have high iron and clay content. These characteristics are a direct result of a good supply of rainfall. In the Southwest where rainfall is less abundant, carbonates are leached less. Furthermore, a higher evaporation rate brings up dissolved salts from lower portions of the soil profile, which adds carbonates to the surface of the A horizon. These soils tend to be alkaline. In large areas of western Texas and some adjoining states, these carbonates have built up a solid, almost impervious, white crust called "caliche."

Many of the characteristics of soil that local land planners should find helpful are those already addressed in national soil surveys prepared by the USDA Soil Conservation Service (SCS), now called NRCS. In fact, the popularity of these soil surveys is the primary reason that soils are often the first consideration in a municipality natural resource inventory. Each soil type identified and mapped by the NRCS has been rated as to its suitability for a variety of construction practices as well as agricultural uses. Limiting soil factors usually evaluated include such things as shallow depth to bedrock, high seasonal water table, perched water table, or restrictive soil horizons called *fragipans*.

Soil factors also discussed, when present, are those layers in the soil horizons that when exposed to air during excavation become highly acidic and subsequently, highly toxic to vegetation (certainly important to those wishing to establish a lawn or plant trees). In addition, such acidity when combined with rainfall or groundwater can cause the release of certain metals from other soil horizons, with subsequent movement into the groundwater or surface water where they can kill or contaminate fish and other aquatic organisms. Mapping these soils (Figure 9.28) and incorporating them into the MMP, therefore, makes good planning sense so that future land development projects are made aware of their existence and measures can be taken to avoid, and if that fails, to limit their excavation and subsequent exposure. Once exposed little can be done to ameliorate the highly acidic conditions produced.

Some interesting and relatively recent discoveries in New Jersey concerning soils at residential construction sites have further dramatized the importance of *soil resource* considerations in both land use and environmental protection. In the first instance, it was discovered that soils at some new residential construction sites harbored unacceptable levels of *arsenic*. This was of great concern to both the prospective residents and local planners and health agencies as well as state environmental regulators. Predevelopment land uses had been composed solely of agriculture, especially fruit orchards and turkey farming operations; and implications were that arsenical compounds could have been used in both activities, and what was being found was a residual of these prior land uses. There was, of course, some speculation that the arsenic could have been naturally derived. To quickly allay the fears of future residents of that particular subdivision the developer was directed to remove the contaminated soils from the site, which was done. This discovery, however, heightened concern among some municipalities, and they quickly proposed and adopted amendments to their local ordinances requiring developers to test the soils for similar contamination during the preliminary subdivision approval process. Another developer, after discovering similar unacceptable levels

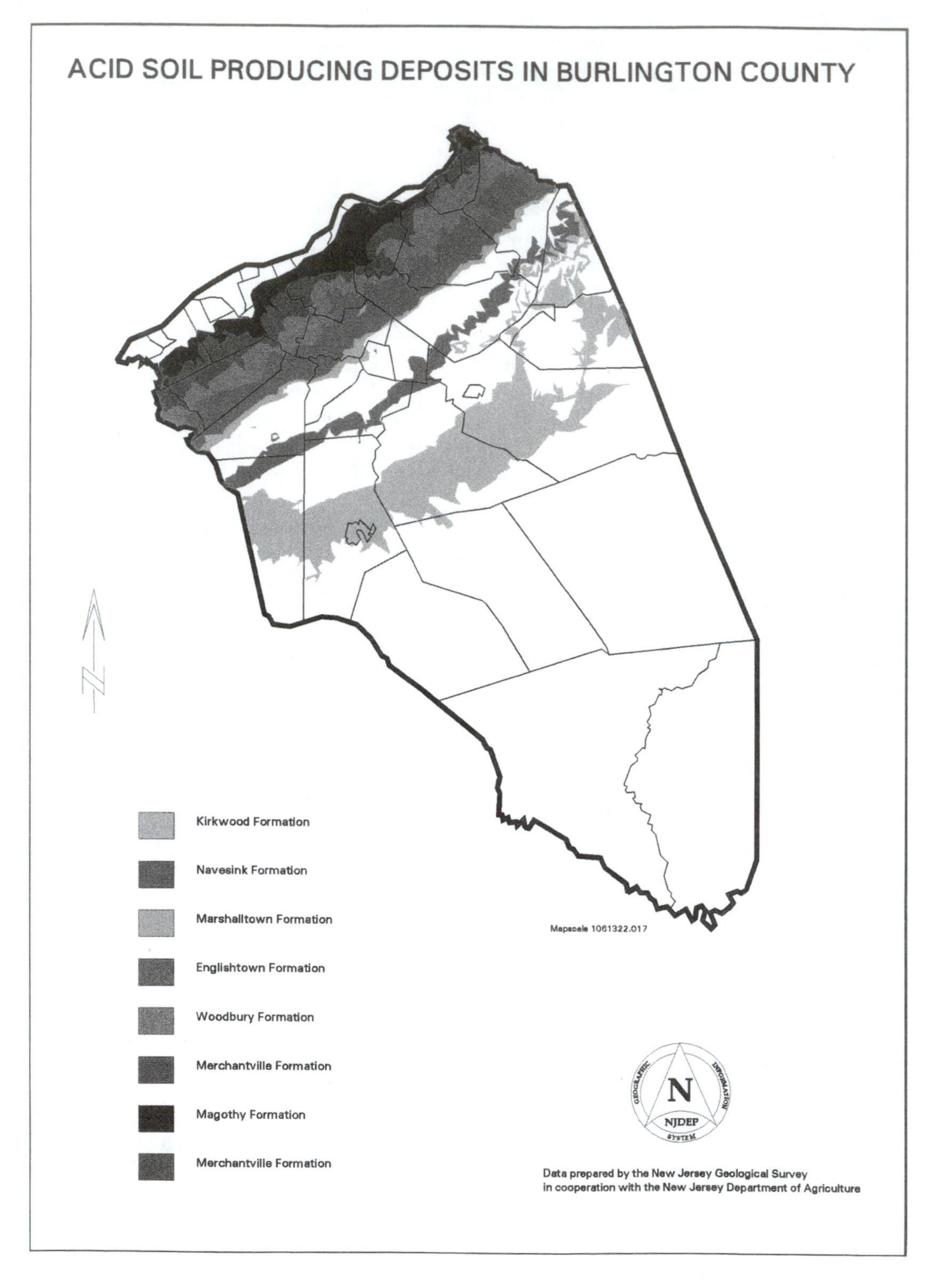

FIGURE 9.28 Exposing acid-producing soils may thwart the ability of homeowners to establish lawns and other vegetation. In addition such soils may threaten nearby surface water bodies and their instream biota. (From New Jersey Department of Environmental Protection, New Jersey Geological Survey.)

of arsenic at his 450 home site, voluntarily removed 20,000 cubic yards of contaminated soil. In this instance, however, the source had been identified as being natural (i.e., indigenous soils). Ironically the developer would not have been under any obligation, under existing state law, to remove the contaminated soil because it was from a natural source. His voluntary removal was commendable and understandable from both a public relations and a liability aspect. At or about the same time the state's geological survey already was conducting a study of native soils in the area to determine whether certain heavy metals such as arsenic were associated contaminants in the soils indigenous to the area, which happened to be the state's coastal plain physiographic province. That study by Dooley[34] determined that certain soils in the coastal plain contained widely varying concentrations of the mineral *glauconite*, or greensand, with some areas consisting of essentially pure concentrations of this material. Dooley found that in certain locations the greensands were moderately or highly enriched in arsenic, beryllium, nickel, vanadium, and zinc — to the extent that in some areas the arithmetic mean concentration of arsenic, beryllium, and chromium *exceeded* the state Department of Environmental Protection (DEP) proposed guidelines for residential soil cleanup at contaminated sites. These findings had particular significance because of the historically extensive use of these greensands from the 1700s to the early 1900s as a fertilizer and soil conditioner. Dooley, quoting Cook (1868), estimated that the greensand used annually in the state, principally within and near the greensand districts was "very near one million tons." General application rates for greensand in the 1860s were 100 to 400 bushels per acre for most forage crops; 5 to 30 tons per acre for potatoes; 2.5 tons per acre for buckwheat; and 5 to 30 tons per acre for wheat, rye, oats, and corn. Dooley then proceeded to estimate the annual metal burden to soil based on this application rate of 1 million tons. By his estimate 16.5 tons of arsenic, 68 tons of barium, 5.4 tons of beryllium, 281 tons of chromium, 3.4 tons of copper, 25.5 lbs of mercury, 25.5 tons of nickel, 12.8 tons of lead, 144 tons of vanadium, and 116 tons of zinc would have been applied annually — a significant loading and certainly one that local land planners in the area should be aware of if they have adopted or are considering adopting mandatory soil sampling ordinances as part of the land subdivision process.

The second intriguing revelation in this state came from a cooperative study between the state Ocean County Soil Conservation District and the local office of the federal NRCS.[35] Again the locale was in the coastal plain physiographic province of the state. Normally soils in the area are very sandy and have good porosity in their natural state with rainfall soaking through them very quickly. However, the agencies have found that the soil disturbances associated with land development practices, such as major cuts and fills and heavy machinery movement, have altered the soil porosity substantially, especially under what would eventually become lawn areas, causing compaction to the extent that engineering calculations of storm water runoff for postdevelopment conditions may be significantly underestimated. For instance, soil bulk density measurements* of these compacted soils ranged from 1.45 to 1.97 g/cm^3. In comparison *solid concrete* has a density of 2.2 g/cm^3 and the normal density of these soils is 1.08 g/cm^3. Local planners have always assumed that grassy yards and like areas are a balance to the impervious surfaces (roadways, parking areas, and rooftops) accompanying site development. That may no longer be a valid assumption, at least in areas with similar soil conditions. Such increased bulk densities also have created a severe obstacle for the reestablishment of vegetation on these disturbed soils, as many new homeowners have quickly found out. Despite the application of tons of fertilizer and lime and thousands of gallons of water and eventually pesticides, newly planted trees, shrubs, and grasses have been withering and dying, simply because their roots could not penetrate these compacted soils. Bulk densities that restrict root growth range from >1.80 g/cm^3 for sands and loamy sands to 1.47 g/cm^3 for clays with >45% clay content. For all these reasons, it would certainly seem reasonable for local land planners to require as-built soil density measurements for newly completed yard areas

* Soil bulk density is the ratio of dried soil (mass) to its bulk volume, including the volume of particles and the pore space between the particles.

as well as for any storm water infiltration basins for which ability to recharge captured storm water could be seriously impaired by compacted soils.

One characteristic not found in the NRCS soil surveys is a calculation of the ability of each soil to effectively capture and recharge precipitation. Methodologies to derive this capability only recently have come into being. The state of New Jersey has developed and published for public consumption such a methodology.[36] (See also Appendix D.) Municipalities dependent on locally derived groundwater for their potable water supply find this methodology helpful in delineating prime recharge areas within the municipality, so that they may be recognized in the MMP and protected.

Finally, one further discouraging feature of our sprawling pattern of land development has been the loss of some of our most productive agricultural soils, buried forever beneath tons of concrete and asphalt roadways, parking lots, house foundations, and massive warehouse complexes and shopping malls. Ironically, much of this loss has occurred in areas well suited for agriculture, where supplemental irrigation is rarely needed and few soil amendments are required. As such good soils are lost we are left with more marginal soils requiring considerably more effort and cost to bring them into production. In some instances this may not be cost-effective. In the meantime, the loss of these localized prime agricultural soils puts more pressure on other areas of the nation, particularly some of the more arid regions of the Midwest and the West (where extensive irrigation is an essential component of agriculture), to provide more and more of our agricultural products. As a consequence, this may exacerbate some of the ecological damages already occurring there because of their extensive use of irrigation, including the leaching of salts and toxic metals and the overdraft of groundwater.* It would seem logical, therefore, for local communities to make much more of an effort to preserve their prime agricultural soils, delineating them and incorporating them into the MMP to be protected at all costs.

9.3.1.7 Geology

Man's manipulation of the earth's surface contours to more desirable configurations is a long-standing tradition in human culture. In the United States, advanced earth-moving technologies developed after World War II greatly enhanced the national ability to tear down, alter, and even obliterate the most durable features of the earth's topography. Land that was once considered marginal for construction suddenly became very viable. Unfortunately, the nation was not as well equipped to deal with the consequences of these often massive disturbances of the earth's contours. We were not particularly concerned about the characteristics of the underlying geologic formations unless, of course, it was composed of bedrock. Then we were only concerned because rock was more expensive to remove than soil was. Perhaps we were naïve, indifferent, or infatuated with our new found prowess to bully the environment.

The detailed examination of subsurface geologic features, however, has not become as widespread or as routine, in spite of over 40 years of documented calamitous consequences for failing to do so. It is somewhat difficult to reconcile the fact that we can demand and be willing to fund, almost without question, expensive, comprehensive, and after-the-fact geologic evaluations at contaminated hazardous waste sites, yet we are reluctant to require such up front analyses as part of the land subdivision process.

There are three geologic structures of which land planners should be especially aware. The first is *cavernous limestone*, a formation of limestone or carbonate rock containing large solution cavities. If these cavities occur below the water table, they are undoubtedly filled with water. Lowering the water table can empty these large voids and therefore remove any supportive buoyancy the water may have been providing to the overlying soil and rock. As a result, the roof of these

* The great Ogallala aquifer, for instance, which runs under six midwestern states (an area of 60,000 square mi) and provides water through 170,000 wells for 15 million acres of land that produces 15% of the nation's crops and 38% of the nation's livestock, has had its water table drop nearly 15 ft. About 24 million acre feet are withdrawn annually, while nature replaces only about 3 million.[37]

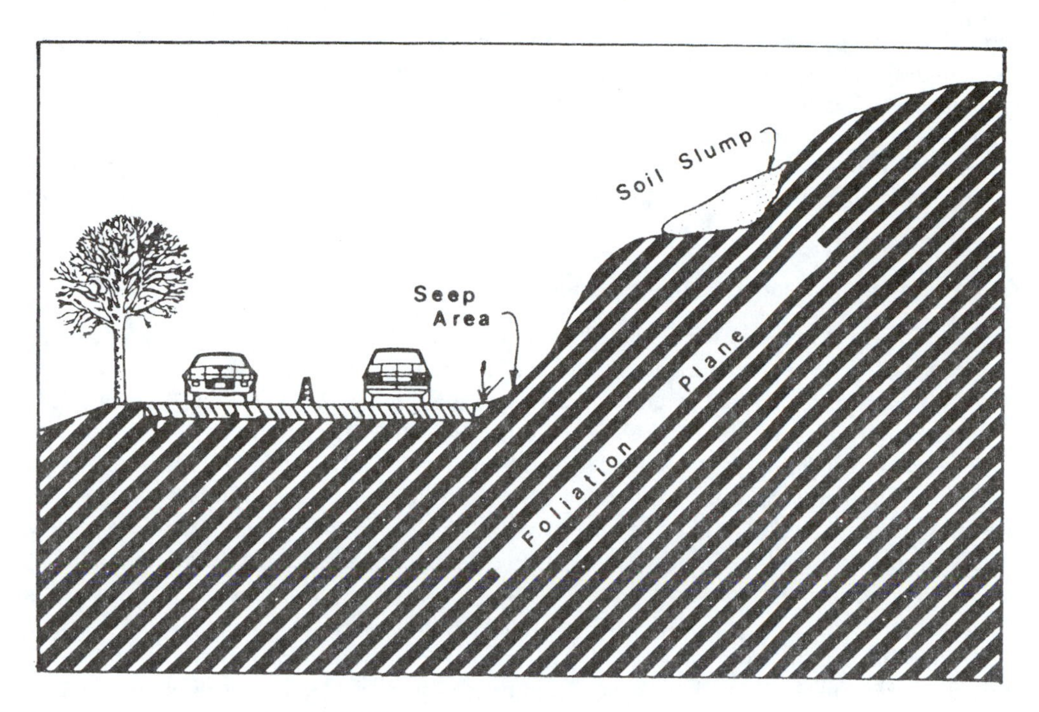

FIGURE 9.29 Construction that disturbs foliated bedrock formations must take into account the orientation of the planes of that rock. Failure to do so often produces chronic maintenance problems in the form of rock slides and soil slumping as well as bottom of slope drainage problems.

cavities may collapse inward, creating a sinkhole at the ground surface that can threaten buildings or structures lying overhead. The state of Florida is one of several states that has experienced this phenomenon. Even limestone cavities located above the water table can pose similar hazards if disturbed by man's activities. Here the cavities often are filled with insoluble clays — the residue of the limestone rock dissolved away by the percolating groundwater. These filled cavities are very vulnerable to disintegration, particularly if part of the overlying soil has been removed or excess amounts of water or wastewater are discharged into or onto the overlying soil. All these can disturb the equilibrium between the rock and the soil. The removal of even part of the overlying soil during land grading allows faster infiltration of rainwater or wastewater, thereby short-circuiting some of the buffering capacity of the soil.

Excessive discharges of wastewater from lagoons or spray irrigation systems, even in the absence of soil removal, can likewise accelerate the chemical weathering of the remaining limestone or wash away the clay contents of the existing cavities — the consequence of which is the acceleration of wastewater to the groundwater system. With sewage treatment plants rarely operating at complete efficiency, this acceleration of wastewater to the groundwater system can represent a hazard to human health.

The second geologic formation worth mentioning is *foliated metamorphic rock*. Foliated or layered metamorphic rocks such as gneiss, slate, or schist are frequently encountered during land development. Figure 9.29 demonstrates some of the common problems that can occur with such formations, including rock slides, soil slump, and areas of seepage. The third geologic formation of concern would seem to be the one that is the most intuitively logical, that of *fault zones* — those areas of the bedrock where the two sides of a fracture have been displaced from each other, and which have a reputation for generating seismic pressure waves some of which are capable of making the earth shake violently enough to produce an earthquake. Despite this recognized hazard we know that in some areas, particularly the West Coast of the United States, local land planners have permitted development close to if not directly over these fault zones.

Finally, as we discussed previously in the section pertaining to soils, delineating those soils in the community having the greatest potential to absorb and recharge fallen precipitation is of importance — the intent being to provide the maximum amount of recharge to any underlying groundwater aquifer. However, it would be of little use if the soil had a tremendous potential to recharge and the underlying geologic formation could not store it. That is why in your calculations of soil recharge capability you must look at the capability of the underlying geologic formation to accept precipitation as well.

9.3.1.8 Threatened and Endangered Species

I admit that this is a resource for which I have a personal interest and bias. Many years ago, as a young teenager I read about the tragic extinction of the *passenger pigeon* (*Ectopistes migratorius*). I was so moved by the story that several months later I made it a point to visit the National Museum of Natural History in Washington, D.C. to see for myself the stuffed body of what is believed to be the last passenger pigeon in America. She had been nicknamed Martha. Despite her many years in a glass coffin, there was still a pink iridescence to her feathers. I stared intently at her eyes, in desperation, hoping that it was not true, that I would somehow see her eyes flicker once more with the spark of life and that she would suddenly leap off the branch and break into flight, but I knew it was just a young boy's fantasy. My sorrow quickly turned to anger because I felt that I had been cheated, and although it has been over 40 years since, I still carry some of that anger with me. It is simply too difficult to comprehend that a species that once was counted in the billions and that formed flocks 240-mi long and 1-mi wide, could be totally eradicated in a span of about 64 years.

Extinction is a natural process that has persisted for billions of years. As conditions in Earth's biosphere changed, species either adapted by forming new species from existing ones (called speciation) or became extinct (e.g., the passenger pigeon) because they could not adapt and reproduce under new environmental conditions. The timetables for both of these processes were previously measured in millions if not billions of years. Humans, however, have altered the timetable for extinction, and have accelerated the rate at which this process occurs. Wilson and Myers* in 1988 estimated the average annual extinction rate of mammal and bird species between 8000 B.C. and 1975. They estimated that between 8000 B.C. and 1600 A.D. there was a loss of *one species every 1000 years*. Between 1600 and 1900 there was a loss of *one species every 4 years*, and between 1900 and 1975 *one species a year* became extinct. Why should we be concerned about any species becoming extinct? The answer is simple. We and they are part of the same complex web of life, clinging precariously to an often hostile biosphere and although we may not know the full extent of their contribution to that web, we should not write off any species as useless. All species, plant and animal including mankind, are part of a giant genetic library that has evolved from successful survival strategies developed over several billion years. The premature raiding of that library may have far-reaching implications that we do not yet fully understand. Therefore, we should not be so quick to write off species tottering on the brink of extinction as useless. They may in fact be our canary in the mine.

9.3.1.9 Steep Slopes

These are generally areas where slopes are 25% or greater. This steepness makes them especially vulnerable to damage and disintegration if disturbed in any way by activities associated with land development.

9.3.1.10 Floodplains

Very few of us have ever taken the time to seriously consider the flow of water through our network of surface waterways. The fact of the matter is that similar to many other parts of our

* This is as cited in Miller.[38]

FIGURE 9.30 Even tiny backyard creeks can become raging torrents when man modifies the landscape to whisk away storm water runoff.

ecological infrastructure, we take that flow for granted, never contemplating its origins or concerning ourselves with its final destination. Only when these surface pathways are filled with excess water or are obstructed in some manner are we forced to take notice. At these times, streams no longer flow quiescently in their normal channels, but rise up to flood adjoining lands (Figure 9.30), erode stream banks and channels, remove surface features, and displace and redeposit tons of silt and rock. It is a process as old as Earth itself and part of the perpetual geologic cycle that continuously sculpts the features of Earth's surface. While these conditions of flooding predate man's existence, they took on new significance with his arrival. Some of the surface features that these floods were now removing or inundating belonged to man. Even man himself became a victim of these excursions.

Man has been attracted to these surface water pathways and their concomitant floodplains for many reasons: for obtaining drinking water, for commerce and transportation, for irrigation of crops or use in industrial processing for agriculture, and finally for recreation. Less quantifiable reasons such as aesthetics or the simple serenity of moving water also were important. The great utility of water subdued our commonsense inclination to avoid such potentially dangerous areas — a characteristic, I'm sorry to say, we seem to have inherited from our ancestors.

Man's encroachment into these flood hazard areas has been less than subtle, perhaps better characterized as brash and reckless. We have diverted, rerouted, filled in, dug out, and covered over these naturally occurring pathways and adjoining floodplains with impunity, placing ourselves confidently in the niches we created. In Earth's closed loop system, however, there is always a price to be paid. In the decade of the 1990s, flood damages have risen to more than $4 billion annually. We have paid dearly for this recklessness.

Even as late as the 1930s, we were still convinced that our structural methods could subdue the nation's continuing flooding problems. Under the federal Flood Control Act of 1938, massive public work flood control projects were undertaken across the United States. By 1960, however,

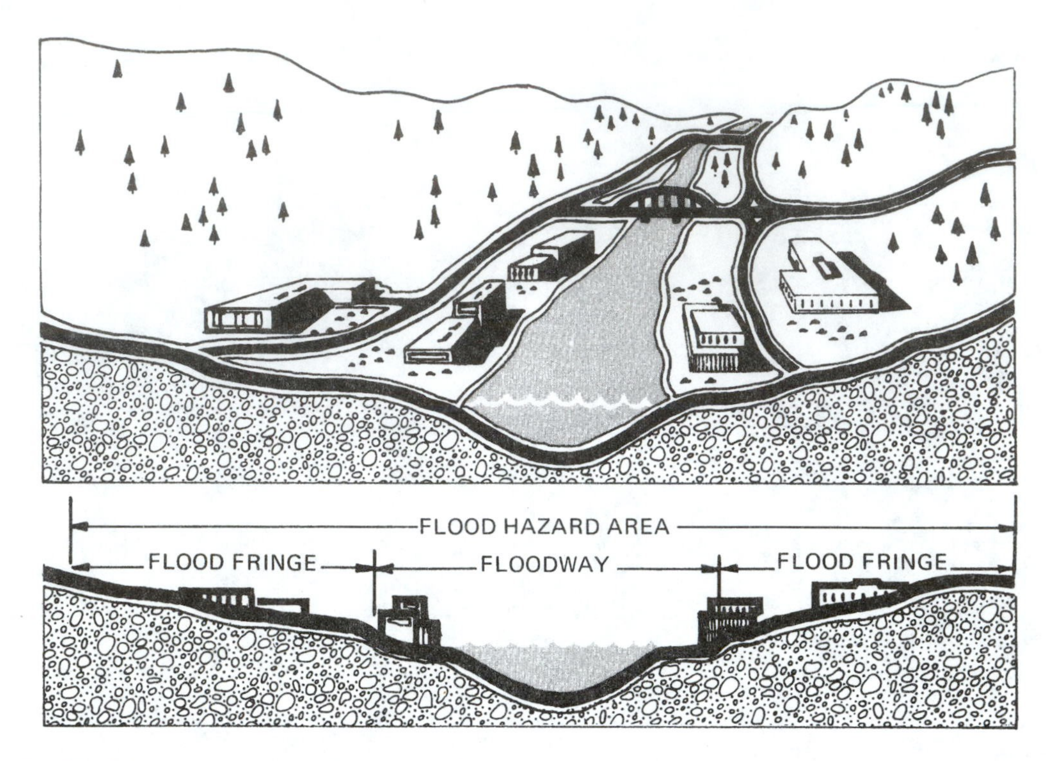

FIGURE 9.31 Typical cross-section view of a flood hazard area showing common man-made encroachments. (From New Jersey Department of Environmental Protection, Flood Hazard Report No. 17, Matchaponix Brook System, New Jersey Division of Water Research, Anderson Nichols Inc., 1973.)

some concession was being made that structural methods alone were not the complete answer.* That year, a new federal Flood Control Act authorized the U.S. Army Corps of Engineers to provide advice and floodplain studies to local governments to assist them in the land use planning process — the primary causal agent of flooding-associated damages. It would be gratifying to say that it has been a steadily improving situation since then, but as most of you know we cannot. While more flood hazard regulations are in place today than in 1960, and we know a great deal more about the characteristics of our national surface water pathways and associated floodplains, we still have difficulty in removing ourselves from and preventing further excursions into these high hazard areas. The fault, however, may be with some of the newer regulations themselves. For example, the federal flood insurance program may actually be encouraging people to remain in flood hazard zones (Figure 9.31) by constantly paying to rebuild houses and other structures that will, in all likelihood, be flooded all over again. A 1997 study by the National Wildlife Federation (NWF)** certainly questions the wisdom of continuing with this practice of subsidizing the habitation of recognized flood hazard areas. The NWF analyzed the National Flood Insurance Program (NFIP) computer databases for repetitive loss properties*** and found that although repetitive loss properties represent only 2% of all properties insured by NFIP, they claimed 40% of all NFIP payments in the period studied (1978 to 1995). The NWF cites one property owner whose total repetitive

* This point is clearly reinforced by the Great Midwest Flood of 1993 where our traditional engineered defenses against flooding, such as levees, flood walls, dams, dikes, and diversion channels, once again proved no match for the powerful onslaught of corseted, moving water that had been denied its traditional floodplains.

** National Wildlife Federation, Higher Ground: A Report on Voluntary Property Buyouts in the Nation's Floodplains, A Common Ground Solution Serving People at Risk, Taxpayers and the Environment, National Wildlife Federation, 1997.

*** The NFIP defines repetitive loss properties as properties that have received two or more flood insurance loss payments of at least $1000 within a 10-year period.

payments on a house valued at $114,880 amounted to $806,591 in the period studied. Not only does this not make good environmental sense but also it does not make good business sense.

Many state floodplain regulations only cover a part of the historic flood hazard area, usually the width of land on either side of the stream expected to be inundated by a 100-year design flood. The problem comes, however (as in 1993 and later in 1995), when floods of a 200- or 500-year magnitude occur. It is highly unlikely that we will ever completely clear our most urbanized floodplain corridors of existing structures or inhabitants, although I strongly support the current trend to buy out inhabited properties in recognized flood hazard areas. The NFIP, according to the NWF report previously cited, believes the buyouts it pursued following the Midwest floods of 1993 in the long run will save $2 for every dollar spent on such acquisitions.

Certainly the value of natural floodplains to strip some of the energy from fast moving flood waters is enormously important; however, we must not forget the other important roles of floodplain areas including the removal and mitigation of various pollutants, and the habitat for an enormous number of species of wildlife (some of which are either threatened or endangered). In a holistic, ecologically based MMP, all these values should be considered.

9.3.1.11 Unique Habitats

Included in this category would be such things as fens or bogs, winter yarding areas for white-tailed deer, snake denning areas, scenic vistas, unique old age forests, areas of unusual vegetation or geologic anomalies, and roosting areas for large populations of native or migratory birds.

In closing it should be recognized that ecological infrastructure does not respect political boundaries; therefore, ecological mapping and inventory may need to be a cooperative undertaking between adjoining municipalities that share parts of the same natural resource or resources. Likewise, the finalized planning strategy in each community to protect these resources, also needs to reflect this same cooperative spirit.

9.3.2 Man-made Infrastructure: Agents of Change, Deterioration, and Destruction

As we inventory our natural resource infrastructure, we also must inventory the agents responsible for its alteration, deterioration, and destruction, namely man-made infrastructure.

9.3.2.1 Storm Drainage Systems

Our proclivity to reconstruct the natural landscape and force it to accommodate our needs often results in the truncation and disruption of naturally occurring drainage patterns. As a result, we have found it necessary to create new, artificial pathways for these altered patterns of water flow. Collectively referred to as *storm drainage systems*, these artificial systems do a commendable job of collecting and whisking away rainfall and snowmelt runoff, propelling it toward its ultimate destination at some downstream waterway. There are, however, deleterious results of taking runoff that once flowed as a broad thin sheet across the landscape, intercepting and isolating it, and then forcing it down unnatural pathways. Ecologically, these systems are a disaster. The smooth cement bottoms (inverts) on the interior of these underground concrete conduits, which work well for ensuring the water is quickly moved through the system, for all intents and purposes are a biological desert. Their darkened interiors, shut off from all sunlight, can hardly be compared with the vibrant, biologically active surface drainage ways they have replaced. To add insult to injury these same conduits are notorious collectors and accumulators of any number of deleterious or polluting materials, including, oils and greases, road salts, toxic heavy metals, and sediment, that are likewise quickly transported through the system and discharged to the receiving waterway. Due to the abiotic nature of these structures, little attenuation or removal of these materials generally occurs as they

FIGURE 9.32 This newly installed storm drain carrying runoff from a recently completed asphalt parking lot has already destroyed a considerable part of the downstream waterway.

move through the system, quite unlike the multiple treatment and attenuation processes of the natural drainage ways they have replaced.

Further environmental damage occurs from the quantity of runoff itself. These abnormally concentrated volumes of runoff are notorious for scouring out the bottom of streambeds at their point of entry (Figure 9.32) and at points well beyond, and also for eroding stream banks and raising the level of flooding along downstream waterways. Figure 9.33 demonstrates the significant change in peak flows that can occur during rainfall events as a result on these concentrated discharges. To partially remedy this situation, the engineering profession has devised add-on structures called detention and retention basins (Figure 9.34). These structures are designed to reduce the peak flow of runoff following rainfall events by capturing part of the runoff, temporarily holding or retaining it for a period of time, and then releasing it at a slower rate of discharge to the stream. In theory the practice works well. In real-time situations, however, some problems have occurred. First, the design and installation of these control structures are done on a project-by-project basis with little accounting as to how the timed releases for one basin influence the discharges from other similar structures installed for other projects, on the same receiving waterway. They can be, and often are, coincidental and additive, thus raising the level of floodwaters in the waterway and sustaining its height for such a protracted period of time that the banks of the stream begin to crumble and fall. A common symptom of this is the widening of the stream channel and the collapse of streamside trees (Figure 9.35). Water quality is a further concern in streams receiving these cumulative discharges. Some consultants have been making claims, albeit somewhat exaggerated, concerning the efficacy of these basins to "treat" storm water runoff. To be sure there is some attenuation of certain waterborne pollutants, particularly those materials attached to particulate matter (sediment and silt) that quickly settles. Dissolved materials such as nitrates and salts, however, are relatively unaffected unless a permanent pool of water is maintained in the basin to encourage biological uptake or further physical and chemical removal. These structures are not capable, despite claims to the contrary, of providing the level of treatment normally associated with a sophisticated secondary sewage treatment plant — a fact local land planners should keep in mind when dealing with water quality issues on sensitive water bodies.

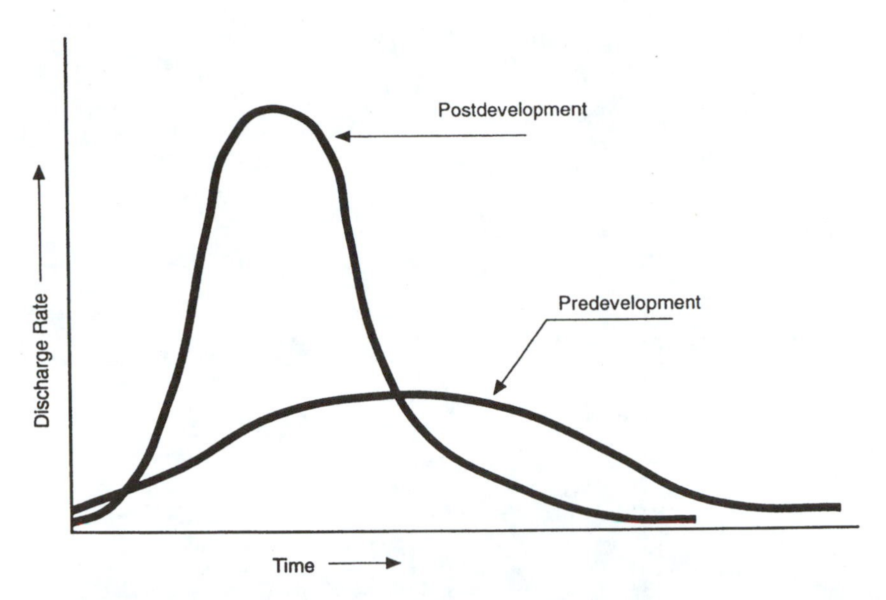

FIGURE 9.33 Typical differences in storm water discharge rates between developed and undeveloped site conditions. (Adapted from U.S. Environmental Protection Agency, Handbook Urban Runoff Pollution Prevention and Control Planning, EPA/625/R-93/004, September 1993.)

FIGURE 9.34 A well-maintained detention basin.

You do not necessarily have to locate every lineal foot of underground storm drain and corollary structures such as curb inlets or manholes, although your public works department may have already done so. Instead, you should focus on locating (again through the use of GPS) the outfalls of each storm drain at their discharge point to the waterway or ditch leading to the waterway. You also should accurately locate all storm water detention and retention basins and similar control structures at the same time.

FIGURE 9.35 The eroding stream banks, widening channel, toppling trees, and excessive sediment bars shown are all indicators of a stream in serious trouble from a hydrological overload due to careless land planning.

FIGURE 9.36 This excavated cross section of a highway undergoing reconstruction clearly demonstrates how impermeable cover (in this case concrete) prevents rainfall from entering the soil and rock fissures below.

9.3.2.2 Impervious surfaces

Impervious surfaces are those surfaces that prevent infiltration of water into the soil. The most common impervious surfaces are transportation related and include roadways (Figure 9.36), parking lots, and driveways. Their prominence on the landscape reflects the dominance of the automobile as the primary mode of transportation in contemporary American society. These transportation-related features also generally comprise the largest proportion of development-related impervious surfaces on a project site, with values ranging from 25 to 60% for single-family residential subdivisions. The remaining impervious surfaces are generally made up of such items as sidewalks, rooftops, and patio surfaces. Why are all these surfaces so important? Basically the reasons are the same as those outlined earlier in our discussion of storm drains. The most obvious, of course, is the *decrease* in *surface infiltration*, which reduces groundwater recharge; this reduction thereby potentially lowers groundwater tables, which can threaten potable water supplies. This reduction in infiltration also reduces the groundwater contribution to stream flow, which can result in intermittent or dry stream beds during episodes of drought and low flow (remember our case example in Chapter 2 involving the disappearing stream). Transportation-related impervious surfaces are intimately connected to our ubiquitous artificial storm drainage systems; and as indicated previously, polluting materials are flushed directly off these impervious surfaces into the storm drainage systems where they are quickly transported, untreated, to the nearest waterway. In addition, the enhanced runoff they produce causes increased erosion both during and after construction, thereby seriously impacting downstream areas and resulting in the loss of riparian habitat and instream habitat that is home to a variety of plants and animals.

Furthermore, the lack of tree cover associated with impervious surfaces leads to greater fluctuations in water temperatures and dissolved oxygen in receiving waterways. (See previous discussion on "heat island" effect discussed earlier in Section 9.3.1.5.) As a result, stream temperatures are warmer in the summer (thereby threatening the survival of such cold water, indigenous species as brook trout, *Salvelinus fontinalis*) and colder in the winter.[39] Impervious surfaces have become synonymous with a human presence, and studies have shown that population density in an area is correlated with the percentage of impervious cover.[40,41] It should come as no surprise then that research from the past 17 years shows a consistent, strong correlation between urbanization and its associated *imperviousness and the overall health* of its watersheds, receiving waterways, and associated biota.[41–48] See Table 9.1 for further details on some of the results of research on impacts caused by urbanization and its associated imperviousness.

Because of this, impervious surfaces are quickly becoming a valuable predictor and surrogate indicator of waterway and water resources degradation, and perhaps of the entire watershed itself. Schueler[48] in his research found that stream degradation begins to occur at relatively low levels of imperviousness (Figure 9.37).

9.3.2.3 Point Source Discharges

Section 402 of the 1972 federal Water Pollution Control Act (later amended as the Clean Water Act) made it *illegal to discharge pollutants* into the nation's navigable waters without obtaining a *national pollutant discharge elimination system* (NPDES) permit. These permits, sometimes derogatorily called, "licenses to pollute," identify and dictate the amount and type of wastes that can be discharged to a surface water body, on a daily, weekly, or monthly basis. It is assumed, albeit erroneously, that at the levels permitted, these waste materials easily can be assimilated by the receiving water body. Assimilation, in some instances, may mean simply dilution.

A majority of the states have been delegated authority to operate the NPDES program, adopting their own regulations and sometimes making them more inclusive than the federal regulations. For instance, the state of New Jersey requires persons who propose to apply their wastewater to either surface waterways or the land (Figure 9.38) to obtain a New Jersey pollutant discharge elimination system (NJPDES) permit. The same applies to those proposing to discharge wastewater by injecting it through a well. The reason for these more inclusive regulations is due to the state's definition of *waters of the state,* which includes both surface and groundwaters.

Permittees out of compliance with their permits can degrade water quality, threaten the public health, and increase potable water treatment costs for downstream water purveyors. As a planner, you may not expect to get directly involved in an enforcement action against either a municipal or

TABLE 9.1
Key Findings Examining the Relationship of Urbanization on Stream Diversity

Ref.	Year	Location	Indicator	Key Finding
Booth	1991	Seattle	Fish habitat channel stability	Channel stability and fish habitat quality deteriorated rapidly after 10% imperviousness
Benke	1981	Atlanta	Aquatic insects	Negative relationship between number of insect species and urbanization in 21 streams
Jones and Clark	1987	North Virginia	Aquatic insects	Urban streams had sharply lower diversity of aquatic insects when human population density exceeded four persons per acre (estimated 15 to 25% imperviousness)
Limburg and Schmidt	1990	New York	Fish spawning	Resident and anadromous fish eggs and larvae declined sharply in 16 tributary streams that were more than 10% urban
Shaver et al.	1994	Delaware	Aquatic insects	Insect diversity at 19 stream sites dropped sharply at 8 to 15% imperviousness
Shaver et al.	1994	Delaware	Habitat quality	Strong relationship between insect diversity and habitat quality; majority of 53 urban streams had poor habitat
Schueler and Galli	1992	Maryland	Fish	Fish diversity declined sharply with increasing imperviousness, loss in diversity began at 10 to 12% imperviousness
Schueler and Galli	1992	Maryland	Aquatic insects	Insect diversity metrics in 24 subwatersheds shifted from good to poor beyond 15% imperviousness
Black and Veatch	1994	Maryland	Fish/insects	Fish, insect, and habitat scores were all ranked as poor in five subwatersheds that were greater than 30% imperviousness
Klein	1979	Maryland	Aquatic insects	Macroinvertebrate diversity declines rapidly after 10% imperviousness
Luchetti and Fuersteburg	1993	Seattle	Fish	Marked shift from less tolerant coho salmon to more tolerant cutthroat trout populations noted at 10 to 15% imperviousness at nine sites
Steedman	1988	Ontario	Aquatic insects	Strong negative relationship between biotic integrity and increasing urban land use/riparian condition at 209 measured stream sites; degradation begins at about 10% imperviousness
Pederson and Perkins	1986	Seattle	Aquatic insects	Macroinvertebrate community shifted to chironomid, oligochaetes, and amphipod species tolerant of unstable conditions
Steward	1983	Seattle	Salmon	Marked shift from less tolerant coho salmon to more tolerant cutthroat trout populations noted at 10 to 15% imperviousness at nine sites
Taylor	1993	Seattle	Wetlands	Mean annual water fluctuation was inversely correlated to plant and amphibian density in urban wetland systems; significant degradation was noted at approximately 10% imperviousness
Garie and McIntosh	1986	New Jersey	Aquatic insects	Drop in insect taxa from 13 to 4 with urbanization shift to collectors such as chironomids
Yoder	1991	Ohio	Aquatic insects	Of 40 sites sampled 100% had fair to very poor IBI scores that had urban runoff and/or CSO impact, compared with only 56% of 52 agricultural sites that had good–exceptional IBI scores

Source: Adapted from Schueler, T. R., Site Planning for Urban Stream Protection, Metropolitan Washington Council of Governments, Center for Watershed Protection, Publication No. 95708, December 1995. With permission.

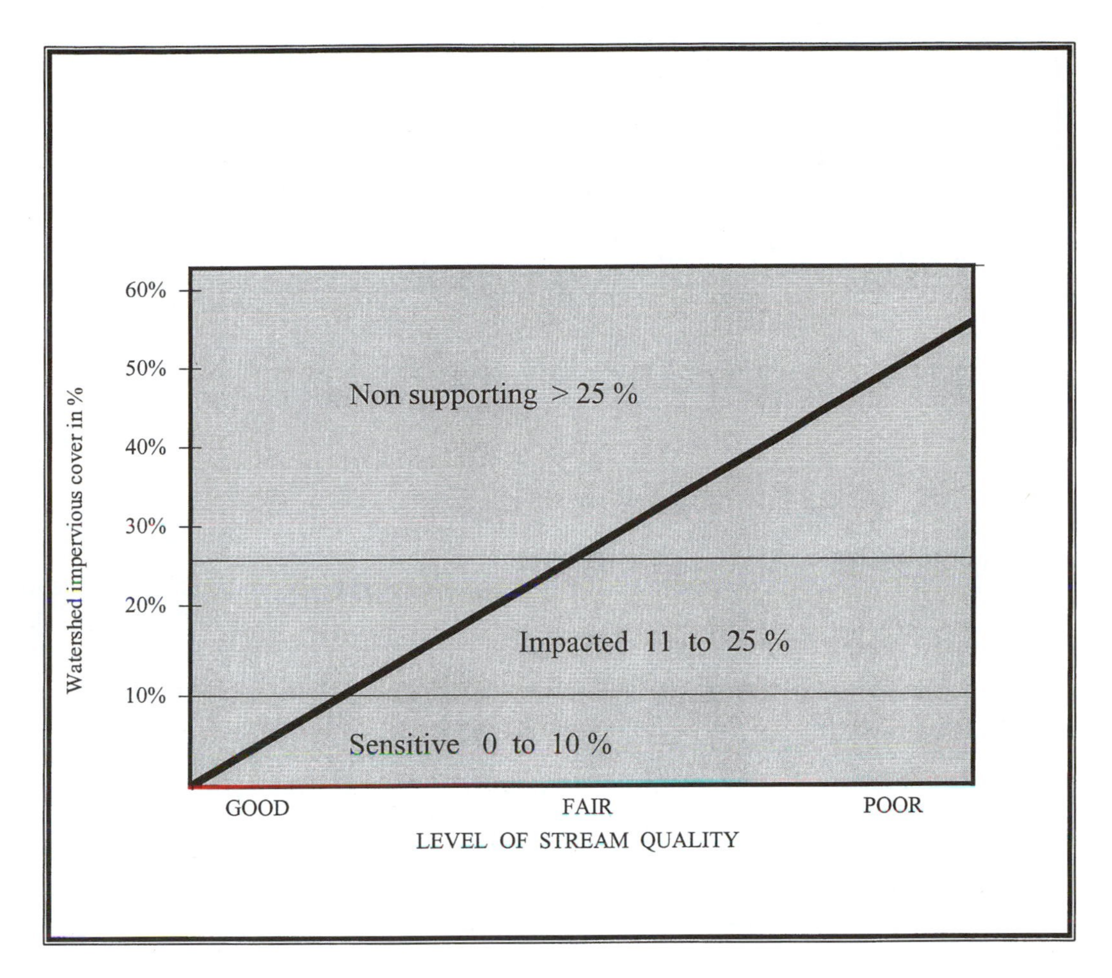

FIGURE 9.37 Impervious cover vs. stream quality. (Adapted from Center for Watershed Protection, *Rapid Watershed Planning Handbook — A Comprehensive Guide for Managing Urbanizing Watersheds,* Ellicott City, MD, October 1998. With permission.)

an industrial discharger. However, an enforcement action may very well involve a land use issue. For instance, a municipal sewage treatment plant may be cited for violating its NPDES permitted flow limits, which in turn causes excursions and violations in other NPDES permitted parameters. Some states, besides imposing fines, also may impose a *sewer ban* on that particular facility. This ban prohibits any further hookups and discharges to the collection system until such time as the wastewater treatment plant no longer exceeds its permitted flows or other permit conditions. In effect a sewer ban becomes a ban on building, an issue in which land planners do have an interest. While the community and the local land planners may not be involved directly in the writing and issuance of permits, they should know, at least on a monthly basis, how well the dischargers in their watershed are doing. They should likewise comment on these permits when they are up for renewal, especially if there has been significant noncompliance. Ambient water quality monitoring throughout the watershed, particularly upstream and downstream of these dischargers, should likewise be routinely conducted; in many instances this is already being done by state, federal, and sometimes county agencies including health departments. Local environmental groups may also be collecting samples. Recent initiatives to more clearly define the assimilative capacity of surface water bodies for which quality remains impaired or degraded is currently underway nationwide. The program is called the total maximum daily load (TMDL) program, and includes extensive

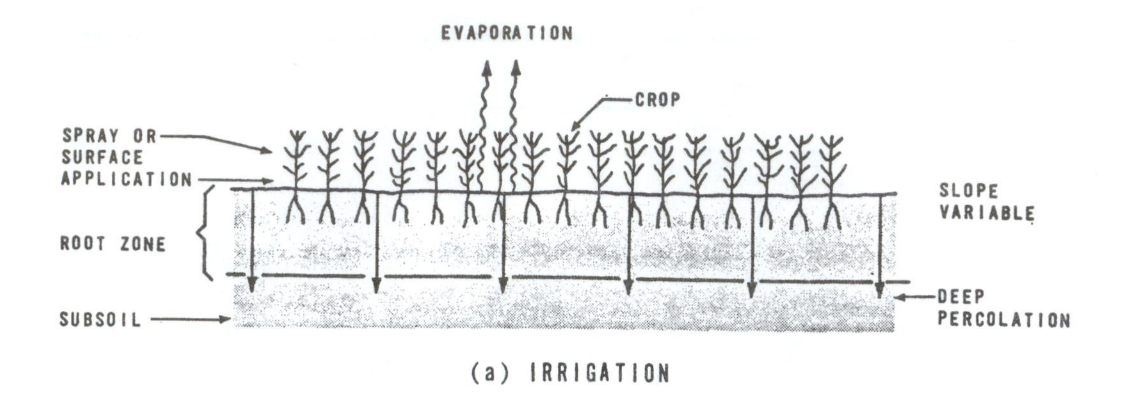

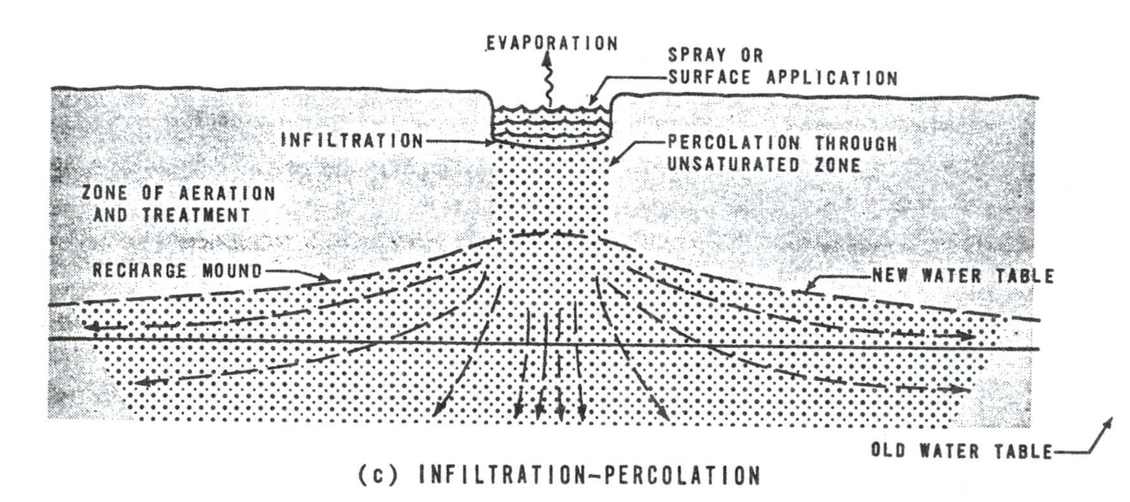

FIGURE 9.38 Methods of land application of wastewater. (From U.S. Environmental Protection Agency, Evaluation of Land Application Systems, EPA 430/9-75-001, March 1975. With permission.)

monitoring and modeling of surface streams to better define what or who is responsible for the continuing degradation. This is primarily a state and federal program but the results could have major impacts on local land use.

Finally, there are some appurtenant components of point source discharger systems, particularly municipal sewage treatment systems, that are worthy of some further discussion. These include *sewage pump stations* and *gravity sewer line collection systems*. Sewage pump stations are little more than temporary collection and transfer points for raw sewage on its way to the sewage treatment plant. They are used wherever gravity sewer lines do not work (i.e., in depressions, valleys, or extremely flat areas). The two biggest problems at pump stations are pump failure and power failure, which can result in back up and bypasses of raw sewage into adjacent waterways or storm drains. Other problems include odors, releases of toxic vapors, and sometimes explosions.

Gravity sewer lines, because of their reliance on gravity to propel raw sewage through the system, are frequently installed in and along stream corridors, sometimes crossing back and forth as the stream wends its way through the watershed. While these collection lines are designed to accommodate a certain amount of infiltrating groundwater and sometimes surface flow, on occasion — due to poor construction — such infiltration and inflow can be excessive. This results in surcharges that cause raw sewage to blow off the top of manholes and flow directly into the adjacent waterway. The same overflows can occur when the municipality or sewerage authority allows too many hookups to its sewer lines. This is especially important to municipalities that desire to expand the areal extent of their service area. Finally, it should not be overlooked that the expansion of wastewater sewer lines into undeveloped areas encourages more extensive development than if individual septic systems only were available. This in turn, as we explained previously, generally promotes a larger amount of impervious surface to be applied.

It is important, therefore, much like our earlier recommendation for storm water outfalls, to accurately locate all these points of discharge (some dischargers may have multiple points of discharge), using GPS technology and downloading their location onto the electronic databases of the municipality's Geographic Information System (GIS).

9.3.2.4 Known Contaminated Sites

You can find most of these sites already identified and located on state and federal GIS databases.

Sites under this category include

1. Old sanitary landfills or dumps whether lined or unlined
2. Federal or state hazardous waste sites as designated under the federal Superfund program
3. Sites with leaking underground storage tanks including gasoline stations
4. Old waste lagoons
5. Major spill sites

9.3.2.5 Agricultural Areas

America is a land of contradictions and a premier example of this is the nation's policies toward its farmland. On the one hand, the federal government, until very recently, encouraged and even financed the drainage and conversion of marginally acceptable land, including wetlands, to agricultural uses. At the same time, elsewhere in the nation, prime agricultural land was being buried, perhaps forever, under miles of concrete and asphalt roadways and beneath the foundations of thousands of dwellings and other buildings.

More than one half of all Americans lived on farms in 1862. By 1920, this percentage had dropped to 30% and by 1940, to about 20%. Today, less than 4% of the American population lives on the nation's 22 million farms and less than 2% of the nation's work force is engaged in farming. As a consequence, few Americans have any firsthand knowledge of modern agriculture or its problems. Furthermore, in my opinion they do not care. Because American agriculture has been so successful in supplying the nation with an abundance of products, such fecundity is often taken for granted. We, however, are biting the very hand that feeds us.

Besides providing food for American households, agricultural land offers other amenities. First, it requires few municipal improvements (infrastructure), thereby reducing municipal operating budgets. Second, it preserves open space and provides aesthetics and habitat for innumerable species of wildlife. This second combination of attributes is what makes agricultural land so attractive to Americans, in general, and urbanites and developers, in particular. In many areas the value of this land for development purposes has surpassed its value for agricultural purposes. As a result, some of the nation's best farmland is rapidly being sold and converted to residential subdivisions and suburbs. Accompanying this land transition is the relocation of a population of Americans unaccustomed to either rural living or normal agricultural practices. Previously routine farming practices, such as the spraying of pesticides and herbicides, the application of animal manures, and the tilling of the soil, must now be carried out in close proximity to numerous human receptors. As a result, farms and residential subdivisions have become incompatible neighbors.

These new rural landholders often complain of such nuisances as excess dust from tilling of the soil, animal noises, and odors from manure spreading. Less frequent, but more serious complaints include actual or perceived personal injury from pesticide or herbicide sprays, injury to property from loose livestock, and contamination of the groundwater. Farmers also are suffering from these encounters, and they are the ones who often suffer the most serious injury. Fences, used to contain livestock, are frequently cut or broken. Hikers, horseback riders, motorbikes, and all-terrain vehicles tear up newly planted crops, compact soils, scare livestock, and cause soil erosion. Free-roaming pets, especially dogs, intrude into pastures to harass and even kill livestock.

These new migrants to rural America come with unrealistic expectations or total insensitivity to farmers and their land. On the one hand, they welcome the open space that farms provide and preserve, but then just as quickly condemn the practices that farmers must carry out to keep their operations profitable. I am not absolving farmers completely. There are cases where farmers have been legitimately criticized and cited for polluting surface streams and groundwater, for improperly storing and applying liquid manures, and for misapplying pesticides. For these transgressions we have included agricultural land as an agent of change, degradation, or destruction. Why so many of these negative environmental impacts from farmland do occur is linked to the value of farmlands as developable property. Farmers under pressure from hostile neighbors and tempted by immediate, large monetary profits are often persuaded to sell out to land developers. These same land developers may not see a need to immediately develop the land, therefore, they often keep it in agricultural production, often to avoid higher taxes. These absentee landlords with no particular devotion or close ties to the land often rent it out to other farmers who also have no particular bond to this particular piece of property, and whose sole intention is to maximize profits. This in turn results in careless land tilling operations often including the cessation of maintenance and destruction of existing soil erosion and sediment control structures such as grassed swales, contour plowing, and runoff control berms. Renting farmers attempt to maximize profits and cut costs any way they can because they realize that the land is soon to be converted to house lots.

9.3.2.6 Potable Surface Water Intakes and Community and Industrial Water Supply Wells

Though not often thought of as agents of change, deterioration, or destruction, potable water surface intakes and community and industrial water supply wells are in fact just that. Surface water intakes, for example, in some parts of the nation, during periods of low flow, can by their withdrawals cause excessive lowering of waterway flows; consequently, water volume is reduced in the stream channel, thereby lessening the chance of diluting concentrations of pollutants, which in turn can induce stress in resident aquatic organisms. Similarly, community water supply wells have been known to draft so much water out of the ground that the surrounding ground surface has subsided by 5 to 6 ft.

We also have other reasons for including these features in our inventory. These components of man-made infrastructure provide a resource (potable water) critical to human survival. Therefore, protection of the waterways, reservoirs, and groundwaters they withdraw from should receive some priority in the MMP.

10 Assessing Community Health, Analyzing the Data, and Setting Objectives and Strategies for the Municipal Master Plan

It is doubtful if many of the nation's municipalities could honestly say that they have a good grasp of the "health" of their community ecological infrastructure and/or natural resources. This has been one of the major omissions and deficiencies of localized land use planning. Fortunately, state, federal, and sometimes county environmental protection and health agencies have been somewhat more diligent, establishing large-scale ambient monitoring networks to collect the scientific data needed to make these types of assessments. The New Jersey ambient monitoring networks shown in Figures 10.1 to 10.3.* for example, are being used to monitor and assess the water quality chemistries of the state's surface and groundwaters and the health of aquatic insect populations so important to the aquatic food chain. Nationwide, other state and federal programs also are tracking the condition of threatened and endangered species, the quality of our air, the health of our forests and woodlands, and the health of the nation's valuable wildlife resources including fish. These ambient data are supplemented by decades of intensive research and other special studies, also carried out under the auspices of many of these same groups as well as academic institutions.

Much of this vital, scientific information while available for some time,** has been languishing in various county, state, and federal repositories. This is not to say that all these data are in an immediately usable form. Some information requires further extraction, analyses, and interpretation (as a number of our figures show) because it may only be available, at least initially, on a spatial scale larger than the boundaries of a single municipality (e.g., watershed scale, Figure 10.4). As we discussed earlier, we also can utilize what are now collectively referred to as *indicators,* which when reviewed and/or compared with or against each other begin to provide us with some holistic sense of a community's ecological well-being. In this regard, three generic categories of indicators have received some recognition and acceptance in general practice and are utilized here. These include *stressor*, *environmental or exposure*, and *response* indicators. Examples of each are given in a prior chapter. However, as we begin to list items under each category, their definitions become much clearer. Also note that we are placing considerably more emphasis on Geographic Information System (GIS) generated maps to display the spatial relationship of these indicators to community boundaries, to individual parcel boundaries, and between the indicators themselves. This is at the heart of our new ecologically based municipal master plan (MMP). See Appendix A for a listing of potential sources (web sites) of these data.

You should not be surprised to find that this list of indicators reveals as much good about your community's natural environment as it does bad (except, perhaps, in areas that are totally urbanized). Healthy and unhealthy conditions will be equally important in setting objectives for your ecologically based MMP.

* All color figures appear after p. 82

** Much, if not all, of the nation's water quality monitoring data, for example, has been stored for decades in the U.S. Environmental Protection Agency (USEPA) national water quality database called STORET, which is accessible to the public including the nation's local land planners.

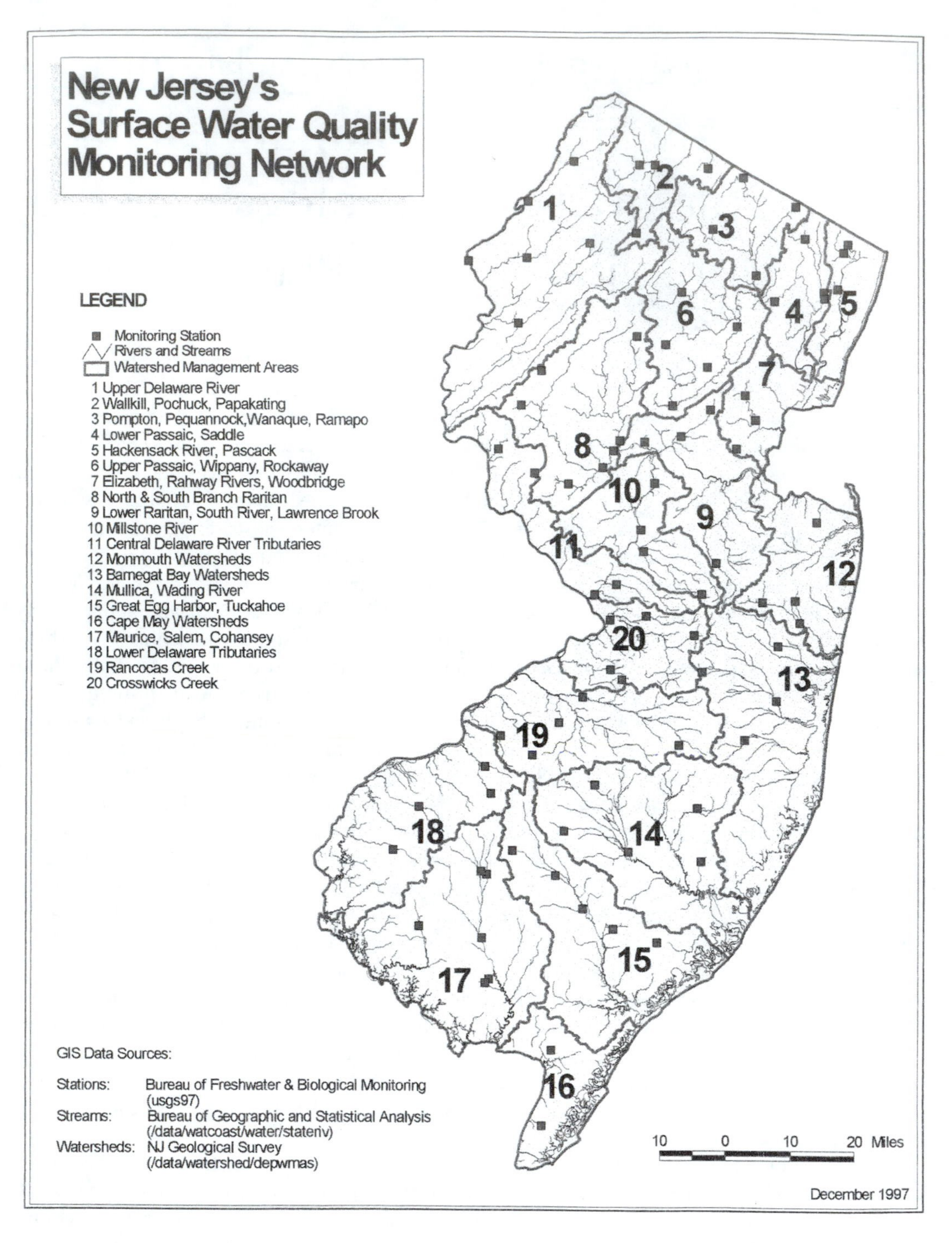

FIGURE 10.1 (From Bureau of Freshwater and Biological Mont., New Jersey Department of Environmental Protection.)

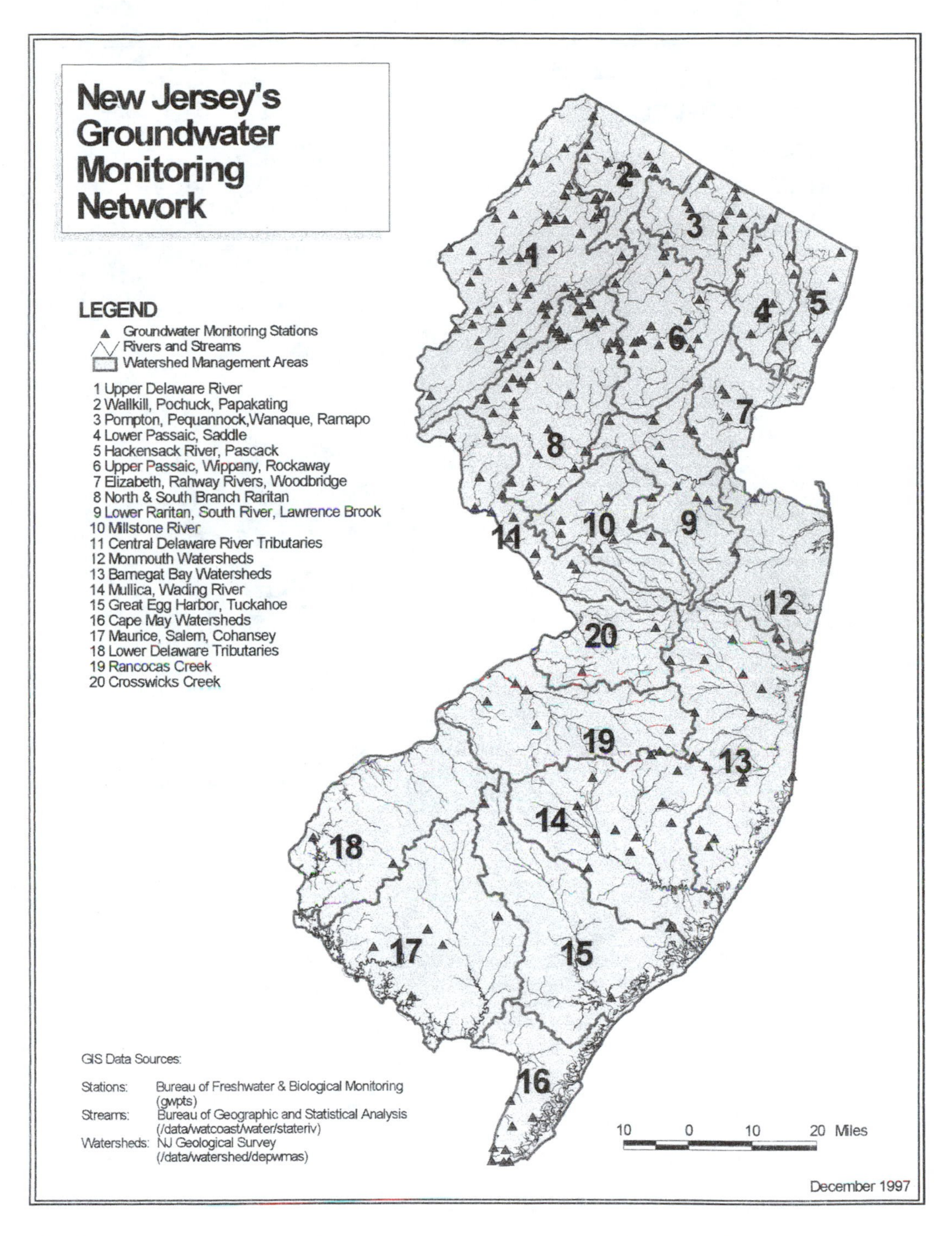

FIGURE 10.2 (From Bureau of Freshwater and Biological Mont., New Jersey Department of Environmental Protection.)

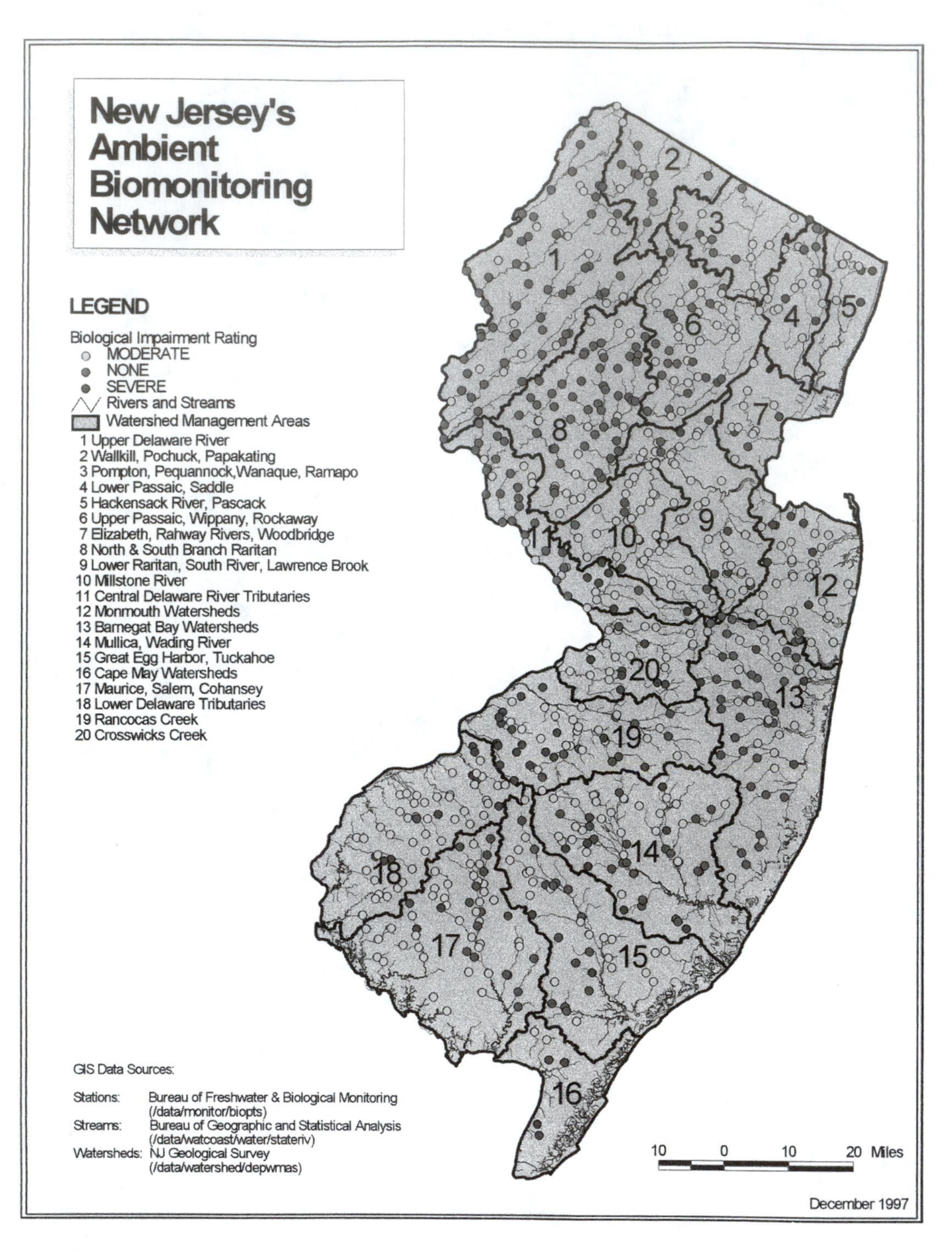

FIGURE 10.3 (From Bureau of Freshwater and Biological Mont., New Jersey Department of Environmental Protection.)

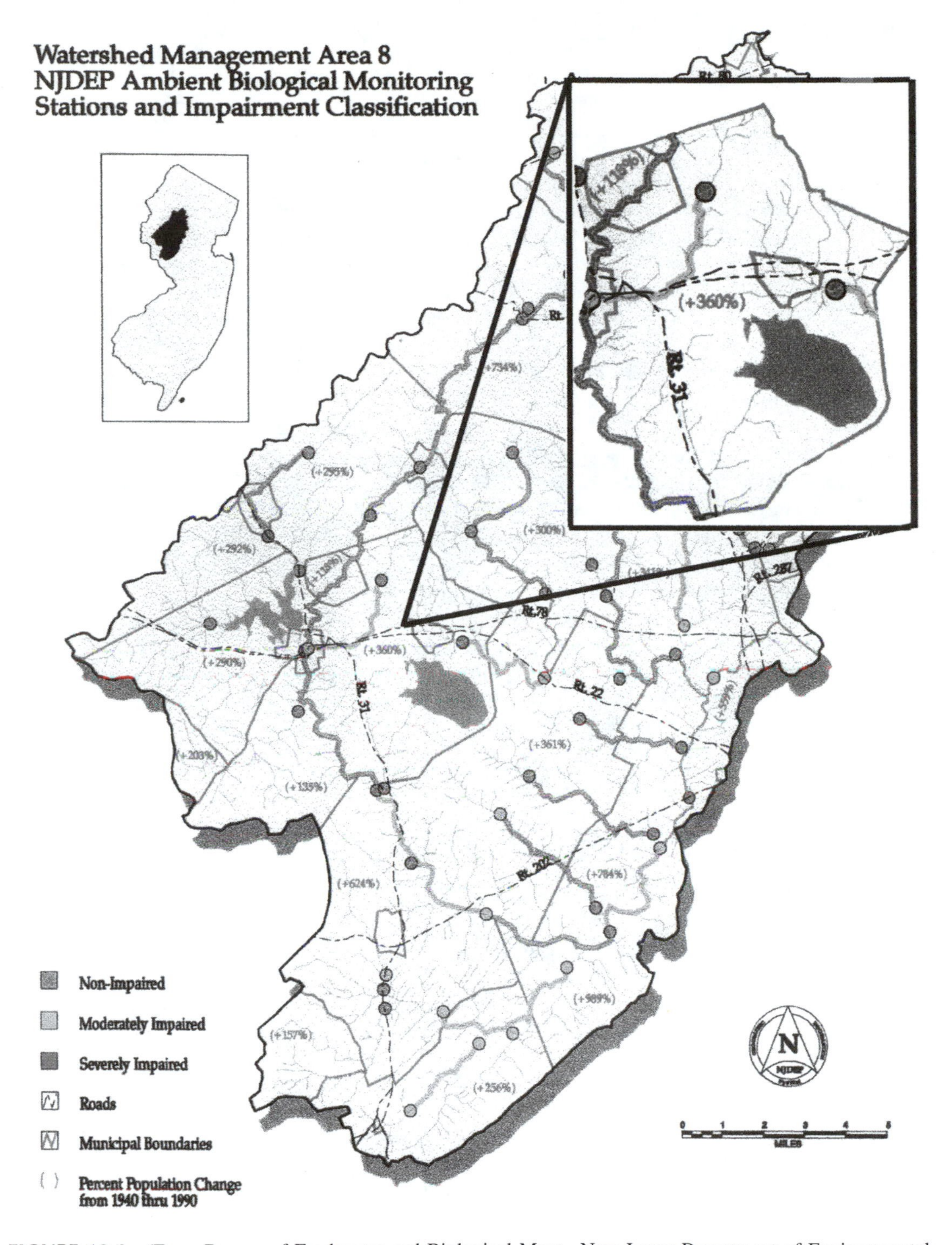

FIGURE 10.4 (From Bureau of Freshwater and Biological Mont., New Jersey Department of Environmental Protection.)

10.1 STRESSOR INDICATORS

Please note that this list of stressor indicators is not meant to be all-inclusive. You may wish to include other indicators not specifically mentioned here. This must be decided on a municipality-by-municipality basis. This also applies to the other categories of indicators that follow:

1. *Number and location of all national pollution discharge elimination system (NPDES) permitted point source dischargers in the community*: Data may also include the number of permit violations for each facility for the preceding 5 years.
2. *Pounds of pollutants discharged per year from each NPDES permitted point source discharger*: Individual pollutants may be selected as shown in Figure 10.5, which tracked the annual load of biochemical oxygen demanding (BOD5) substances from multiple dischargers into various waterways in a large watershed. Such information gives some idea of the loading or demand on these surface waterways to assimilate or dilute pollutants.
3. *Number, size, and location of all storm water outfalls in each subwatershed.**
4. *Percentage of impervious cover existing in each subwatershed*: Imperviousness is a particularly useful indicator to estimate the impacts of land development on aquatic ecosystems. Studies have linked the amount of impervious ground cover in a watershed directly to substantial changes in the hydrology, habitat structure, water quality, and biodiversity of aquatic ecosystems. Increased imperviousness seriously alters the hydrological dynamics of the landscape, thereby increasing runoff volume, and the frequency, duration, and extent of flooding along downstream waterways.
5. *Location and size of all storm water retention and detention facilities in each subwatershed.*
6. *Pounds of pesticides applied annually to agricultural land* (Figure 10.6).
7. *Pounds of pesticide applied annually to lawns and gardens.*
8. *Number and location of sanitary landfills and contaminated sites including those undergoing remediation* (Figures 10.7 and 10.8).
9. *Location of sanitary sewers with excessive infiltration and inflow.*
10. *Location of areas of failing septic systems.*

10.2 ENVIRONMENTAL OR EXPOSURE INDICATORS

1. *The concentration of pollutants measured in the surface waters* (Figure 10.9) *and sediments in each subwatershed, and plotted trends shown over at least a 10-year period*: Concerning sediments, in particular, certain chemicals in water tend to bind to particles and collect in bottom sediments. When present at elevated levels, these chemically contaminated sediments can kill or otherwise harm bottom dwelling organisms such as benthic macroinvertebrates. Pollutants in sediments also can accumulate in aquatic organisms and move up the food chain to fish, shellfish, and eventually humans. Because of these effects, the presence of contaminated sediment is a good indicator of the current health of a waterway.
2. *The number and location of contaminated community potable water supply wells* (Figure 10.10).
3. *The number and location of contaminated residential and/or private potable water supply wells.*
4. *The location and areal extent of identified plumes of contamination in the groundwater* (Figure 10.11).

* Identifying, measuring the size of, and accurately locating (by Global Positioning System [GPS]) all storm water outfall pipes along all of a municipality's waterways is an excellent field project for citizen volunteer groups who want to have a larger role in protecting the environment.

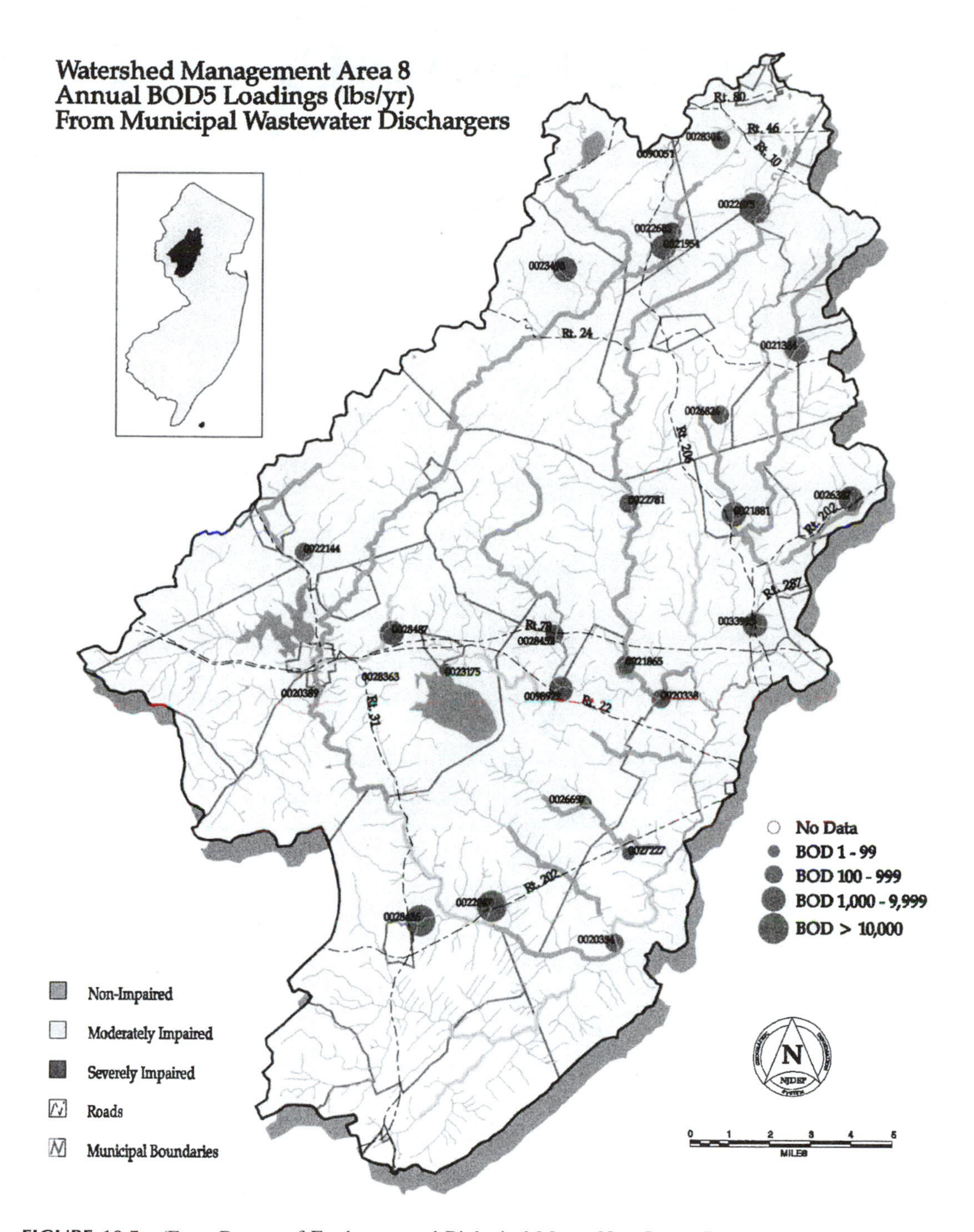

FIGURE 10.5 (From Bureau of Freshwater and Biological Mont., New Jersey Department of Environmental Protection.)

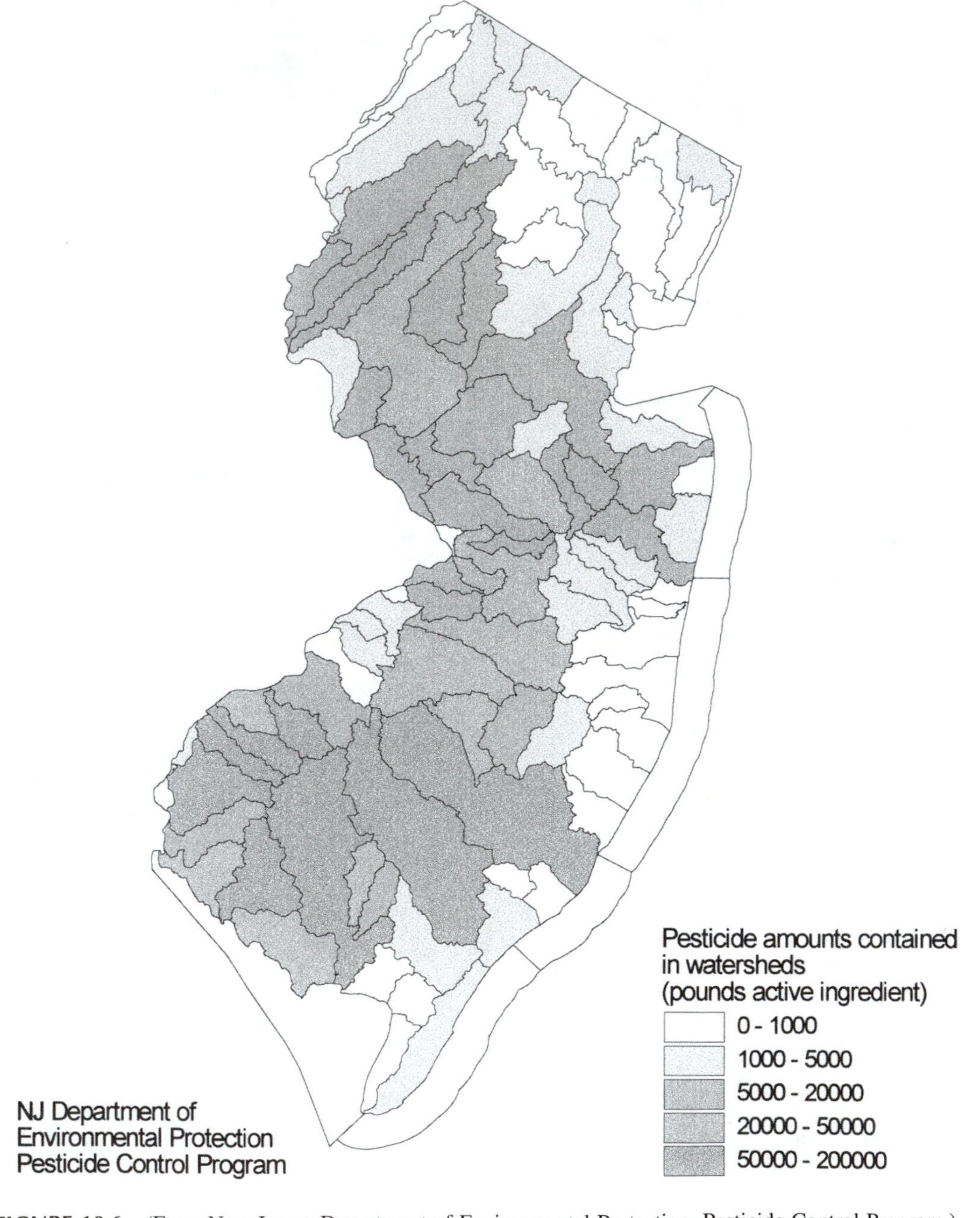

FIGURE 10.6 (From New Jersey Department of Environmental Protection, Pesticide Control Program.)

FIGURE 10.7 Most abandoned hazardous waste sites cannot be entered without special protective clothing. This site once housed a pesticides manufacturing facility and the soils on and around the site are heavily contaminated. The author is shown preparing to collect soil samples.

5. Areas of identified surface soil contamination (i.e., residual pesticides or other materials left over from prior agricultural practices or spills from other sources).
6. Location of sites where soil contamination exceeds soil cleanup criteria.

10.3 RESPONSE INDICATORS

1. *Impaired or healthy benthic macroinvertebrate population data for each subwatershed including deformities identified* (Figure 10.12)*: Benthic macroinvertebrates are macroscopic invertebrate animals inhabiting aquatic habitats. In freshwater, common forms are aquatic insects (Figure 10.13), worms, snails, and crustaceans. Macroinvertebrates are ubiquitous, fulfilling an important role in the aquatic food web. Species comprising the instream macroinvertebrate community occupy distinct niches governed by environmental conditions and their tolerance to pollution. Changes in environmental conditions are reflected by commensurate shifts in macroinvertebrate community structure. Some assessment of ambient water quality can then be based on standardized measures of said changes in community structure (Figures 10.14 and 10.15). These evaluations, called *bioassessment ratings*, are based on a statistical analysis of multiple subsets of 100 organisms collected at varying locations along each waterway. The standard assessment protocol was developed by the U.S. Environmental Protection Agency (USEPA) and is termed the *rapid bioassessment protocol*,[49] of which there are several levels. The advantages of using benthic macroinvertebrates are further elaborated in Table 10.1. Interestingly, an increasing number of studies also reveals that benthic macroinvertebrate

---

* **Nonimpaired** means an instream condition comparable to the best that can be expected within a given ecoregion, or the station exhibiting an optimum macroinvertebrate community structure for that size and type of stream.

Moderately impaired is a condition where the stream is exhibiting signs of biological stress, such as the loss of pollution intolerant macroinvertebrate species and/or a reduction in macroinvertebrate diversity.

Severely impaired means an instream condition where only a few pollution-tolerant macroinvertebrate species are present, though a few kinds may be present at high densities. A significant loss of species diversity is observed.

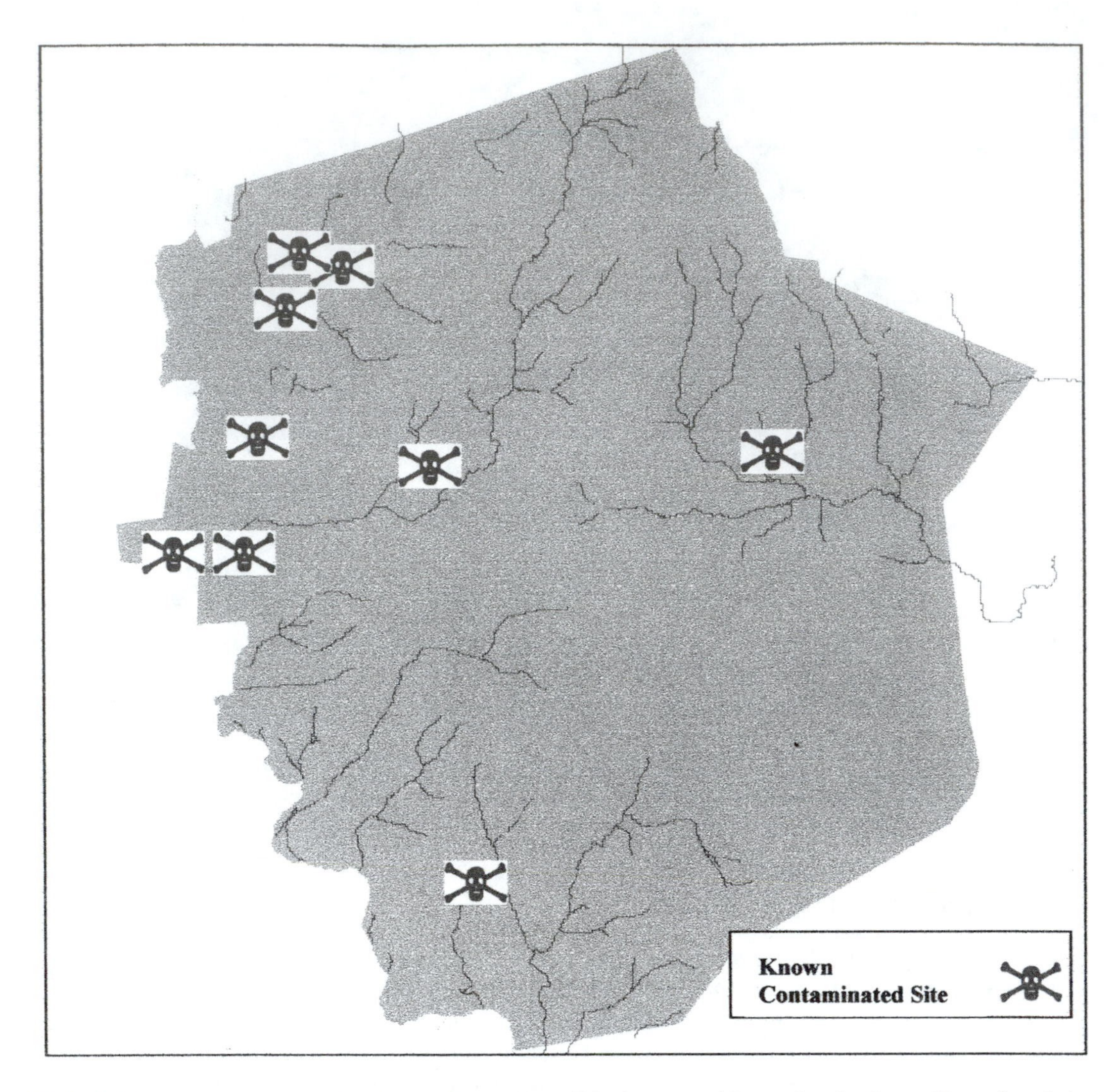

FIGURE 10.8 (Adapted from New Jersey Department of Environmental Protection databases. Example only.)

deformities (Figure 10.16) may, by themselves, be a good indicator of chemical contamination, particularly of stream bottom sediments.[50–53]

2. *Stressed and healthy fish population data for each subwatershed, including lakes* (Figure 10.17): There are a number of advantages of using fish as indicators, including the following:
   a. Fish are good indicators of long-term (several years) effects (Figure 10.18) and broad habitat conditions because they are relatively long-lived. Ross et al.[54] and Matthews[55] found that stream fish assemblages were stable and persistent for 10 years, recovering rapidly from droughts and floods, indicating that large population fluctuations are unlikely to occur in response to purely natural environmental phenomena.
   b. Fish communities generally include a range of species that represent a variety of trophic levels (omnivores, herbivores, insectivores, planktivores, and piscivores). They tend to integrate effects of lower trophic levels; thus, fish community structure is reflective of integrated environmental health.
   c. Fish are at the top of the aquatic food chain and are consumed by humans, making them important subjects in assessing contamination.

WATERSHED MANAGEMENT AREA # 8

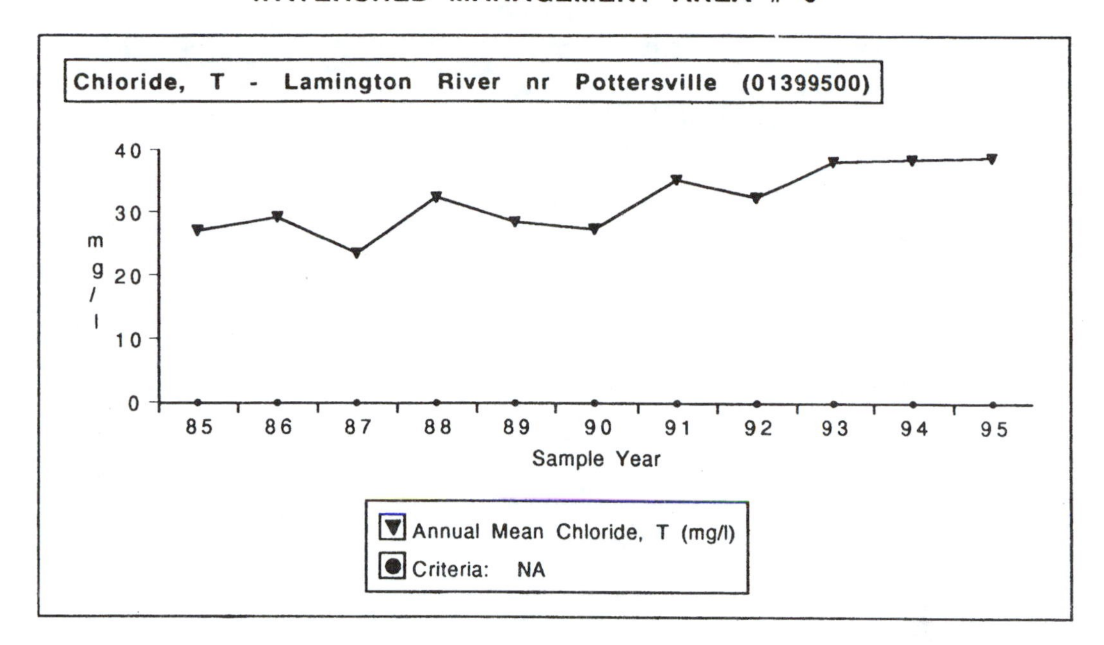

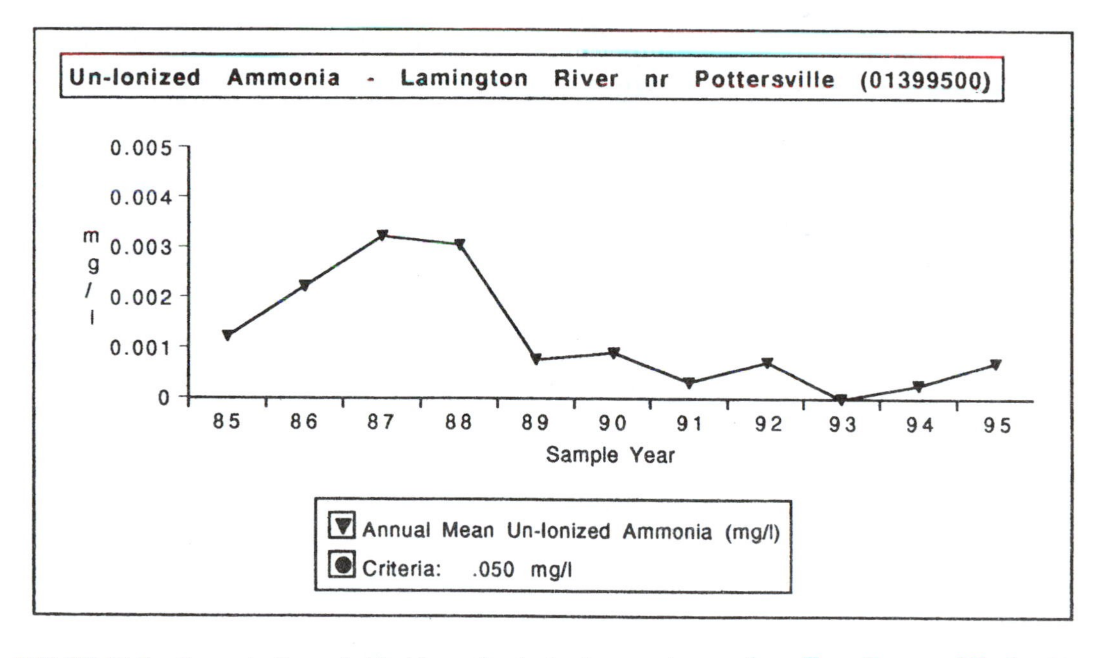

FIGURE 10.9 Concentrations of chlorides and unionized ammonia over time. (From Bureau of Freshwater and Biology Mont., New Jersey Department of Environmental Protection.)

   d. Monitoring fish communities provides direct evaluation of "fishability," which emphasizes the importance of fish to anglers and commercial fisherman.
   e. Fish account for nearly half of the endangered vertebrate species and subspecies in the United States.

3. *Number and location of fish consumption bans*: Fish consumption advisories are a good indicator of bioaccumulation of toxic substances in fish and shellfish. Bioaccumulation

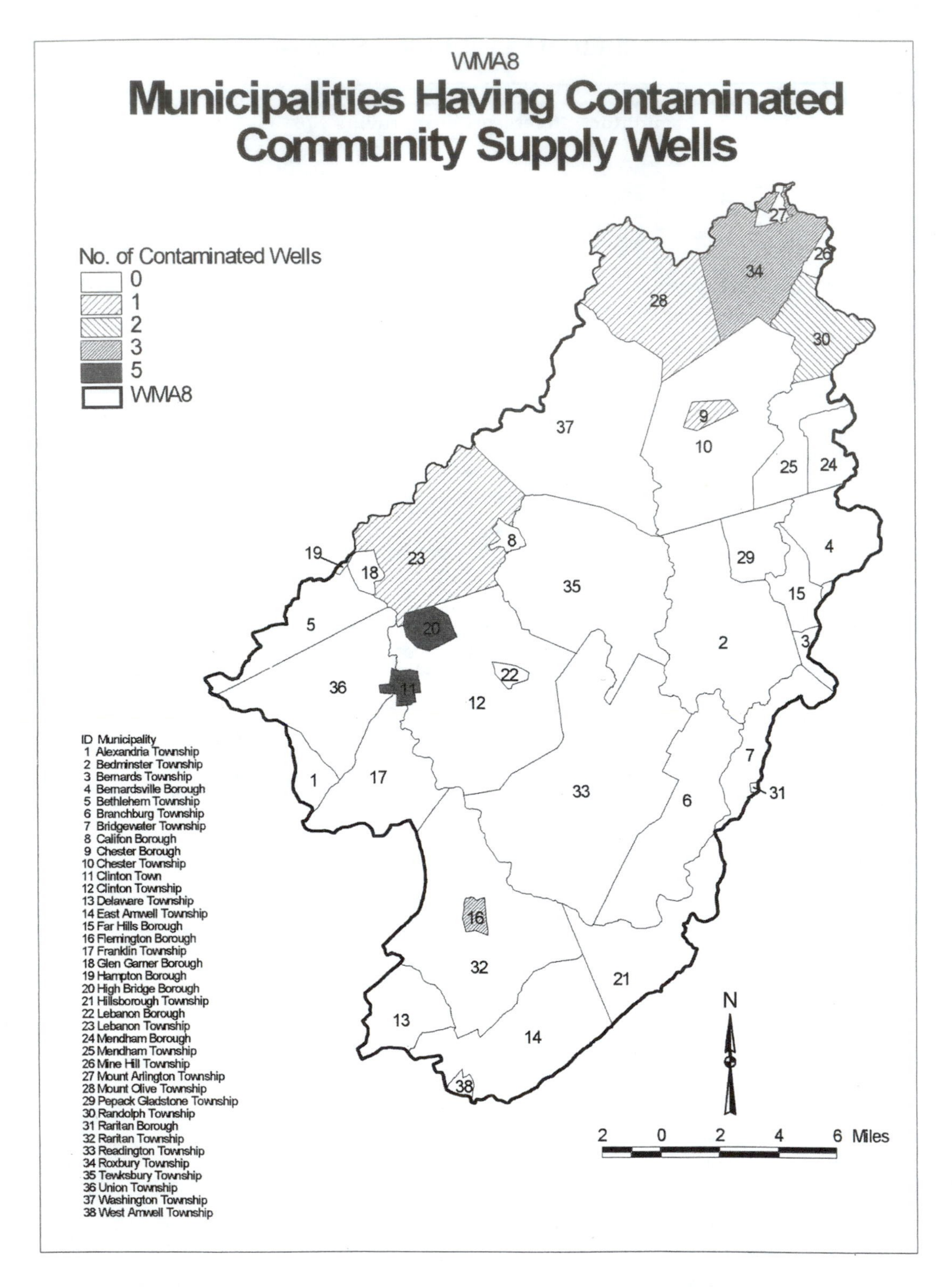

FIGURE 10.10 (From New Jersey Department of Environmental Protection, GIS program and other databases. Example only.)

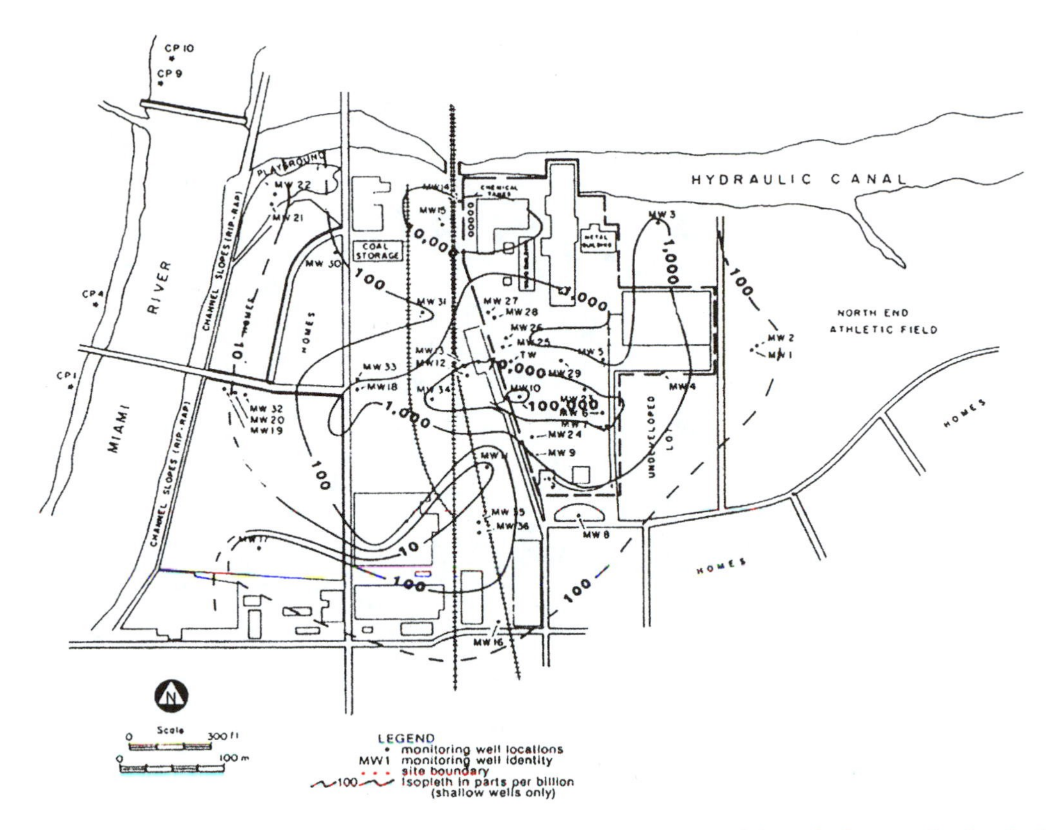

FIGURE 10.11 Example Superfund site with groundwater contaminated by volatile organic chemicals. (Adapted from U. S. Environmental Protection Agency, *Handbook Ground Water*, EPA/625/6-87/016, 1987. With permission.)

is the process by which fish accumulate pollutants in their tissues by eating smaller organisms already contaminated with the pollutant. Pollutants also can enter fish and shellfish directly from the surrounding water through the gills during respiration or feeding or by absorbance through the skin. These pollutants may cause fish and shellfish to be unsafe for human consumption. State environmental protection agencies monitor fish to determine whether levels of contamination in fish pose a threat to the health of the people who eat them. Where fish contamination levels exceed safe levels, states often issue advisories to the public recommending some limitation on fish consumption or no consumption of fish at all (Figure 10.19). Advisories also may target a specific subpopulation at risk, such as children, pregnant women, and nursing mothers.

4. *Number and location of deformed amphibians*: In 1995, some middle school students on a field trip to a farm pond in southern Minnesota discovered large numbers of frogs with misshapen, extra or missing limbs (Figure 10.20). About 50% of the northern leopard frogs they caught were malformed. Since then, amphibian malformations from other parts of North America have been reported. Species that have been reported with malformations include northern leopard frogs, wood frogs, bullfrogs, green frogs, mink frogs, gray tree frogs, Pacific tree frogs, spring peepers, American toads, long-toed salamanders, tiger salamanders, and spotted salamanders. Malformed amphibians are not a new phenomena, but reports were only infrequent until very recently. Since 1995, reports have become increasingly common. Hypotheses as to causes abound. It is possible that such

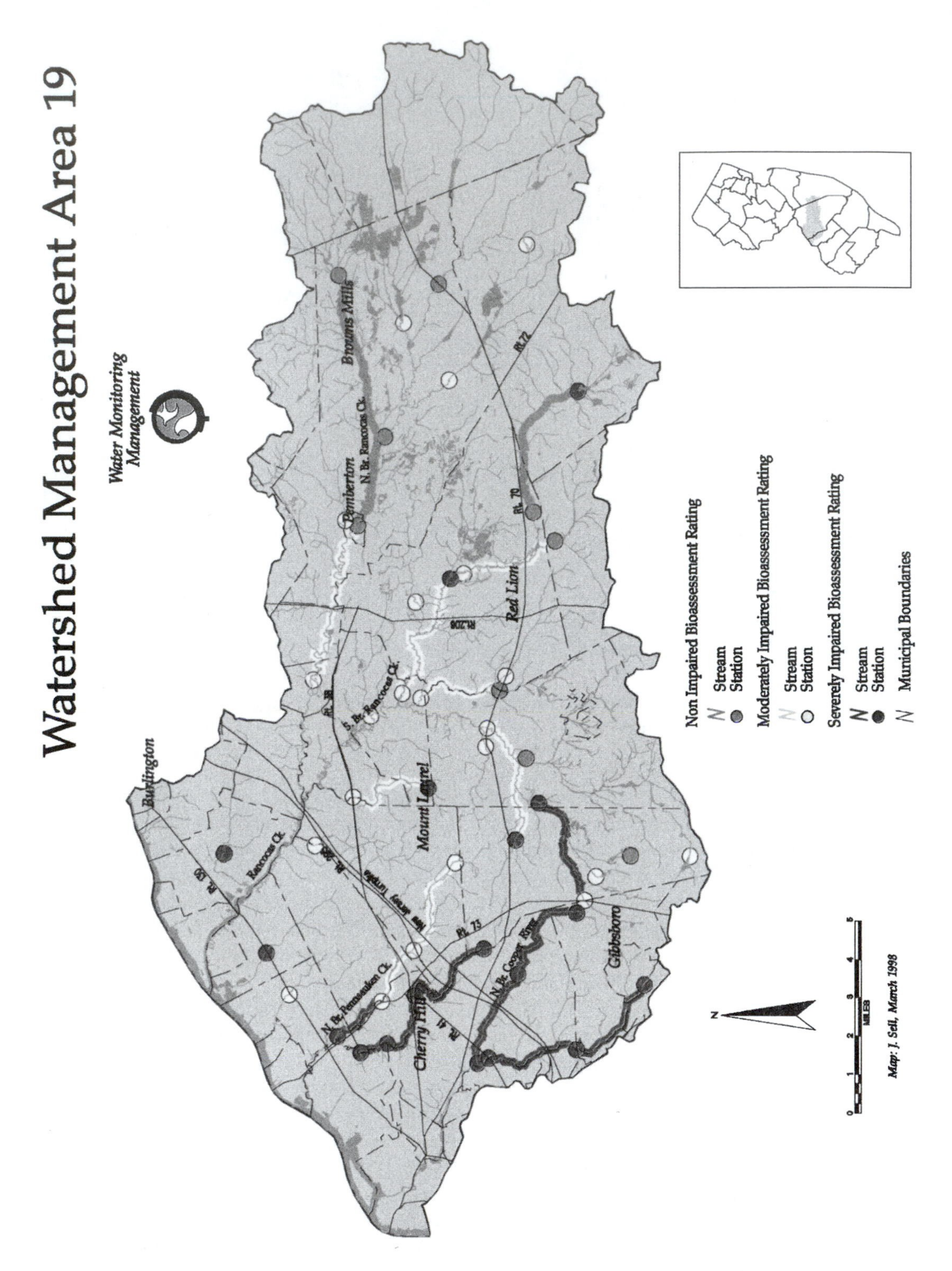

FIGURE 10.12 Biological impairment in a watershed as measured by the benthic macroinvertebrates. (From Bureau of Freshwater and Biological Mont., New Jersey Department of Environmental Protection. With permission.)

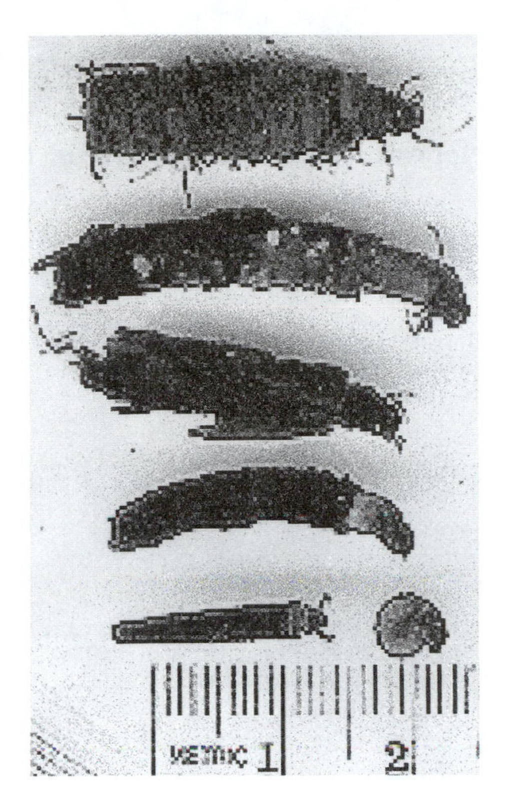

FIGURE 10.13 Some typical benthic macroinvertebrates (caddis fly larvae). (From Bureau of Freshwater and Biological Mont., New Jersey Department of Environmental Protection. With permission.)

malformations occur due to chemicals that humans add to the environment — such as herbicides, insecticides, or fungicides. For example, some insecticides mimic a growth hormone that may cause amphibian embryos to develop abnormally. Biologists in Quebec found a higher rate of malformed amphibians in areas with a history of pesticide use than in sites not known to be exposed to pesticides.[56] Others are suggesting that chemicals in their breeding habitat may make these amphibians more susceptible to diseases or parasites.* Some scientists are even hypothesizing that increased ultraviolet radiation (a result of Earth's thinning ozone layer) is breaking down supposedly nontoxic chemicals into toxic ones. Without a clearer understanding of the cause it is difficult to determine if human health is also at risk. The possibility that these malformed amphibians indicate a greater environmental problem is a very real concern.

5. *Density of white-tailed deer populations* (Figure 10.21).
6. *Number of threatened and endangered species actually identified and their location.*
7. *Waterway habitat quality assessment by subwatershed*: Waterways with channel widening, bank undercutting, and chronic tree fall are of special concern (see previous Figure 9.35).
8. *Identification and location of all hypereutrophic lakes and ponds, with identified episodes of massive algae blooms, fishkills, excessive weed growth, and heavy sedimentation.*
9. *Identification of those reaches of the community's surface waterways that* fail *to meet adopted chemical* (Figure 10.22) *or biological water quality standards; conversely, the*

* Some very recent research appears to link parasites to at least some of these deformities. Those studies indicate that trematodes, simple parasitic flatworms, can infect the developing legs of tadpoles, causing them to grow multiple hind legs. However, stress due to some other environmental conditions may still be the main contributing factor, making the tadpoles more susceptible to infestations by such parasites.

Aquatic Organisms as Environmental Indicators

Benthic Macroinvertebrates Usually Indicative of *Good* Water Quality

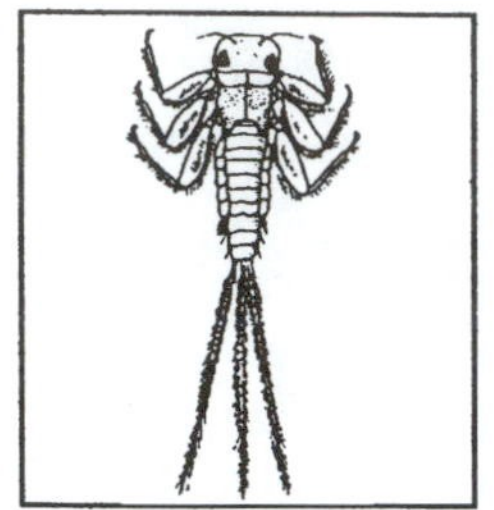

Figure 1 - Mayfly

Mayfly nymphs are often abundant wherever the water is clean. They are sensitive to various types of water pollution, including low dissolved oxygen, ammonia, biocides, and metals.

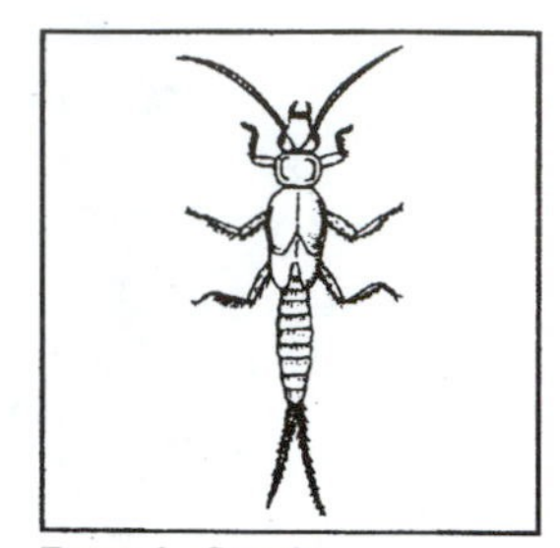

Figure 2 - Stonefly

Stonefly nymphs are usually found only in cool, well-oxygenated waters free of pollution. Though not usually found in the numbers characteristic of mayflies, the presence of even a few stoneflies is indicative of good water quality.

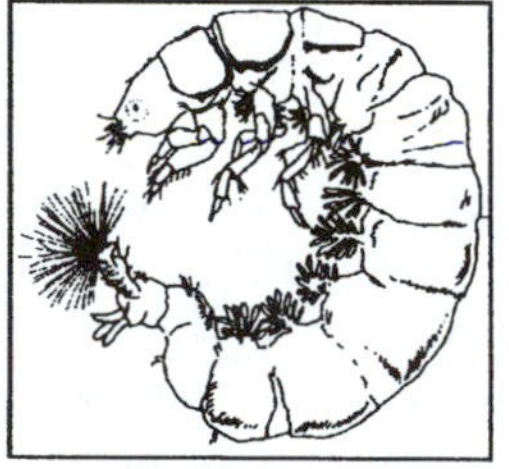

Figure 3 - Caddisfly

Most caddisfly larvae, many of whom build portable cases of stones, sticks, sand and other detritus, are intolerant of water pollution.

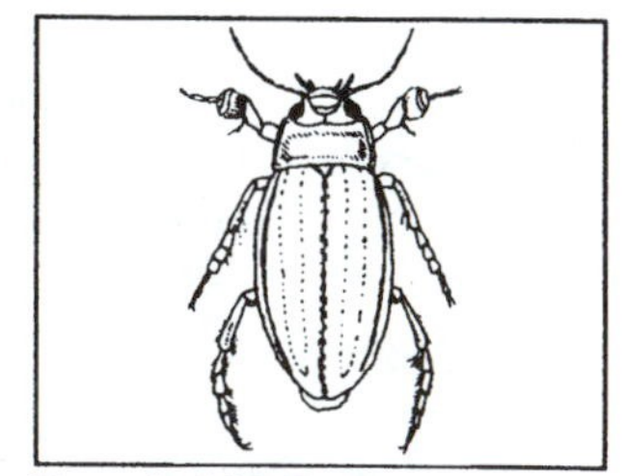

Figure 4 - Adult Beetle

Aquatic beetles are common in well oxygenated, swiftly running waters; many species are referred to as "riffle beetles." They are usually indicative of clean water since they are sensitive to wetting agents (soaps and detergents) and other pollutants.

Illustrations modified from W.B. Clapham, Jr., "NATURAL ECOSYSTEMS," The Macmillan Company, New York, 1973.

FIGURE 10.14 (From Bureau of Freshwater and Biological Mont., New Jersey Department of Environmental Protection.)

identification of those reaches of the community's surface waterways that do meet *all adopted water quality standards.*

10. *Identification of all surface water bodies, including lakes, ponds, and reservoirs exhibiting sediment toxicity.*
11. *Number and location of noise complaints due to highway traffic.*

Aquatic Organisms as Environmental Indicators

Benthic Macroinvertebrates Usually Indicative of *Poor* Water Quality

Figure 5 - Midge Larvae

Midges (chironomids) are among the most common of aquatic invertebrates. They occupy a variety of aquatic habitats, including lakes, ponds, bogs, rivers, creeks, and marshes. They even exploit manmade habitats such as sewage treatment plants, water treatment plants, fish pools, irrigation ditches, and bird baths. Many species are very tolerant of pollution.

Aquatic sowbugs are an important freshwater isopod, abundant in waters enriched with organic nutrients and low in dissolved oxygen. They are commonly observed in the recovery areas below sewage treatment plants.

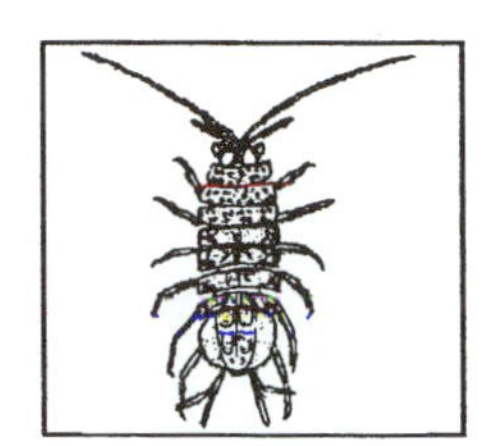

Figure 6 - Sowbug

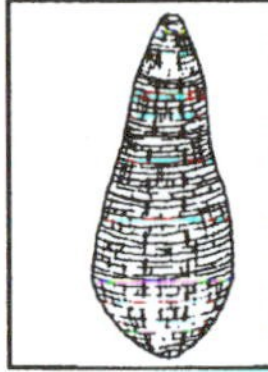

Figure 7 - Leech

Leeches and other segmented worms are very common in our lakes and streams, though not often noticed. They are tolerant of poor water quality and severe pollution.

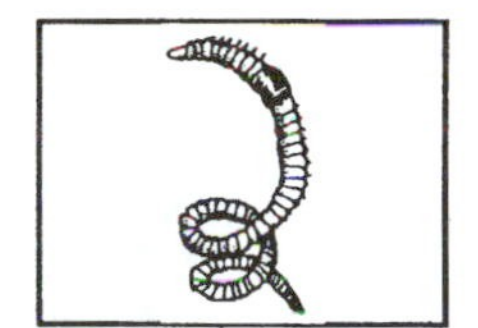

Figure 8 - Tubifex Worm

Black fly larvae are filter feeders, capturing and ingesting plankton and bacteria from the surrounding water with specialized antennae. Some species are very tolerant of poor water quality and thus can be used as indicators of pollution.

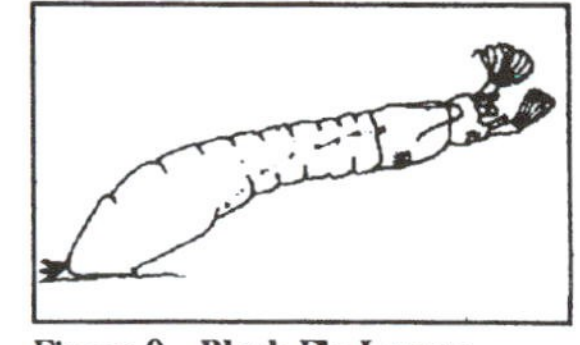

Figure 9 - Black Fly Larvae

Illustrations modified from W.B. Clapham, Jr., "NATURAL ECOSYSTEMS," The Macmillan Company, New York, 1973

FIGURE 10.15 (From Bureau of Freshwater and Biological Mont., New Jersey Department of Environmental Protection.)

12. *Location of dead and dying vegetative communities*: In addition to the attributes afforded by vegetation in general and discussed previously, vegetation also can be a useful indicator, both of its own health and that of the surrounding environs. For instance, on land surfaces where excessive amounts of liquid food processing waste and partially treated domestic wastewater have been applied (Figure 10.23), it is not uncommon to

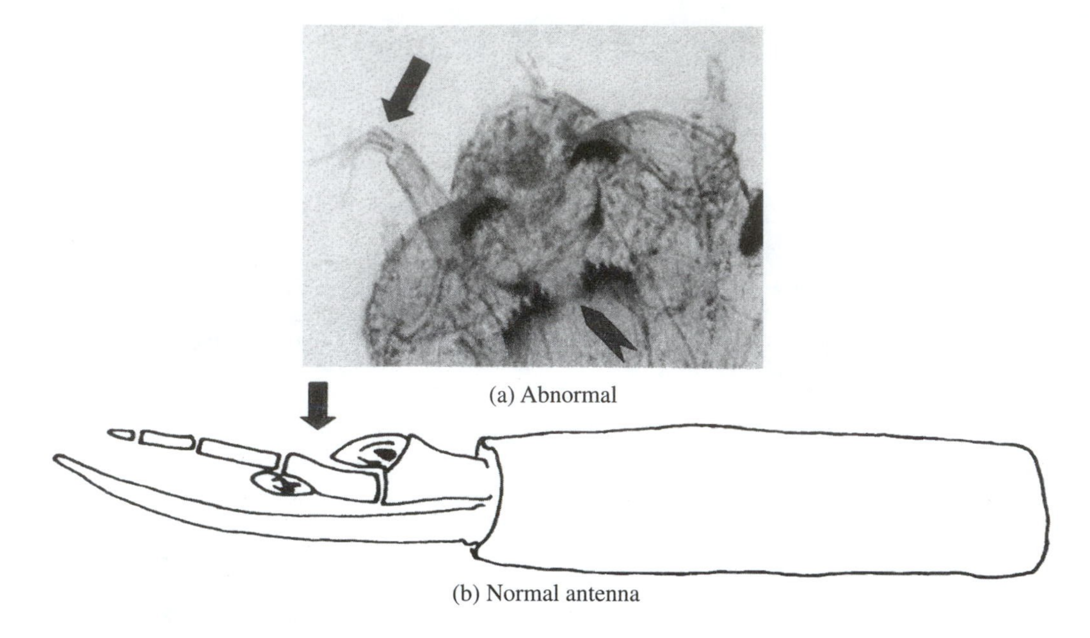

(a) Abnormal

(b) Normal antenna

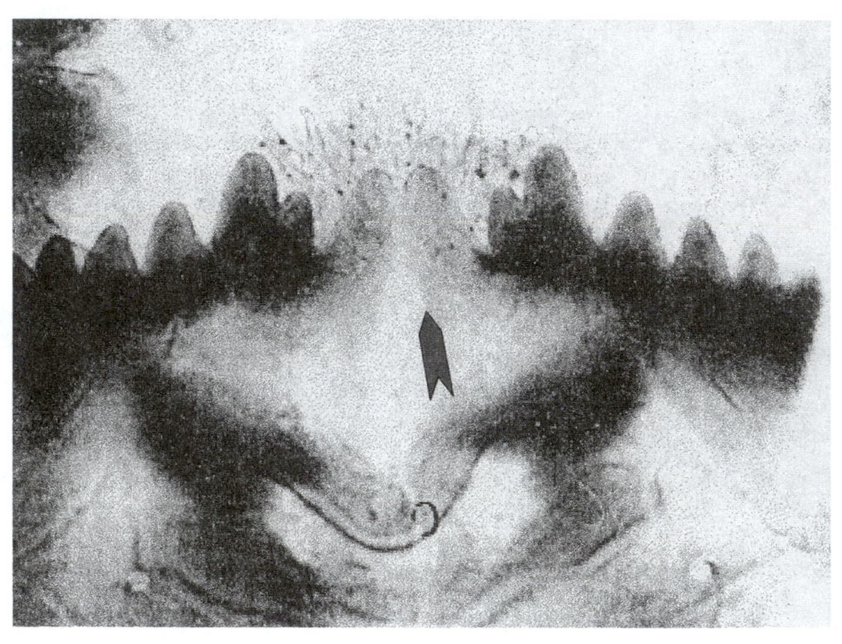

Normal teeth

(c) Normal teeth

FIGURE 10.16 Note the abnormal teeth and antenna in the top picture compared to the bottom pictures which depict normal antenna and teeth. The normal pictures on the bottom (b and c) are magnified to show detail. (From: (a) J. Kurtz, NJDEP; (b) *An Introduction to the Aquatic Insects of North America*, 2nd ed., R. W. Merrit & K. W. Cummins, Kendall/Hunt Publ. Co., 1988; (c) *A Key to Some Larvae of Chironomidae (Diptera) From the Mackenzie & Porcupine River Watersheds*, D. R. Oliver, D. McClymont, & M. E. Roussel, Fisheries & Marine Service Technical Report #791, 1978.)

see on-site and downslope off-site vegetation exhibiting dramatic shifts in plant community diversity and taxa, including at times the wholesale death of an entire community of indigenous plants. Where mesic perennial shrubs and herbs and a mature forest canopy once stood, sedges, rushes, and other water-loving and water-tolerant plants now prosper

TABLE 10.1
Advantages of Using Benthic Macroinvertebrates

1. Are good indicators of localized conditions of water quality due to their limited mobility; as such, are well suited for the assessment of site-specific pollution impacts
2. Are sensitive to environmental impacts from both point and nonpoint sources of pollution
3. Integrate the effects of short-term environmental variations, such as oil spills and intermittent discharges
4. Sampling relatively easy and inexpensive
5. Are holistic indicators of overall water quality, even for substances at lower than detectable limits
6. Are normally abundant in New Jersey waters
7. Serve as the primary food source for many species of fish important commercially and, recreationally
8. Are a direct measure of water quality degradation in a manner closely aligned with the goals of the Clean Water Act, unlike chemical monitoring, where impacts to the environment are by inference, not direct measurement
9. Can be used to assess nonchemical impacts to the benthic habitat, such as by thermal pollution or excessive sediment loading (siltation)
10. To the general public, impacts to resident benthic macroinvertebrate communities offer more tangible measurements of water quality than more esoteric listings of chemical analysis results
11. When used together with chemical/physical parameter monitoring, can be used to identify sources of impairment

Note: Biological monitoring cannot replace chemical monitoring, toxicity testing, and other environmental measurements; each of these tools provides the analyst with specific information only available by that procedure.

Source: Adapted from New Jersey Department of Environmental Protection, Ambient Biomonitoring Network, Raritan River Drainage Basin, 1993-1994, Benthic Macroinvertebrate Data, Executive Summary, Trenton, NJ, July 1995.

FIGURE 10.17 These biologists are collecting samples of fish using a technique known as electroshocking, which allows the fish, if not sacrificed for tissue analysis, to be returned to the water unharmed after measurement and species identification. (Courtesy New Jersey Division of Fish, Game, and Wildlife, New Jersey Department of Environmental Protection.)

amid the skeletons of water-logged oaks and pines. Unfortunately, it is not uncommon also to see the eventual death of even water-tolerant plant communities due to toxicity of the pollutants in the wastewater being discharged. Such signs of plant community shifts and outright death should be a warning sign to community planners that the groundwater might also be in jeopardy, as well as any downstream water bodies that might be receiving this polluted runoff. Vegetation is also a good indicator of air pollution.

FIGURE 10.18 These adult channel catfish, collected as part of an annual fish survey, are showing abnormally high rates of bacterial and fungal infections. Such unusually high instances of infection may be a warning sign and certainly would warrant further investigation.

Vegetation, sometimes a thousand feet or more downwind of some industrial operations, even under today's tough restrictions, is often little more than a forest of lifeless, blackened tree trunks. If these operations can do that to the vegetation, they might also be doing the same to the lungs of community residents at the same time. Finally, vegetation also can be an excellent indicator of wildlife populations, particularly large mammalian herbivores, such as elk and deer, that may be exceeding their habitat's carrying capacity (Figure 10.24).

13. *Density and number of deer-related automobile accidents by lane mile*: These statistics may help corroborate deer density data and serve as further proof that populations of these large mammals are in need of reduction in certain areas.
14. *Number and location of areas of crop damage or damage to landscape shrubbery caused by deer or other wildlife.*
15. *Location of sites where wildlife populations, particularly large herbivores, show signs of starvation and/or overpopulation.*
16. *Number and location of streams on which episodes of flooding have been increasing.*
17. *Number and location of subdivisions or groups of homes where individual wells have dried up due to overwithdrawal of groundwater.*
18. *Acres of prime agricultural soils lost due to coverage by impervious surfaces.*
19. *Number of acres of forested land lost due to fragmentation by development.*

10.4 ANALYZING THE DATA AND SETTING THE OBJECTIVES AND STRATEGIES

At this point, one of the major complaints heard from local land planners is that they are simply overwhelmed by the amount of data, scientific and otherwise, that is available to them. They are often are confused as to what to do with all the data. This is good, not the confusion part, but the fact that the amount of information is so overwhelming. It means that land planners have done their homework. Conversely, they often commend the inventory and information-gathering phases

DO YOU LIKE TO FISH?

B-55

Good! Fishing is fun! But, be careful. Some fish in Minneapolis/St. Paul lakes have poisons that make the fish unsafe for you to eat. The fish may not taste, smell or look bad but the poisons in the fish may make you or your child sick after eating them for a long time.

Learn how to protect yourself!

1. **Eat fish that are <u>SAFE</u>.**

These fish have very few poisons in them and are safe to eat. Most people can eat one meal of these fish 1 time a week. But pregnant or nursing women and children under age 6 should eat these fish no more than once a month.

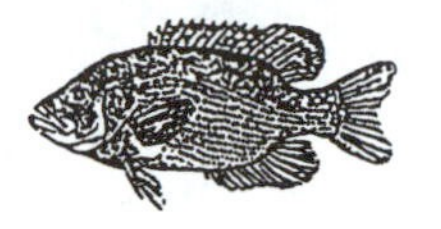

bluegill crappie rock bass perch

2. **Eat fish that are <u>NOT SAFE</u> less often.**

These fish have the most poisons in them. Pregnant or nursing mothers and children under age 6 should never eat these fish. Others should eat these fish once a month or less. It is best to eat the smallest and youngest of these fish.

carp sucker white bass catfish buffalo

3. **Do not eat the fatty parts of fish.**

Some poisons build up in the fatty parts of fish. It is best if you cut off the fatty parts before you cook the fish. Throw away the water that fish have been cooked in. Do not make fish soup.

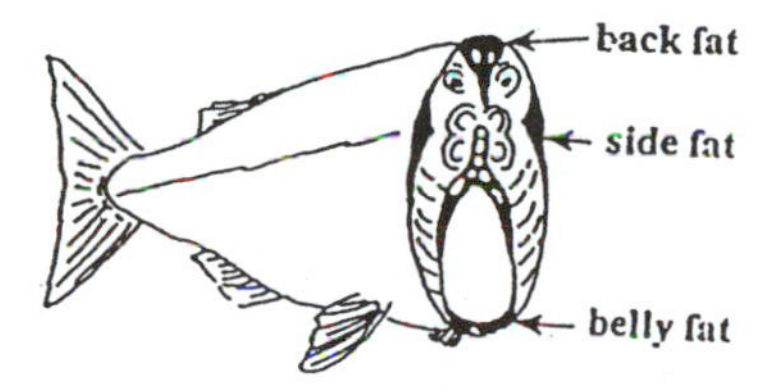

4. **Eat fish from safe lakes.**

A. Bluegill, sunfish and crappie from these lakes are SAFE to eat every day.

Big Carnelian (May Township)
Byllesby Reservoir (Randolph Township)
Coon (Columbus Township)
Elmo (Lake Elmo)
Forest (Forest Lake)
Long (New Brighton)
Minnetonka (Minnetonka)
Parley (Laketown)
Rebecca (Hastings)
Snelling (Fort Snelling State Park)
Wasserman (Laketown)
Wirth (Minneapolis)
Big Marine (New Scandia)
Christmas (Shorewood)
Crystal (Burnsville)
East Vadnais (Vadnais Heights)
Harriet (Minneapolis)
Medicine (Plymouth)
O'Dowd (Shakopee)
Pickerel (Lilydale)
Rebecca (Greenfield)
Waconia (Waconia)
White Bear (White Bear Lake)

B. Bluegill, sunfish, crappie, pike and small walleye (less than 20 inches) from these lakes are SAFE to eat every day.

Big Carnelian (May Township)
Big Marine (New Scandia)
Byllesby Reservoir (Randolph Township)
Christmas (Shorewood)
Coon (Columbus Township)
Crystal (Burnsville)
Elmo (Lake Elmo)
Harriet (Minneapolis)
Long (Long Lake)
Pickerel (Lilydale)
Snelling (Fort Snelling State Park)
Wasserman (Laketown)
White Bear (White Bear Lake)

Most fish from the Mississippi, Minnesota and St. Croix Rivers in the Minneapolis/St. Paul area, are **NOT SAFE** to eat.

MINNESOTA DEPARTMENT OF HEALTH • SECTION OF HEALTH RISK ASSESSMENT • 925 S.E. DELAWARE ST., MINNEAPOLIS, MN 55414 • 612/627-5046

FIGURE 10.19 Typical fish advisory. (From U.S. Environmental Protection Agency, Guidance for Assessing Chemical Contaminant Data For Use in Fish Advisories, EPA 823-R-95-001, March 1995.)

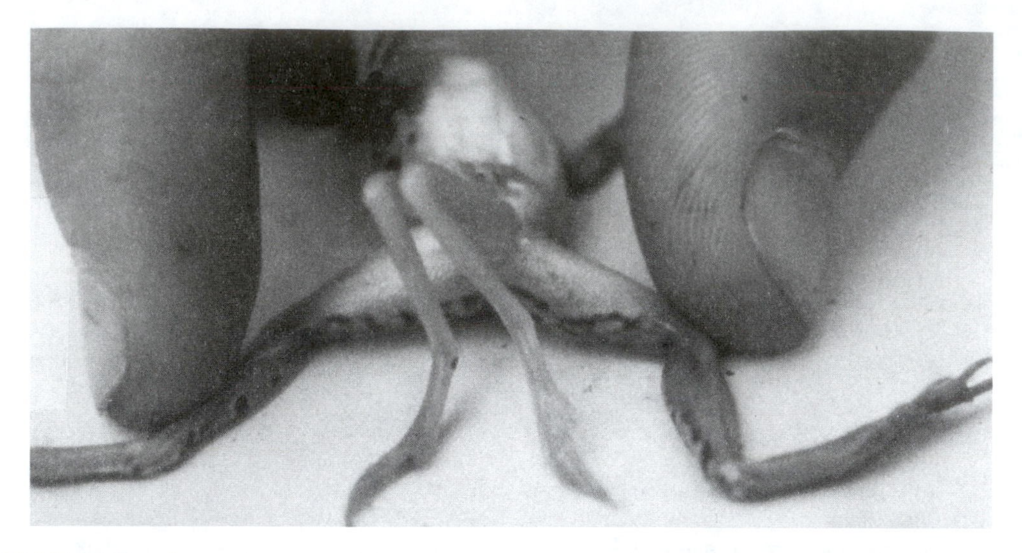

FIGURE 10.20 Can such abnormalities be a warning sign for us? (From Minnesota Pollution Control Agency. With permission.)

as worthwhile undertakings, with most admitting that they know more about the overall health of their community's natural resources than they did before they started. This new knowledge, however, often makes it difficult for planners to return to the status quo of myopic, site-by-site plan review, because they now know that there are a substantial number of ecological implications that extend far beyond the borders of individual site plans. I consider this to be a plus because that is exactly the mindset we are trying to encourage among the nation's local land planners.

10.4.1 Example Applications

In any event, we are now ready to proceed to the final step in the process — that of analyzing the data and establishing objectives and/or strategies for incorporation into the MMP. Where to begin is undoubtedly one of the initial questions most commonly asked. For this we must fall back on the single most important premise of the ecologically based MMP — that of preventative planning. Thus, we should be looking first at those portions of the ecological infrastructure and natural resources that, to date, remain relatively unimpacted by the activities of man. Our intent, of course, is to devise strategies to keep them that way. Regretfully, due to space constraints, we are unable to analyze and develop objectives and strategies for all the potential ecological issues and concerns that have been identified in the previous sections of this chapter and in Chapter 9. We do, however, present some example applications as they pertain to some of the key elements of our ecological infrastructure, notably water resources, wildlife, wetlands, and soils. This representative cross section of examples provides a sufficient number of templates and concepts that local land planners can then easily copy or adapt to fit the needs of their particular community.

10.4.1.1 Surface Waterways

At this point we begin our query of the data by examining our list of *response indicators*, in particular, the instream quality assessments provided by the collection and analyses of bottom-dwelling benthic macroinvertebrates. No one indicator, however, whether response, *environmental* or exposure, or *stressor*, should be viewed separately. Each indicator holds clues that ultimately need to be viewed collectively for proper assessments to occur and appropriate decisions to be made. Although we are beginning with an individual indicator, we quickly find the process evolving into more collective evaluations. Benthic macroinvertebrates, as we discussed earlier, are macroscopic invertebrate animals inhabiting aquatic habitats. In freshwater, the most common forms are

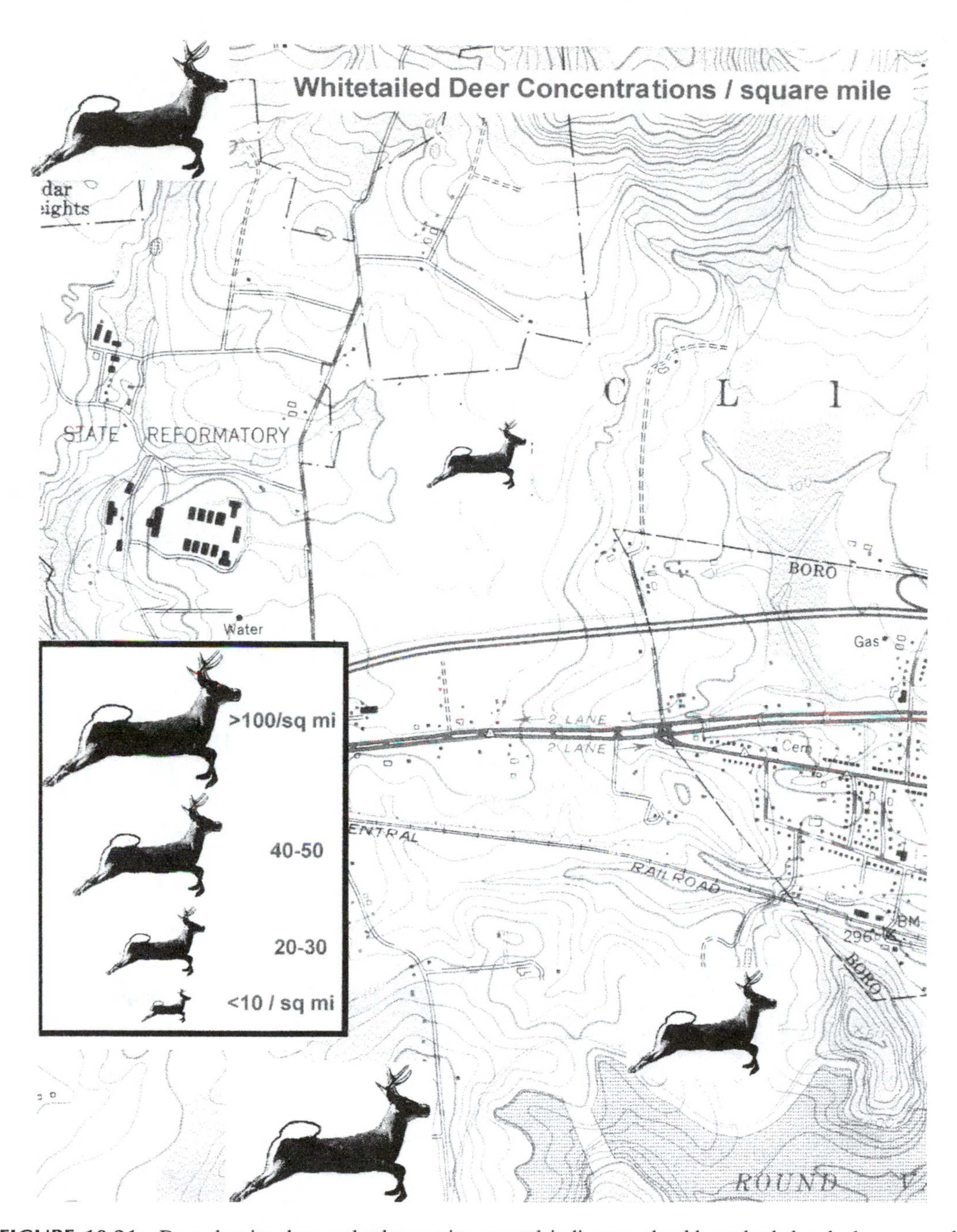

FIGURE 10.21 Deer density data and other environmental indicators should put both local planners and developers on notice that the management of such large mammals must become an important consideration in every land development project.

insects, worms, snails, and crustaceans. These organisms play an important role in the *aquatic food web* and occupy distinct ecological niches governed by ambient environmental conditions and their *tolerance* to *pollution.* When ambient environmental conditions change, as a result of introduced pollution and sometimes other physical disturbances, there are commensurate and observable *changes* and *shifts* in macroinvertebrate community structure, although these communities of organisms often display a surprising resiliency before they finally succumb. Pollution-tolerant macroinvertebrates thrive and pollution-intolerant macroinvertebrtaes decrease and even disappear. Some approximations can, therefore, be made of *ambient water quality,* based on standardized

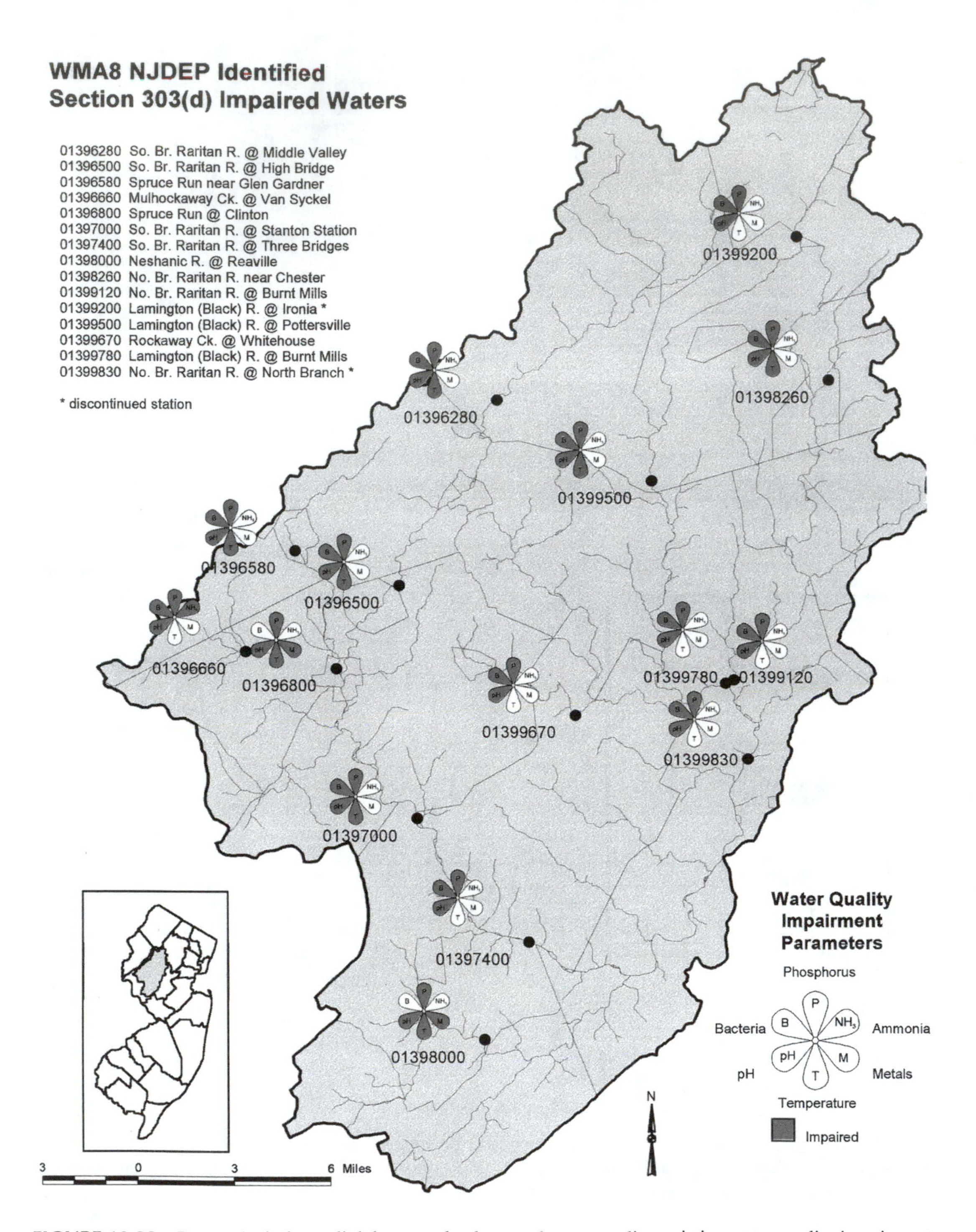

FIGURE 10.22 Due to the intimate link between land use and water quality, existing water quality impairments of local streams should be clearly identified in the municipal master plan. Such data can significantly bolster community arguments for wider buffer zones or enhanced wastewater treatment. (From Bureau of Freshwater and Biological Mont., New Jersey Department of Environmental Protection.)

FIGURE 10.23 The owner of this land application (spray irrigation) site has applied more wastewater to the land than the soil can tolerate. As a result, indigenous vegetation has been killed and potable water supply wells in an adjacent residential area have been contaminated.

FIGURE 10.24 The high browse line on these red cedar trees indicate that the whitetailed deer population in the area is at or near the carrying capacity of the vegetative cover to sustain them. (Courtesy New Jersey Division of Fish, Game, and Wildlife, New Jersey Department of Environmental Protection.)

FIGURE 10.25 This small brook trout has never known the inside of a commercial hatchery. His presence is an excellent indicator of relatively clean water quality.

measures of these surrogate populations. Please refer to previous Figures 10.14 and 10.15, which illustrate some of the representative benthic macroinvertebrates and their relationships to good and poor water quality. These assessments are based largely on a methodology developed by the USEPA in 1989 that is entitled, "*Rapid Bioassessment Protocols for the Use in Streams and Rivers*."[49] Some states have modified the EPA protocols somewhat to accommodate localized conditions.

Such assessments historically have been carried out by state, and sometimes federal agencies, but due to the relative simplicity of the protocol, local environmental and citizen groups are now supplementing state benthic macroinvertebrate monitoring networks with their own collected data and analyses, filling in gaps that may exist due to the spatial scale of the much larger state networks.

Thus, how do we use this particular data? Because *preventative planning* is a major premise of our ecologically based MMP, we need to look first at those areas where stream segments have been rated as *unimpaired* (see previous Figure 10.12). This is not to say, however, that unimpaired also could not be represented by the presence of a thriving population of wild and naturally reproducing trout (Figure 10.25) and salmon. Because such unimpaired stream segments are becoming an increasingly scarce commodity, a major *objective* of the MMP should be to *prevent their degradation*. These areas, including their contributing watersheds, (if the community deems their preservation important) can be identified in the MMP as *aquatic life preservation areas* or in the case of the presence of wild trout (Figure 10.26), as *wild trout protection areas*. Creating these area designations administratively is a relatively simple process. The major question, however, is how do we actually accomplish this objective physically on the ground. Intuitively, we know that keeping the stream channel intact and adjoining riparian corridor undisturbed is imperative, and a common first step to accomplish this has been to establish parallel buffer zones (sometimes referred to as stream management zones* [SMZs]) on both sides of the stream.** Some local governments have been prone to establish a one size fits all, fixed buffer width for the entire length of a stream

* It should be noted that the buffer or SMZ described here does not necessarily equate to what has been described by scientists as the riparian zone — a specific ecosystem type that lies adjacent to surface water and reflects the influence of its proximity to surface water. Riparian zones are also not wetlands, although both are subject to saturation and inundation. The difference between the two is that riparian lands are not generally saturated at the frequency or for the prolonged intervals that wetlands are. Riparian zones can, however, include wetlands. Riparian zones almost always include the floodplains of a waterway.

** Buffer zones alone, however, may not completely protect the integrity of a waterway or the biodiversity and viability of the organisms that inhabit it (Harding, J. S., Benfield, E. F., Bolstad, P. U., Helfman, G. S., and Jones, E. B. D., III, Stream biodiversity: the ghost of land use past, *Proc. Natl. Acad. Sci. U.S.A.*, 95, 14843, 1998.) A combination of preventative measures applied on a whole watershed basis would be the best strategy.

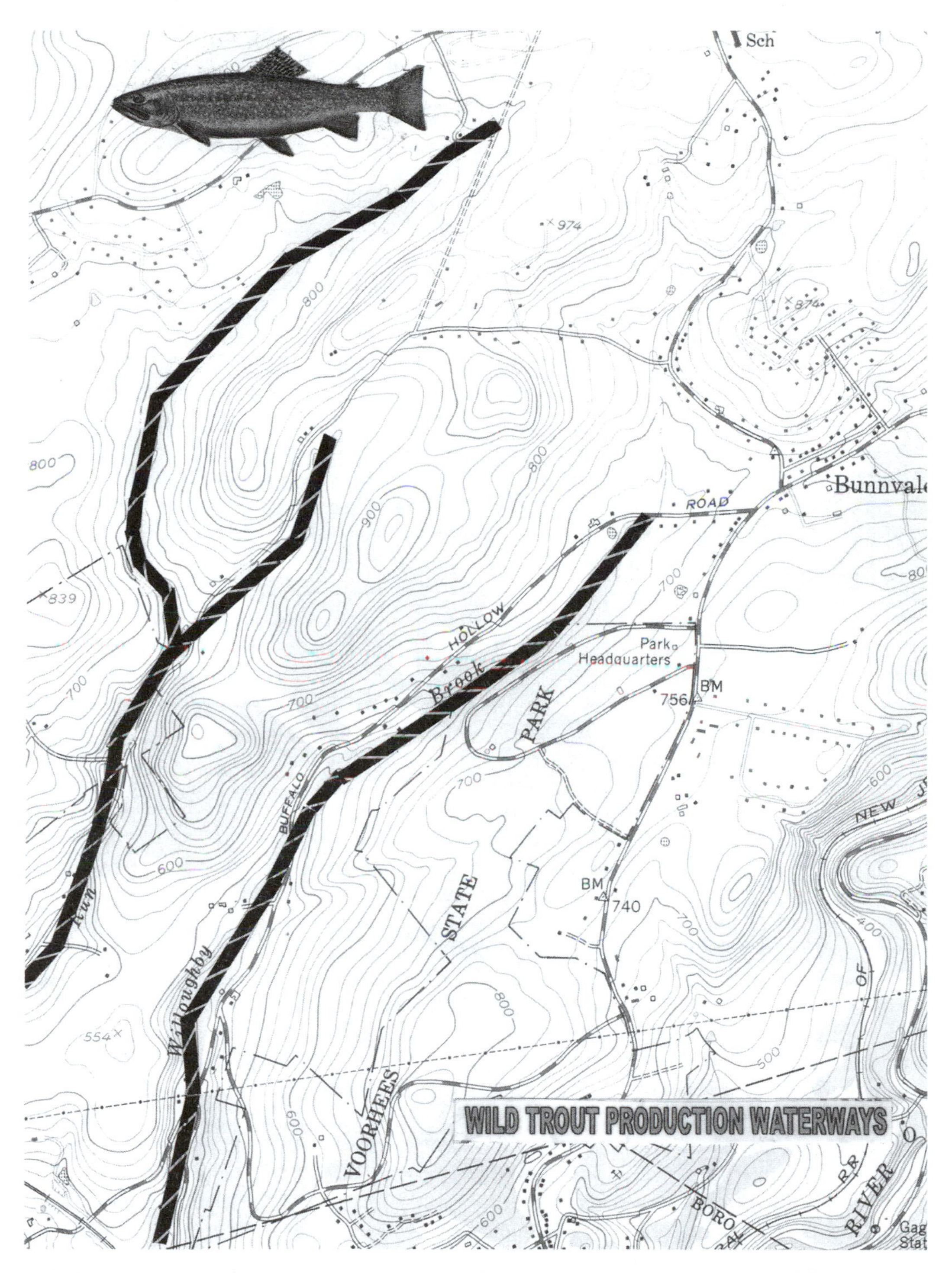

FIGURE 10.26 Wild trout populations can be an excellent barometer of watershed health and their protection should receive considerable priority in the municipal master plan.

and sometimes for every stream in the community irrespective of individual stream size, character of its contributing watershed, and varying physiography along each stream corridor and between streams and varying objectives. Some state governments have been similarly prone to establish fixed width management or protection zones along their waterways. Two northeastern states, in particular, are worth noting because they include the participation of local governments in their strategies. In 1996, the state of Massachusetts enacted what is now commonly referred to as the Massachusetts Rivers Protection Act, which established riverfront management areas, 200 ft wide along each side of all the state's 9000 mi of waterways (25 ft wide in municipalities with large populations and in densely developed areas). The act does not prohibit activities near rivers. Applicants, however, proposing work in the riverfront area must obtain a permit (called an order of conditions) from the *local conservation commission* [emphasis added] or the Massachusetts Department of Environmental Protection (DEP) on appeal. Applicants must demonstrate that their projects meet two performance standards — there are no practicable and substantially equivalent economic alternatives and no significant adverse impacts on the riverfront area to protect wildlife habitat, fisheries, shellfish, groundwater, and surface water supplies; and they must prevent flooding, storm damage, and pollution. Whether this approach can achieve the desired effect, that of preventing the degradation of the state's 9000 mi of waterways, remains to be seen. Our second state, the state of New Hampshire, in 1994 passed a Comprehensive Shoreland Protection Act RSA 483-B, which provided protection to all land within 250 ft of the "reference line" of public waters. The New Hampshire Department of Environmental Services (DES) is responsible for enforcing the standards within the protected shoreland areas, *unless a community adopts an ordinance or shoreland provisions that are equal to or more stringent than the state act* [emphasis added]. Two notable provisions are included. First, no fertilizer, except limestone, can be used within 25 ft of the reference line of any property; 25 ft beyond the reference line, low phosphate, slow release nitrogen fertilizer, or limestone may be used on lawns or areas with grass. Second, where existing, a natural woodland buffer shall be maintained within 150 ft of the reference line for the purpose of minimizing erosion, preventing siltation and turbidity, stabilizing soils, preventing excess nutrients and chemical pollution, maintaining natural water temperatures, maintaining a healthy tree canopy and understory, preserving fish and wildlife habitat, and respecting the overall natural condition (ecological integrity?) of the protected shore land. Figure 10.27 elaborates more fully the restrictions and permitted activities. Unfortunately, the act only applies to streams of fourth-order size (Strahler method) or higher. Because the quality of fourth-order streams is highly dependent on what happens to the quality of first- through third-order streams farther up in the watershed, this strategy may not be sufficiently protective in the long term based on our current knowledge. This author strongly recommends, especially if a community expects its delineations to survive judicial scrutiny, the use of variable buffer widths that are in concert with the variability of localized conditions.

Tremendous amounts of research have been conducted over the past several decades concerning the efficacy of stream buffers as mitigators of storm waterborne pollutants and the physical degradation of stream channels. In addition, the utility of these riparian corridors as both habitat and passageways for a variety of wildlife also has been extensively studied. Let us put this latter use aside for the moment, and look at the results and recommendations of those studies as they pertain to the function of riparian buffers as mitigators of stream pollution and degradation. Much of the data on buffer effectiveness has been derived from studies of buffers applied in an agricultural, silvicultural, or wastewater treatment setting plus studies of forest ecosystems and riparian forest buffers.[57] Regardless of where they have been applied and studied, the consensus of opinion among researchers is quite clear — adequately sized buffers can provide substantial protection to surface water bodies and the organisms that inhabit them. Summarized next are additional findings and recommendations from those same buffer studies that should be considered in buffer design:

1. Pollutant removal efficiency varies with the pollutant, soil type and condition, vegetative cover and condition, retention time, depth to water table, and slope.
2. Buffers are capable of removing the following pollutants: total suspended solids (TSS) with removal rates ranging from 45 to 99%; phosphorus with reported removal rates ranging from 23 to 96%; nitrogen with removal rates reported as high as 90%; petroleum

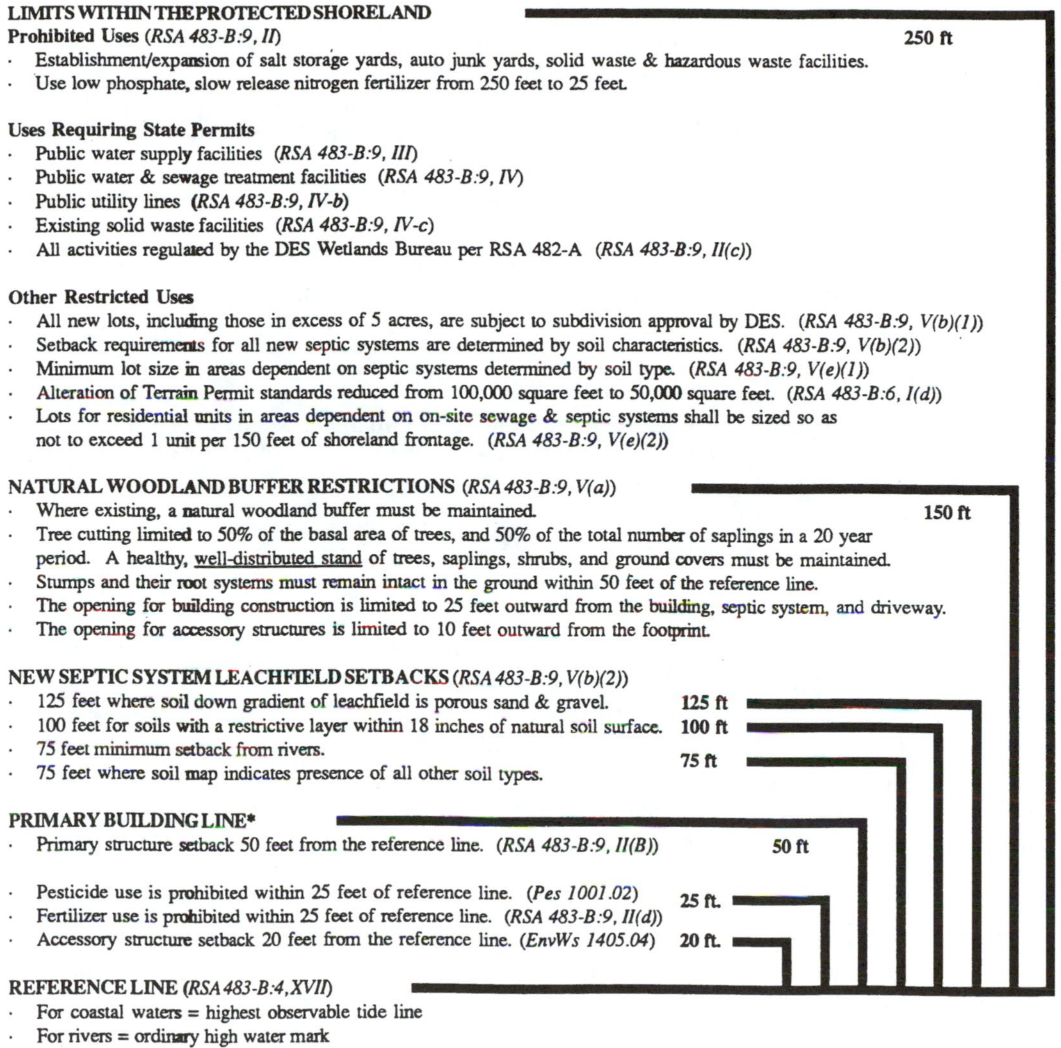

Minimum Shoreland Protection Standards, RSA 483-B

LIMITS WITHIN THE PROTECTED SHORELAND — 250 ft

Prohibited Uses (*RSA 483-B:9, II*)

- Establishment/expansion of salt storage yards, auto junk yards, solid waste & hazardous waste facilities.
- Use low phosphate, slow release nitrogen fertilizer from 250 feet to 25 feet.

Uses Requiring State Permits

- Public water supply facilities (*RSA 483-B:9, III*)
- Public water & sewage treatment facilities (*RSA 483-B:9, IV*)
- Public utility lines (*RSA 483-B:9, IV-b*)
- Existing solid waste facilities (*RSA 483-B:9, IV-c*)
- All activities regulated by the DES Wetlands Bureau per RSA 482-A (*RSA 483-B:9, II(c)*)

Other Restricted Uses

- All new lots, including those in excess of 5 acres, are subject to subdivision approval by DES. (*RSA 483-B:9, V(b)(1)*)
- Setback requirements for all new septic systems are determined by soil characteristics. (*RSA 483-B:9, V(b)(2)*)
- Minimum lot size in areas dependent on septic systems determined by soil type. (*RSA 483-B:9, V(e)(1)*)
- Alteration of Terrain Permit standards reduced from 100,000 square feet to 50,000 square feet. (*RSA 483-B:6, I(d)*)
- Lots for residential units in areas dependent on on-site sewage & septic systems shall be sized so as not to exceed 1 unit per 150 feet of shoreland frontage. (*RSA 483-B:9, V(e)(2)*)

NATURAL WOODLAND BUFFER RESTRICTIONS (*RSA 483-B:9, V(a)*) — 150 ft

- Where existing, a natural woodland buffer must be maintained.
- Tree cutting limited to 50% of the basal area of trees, and 50% of the total number of saplings in a 20 year period. A healthy, well-distributed stand of trees, saplings, shrubs, and ground covers must be maintained.
- Stumps and their root systems must remain intact in the ground within 50 feet of the reference line.
- The opening for building construction is limited to 25 feet outward from the building, septic system, and driveway.
- The opening for accessory structures is limited to 10 feet outward from the footprint.

NEW SEPTIC SYSTEM LEACHFIELD SETBACKS (*RSA 483-B:9, V(b)(2)*)

- 125 feet where soil down gradient of leachfield is porous sand & gravel. — 125 ft
- 100 feet for soils with a restrictive layer within 18 inches of natural soil surface. — 100 ft
- 75 feet minimum setback from rivers. — 75 ft
- 75 feet where soil map indicates presence of all other soil types.

PRIMARY BUILDING LINE* — 50 ft

- Primary structure setback 50 feet from the reference line. (*RSA 483-B:9, II(B)*)

- Pesticide use is prohibited within 25 feet of reference line. (*Pes 1001.02*) — 25 ft.
- Fertilizer use is prohibited within 25 feet of reference line. (*RSA 483-B:9, II(d)*)
- Accessory structure setback 20 feet from the reference line. (*EnvWs 1405.04*) — 20 ft.

REFERENCE LINE (*RSA 483-B:4, XVII*)

- For coastal waters = highest observable tide line
- For rivers = ordinary high water mark
- For natural fresh waterbodies = natural mean high water level
- For artificially impounded fresh waterbodies = water line at full pond

* If a municipality establishes a shoreland setback for primary buildings, whether greater or lesser than 50 feet, that defines the Primary Building Line for that municipality.

FIGURE 10.27 (From New Hampshire Department of Environmental Services, Environmental Fact Sheet WD-BB-35. With permission.)

hydrocarbons and metals with removal rates ranging from 40% for lead, as well as greater than 60% for copper, zinc, aluminum and iron.[57]

3. Surface runoff across the buffer should provide, wherever possible, an average residence time of 9 min[58] so that there is sufficient time for settling and filtering — the two primary mechanisms for removal of particulate-related pollutants. It should be further noted that these mechanisms work best when the velocity of the runoff across the buffer is less than 1 ft/s.

4. Buffers have a finite capacity to remove pollutants; when these thresholds are exceeded, pollutants may not be attenuated and previously sequestered pollutants may be remobilized.
5. Buffers on slopes greater than 10% are less effective in pollutant removal. That is to say that given equal buffer widths, less steep slopes provide more effective treatment. Consequently, steeper slopes require wider buffer widths.
6. Grassed buffers when properly designed and maintained can remove particulate-associated pollutants; however, they cannot generally recycle pollutants as effectively as forested buffers can.
7. Most suspended solids in urban runoff are composed of silt-sized particles.

A number of studies indicates that for smaller first-order or headwater stream, a 25 ft wide heavily wooded buffer (about the width of two mature trees), on either side of the stream banks, can provide some minimal protection and three items critical to stream homeostasis:

1. Protection of the stream bank from erosion
2. Sufficient supply of food (mostly in the form of leaf litter and woody debris) to support a healthy community of benthic macroinvertebrates
3. Shading* for the stream in the summertime to prevent extreme variations in stream temperature caused by incoming solar radiation (sunlight) — an attribute of special significance to waterways that also harbor populations of cold water fish, such as brook trout.

It should be noted, however, that as we move farther downstream and stream channel width increases, tree canopy has less of an influence on regulating stream temperatures.

A 25 ft wide buffer, however, is generally not adequate to ensure that potentially harmful pollutants converging on the waterway, primarily through storm water runoff, can be effectively trapped, removed, or rendered harmless — a concern not only to macroinvertebrates and fish life but also to man, especially if the stream is tributary to a downstream public water supply reservoir or intake, or is used for contact recreation such as swimming. In this instance, buffer zones, again well vegetated and preferably heavily wooded, would need to be generally on the order of 150 to 200 ft wide.[59,60]** The Metropolitan Washington Council of Governments (MWCOG) has developed a somewhat more specific buffer width guidance[57] that is based on a formula that incorporates consideration of land slope, vegetative cover, and particle size of transported sediment (Table 10.2). Their recommended widths are based on design guidance provided by the works of Wong and McCuen,[60] Trimble and Sartz,[61] and Heniger and Ray Engineering Associates Inc.[62] To use the design width table you must first determine the slope of the area. Following this, you must then decide whether vegetation in the proposed buffer area is of a high or moderate density. High density as defined here means that the vegetative cover, whether grassed, forest, or transition forest is thick enough such that in 80% or more of the buffer area, bare ground cannot be seen through the blades of grass or ground litter. Medium density means it is possible to see bare ground between the blades of grass and surface litter in more than 20% of the buffer area. Next, you must determine whether the buffer has a potential for high or low human disturbance. Riparian buffers next to commercial, residential, or industrial uses or close to roads, sidewalks, footpaths, or backyards would be considered as having a high potential for human disturbance. Finally, you must determine the average particle size expected to be carried in the storm water runoff. The MWCOG indicates that

* It should be noted that the Metropolitan Washington Council of Governments (MWCOG)[57] recommends a buffer width of up to two times the height of the dominant tree species to provide adequate shading of the stream.

** Verry suggests a riparian management zone width of 10 times the bankfull width of the stream on either side (Verry, E. S., Effects of forestry practices on physical and chemical resources, Conf. Proc., Water's Edge: Sci. Riparian For., BU-6637-S, Minnesota Extension Service, University of Minnesota, January 1996.)

TABLE 10.2
Buffer Width Selection Based on Vegetation Type

Slope (%)	Type of Sediment Load (width in ft)		
	Sandy	Silty	Clay
	High-Density and Low-Disturbance Potential		
0	50	110	150
1	54	119	162
2	58	128	174
3	62	136	186
4	66	145	198
5	70	154	210
6	74	163	370
7	78	172	390
8	82	180	410
9	86	189	430
10	90	198	450
	Medium-Density or High-Disturbance Potential		
0	150	350	735
1	162	356	794
2	174	383	853
3	186	409	911
4	198	436	970
5	210	462	1,029
6	370	814	1,813
7	390	858	1,911
8	410	902	2,009
9	430	946	2,107
10	450	990	2,205

Source: Adapted from Metropolitan Washington Council of Governments (MWCG), Riparian Buffer Strategies for Urban Watersheds, Publication No. 95703, Washington, D.C., 1995. With permission.

buffers in urban areas that are designed for silt-sized particles generally trap the majority of sediment. Once these determinations are made you may then go to the table and based on vegetation density or disturbance potential, select the appropriate width by choosing slope and the type of sediment.

Some researchers[63] (Figure 10.28[64,65] and Figure 10.29) have proposed a tiered system of buffers zones. Despite different authorships there is considerable similarity between these tiered buffer zone approaches. In all instances the tier closest to the water body is intended to be an area of *no disturbance*, primarily to protect the near channel ecosystem integrity, although there is some variance of opinion as to how wide this initial tier should be. Similarly, the second tier is designed to keep upland disturbances to a minimum, mostly by severe restrictions on activities and uses and implementation of extraordinary pollution prevention measures for any activities that might be permitted already. Finally, the third or outermost tier is intended to be an area where a standard array of best management practices (BMPs) would accompany all proposed development. It also should be noted that all these proposals recognize and recommend the necessity of having some flexibility in buffer boundaries, expanding as necessary to incorporate any or all the following: the 100-year floodplain or historic floodplain of record; areas of steep slope (Figure 10.30) (Schueler[63]

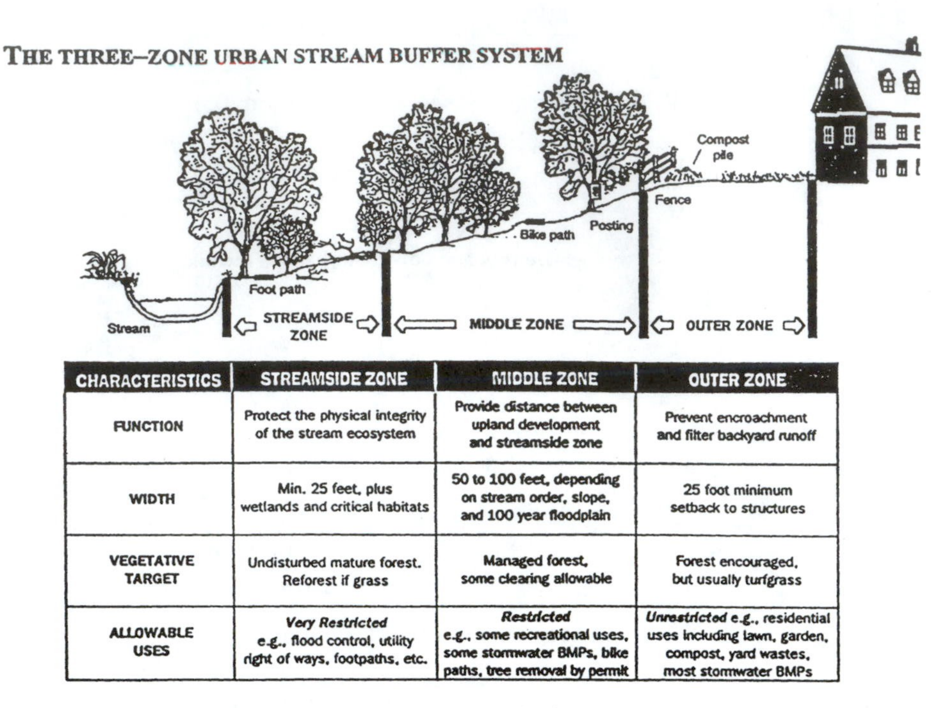

CHARACTERISTICS	STREAMSIDE ZONE	MIDDLE ZONE	OUTER ZONE
FUNCTION	Protect the physical integrity of the stream ecosystem	Provide distance between upland development and streamside zone	Prevent encroachment and filter backyard runoff
WIDTH	Min. 25 feet, plus wetlands and critical habitats	50 to 100 feet, depending on stream order, slope, and 100 year floodplain	25 foot minimum setback to structures
VEGETATIVE TARGET	Undisturbed mature forest. Reforest if grass	Managed forest, some clearing allowable	Forest encouraged, but usually turfgrass
ALLOWABLE USES	*Very Restricted* e.g., flood control, utility right of ways, footpaths, etc.	*Restricted* e.g., some recreational uses, some stormwater BMPs, bike paths, tree removal by permit	*Unrestricted* e.g., residential uses including lawn, garden, compost, yard wastes, most stormwater BMPs

Three lateral zones comprise the foundation of an effective urban stream buffer zone. The width, function, management and vegetative target vary by zone.

FIGURE 10.28 (From Schueler, T., Site Planning for Urban Stream Protection, Metropolitan Washington Council of Governments (MWCG), Center for Watershed Protection, Publication No. 95708, 1995. With permission.)

recommends an additional 4 ft of buffer for each 1% in slope grade above 5%); and adjacent wetlands and critical habitats.*

Wooded buffer zones have some distinct advantages over grassland-dominated ones. First, of course, is the hydraulic resistance to the flow of water offered both by the trunks of trees as well as the leaf and debris litter that covers the forest floor. This lowers flow velocity, whether the flow is perpendicular to the stream (runoff) or parallel to it (flooding), allowing sediment deposition, possible increases in soil infiltration, and a greater surface area for attenuation and removal of pollutants.**, ***

* The U.S. Department of Agriculture (USDA) Forest Service[65] further recommends that the width of tier 2 or zone 2 be increased, if necessary, to incorporate any soils designated by the Natural Resource Conservation Service (NRCS) as Hydrologic Soil Group D and those soils of Hydrologic Group C that are subject to flooding.

** Following the Midwest flood of 1993, a study was initiated (Dwyer, J. P., Wallace, D., and Larsen, D. R., Value of woody river corridors in levee protection along the Missouri River in 1993, *J. Am. Water Resour.*, 33, No. 2, April 1997) along a 39-mi segment of the Missouri River to determine if there was an association between woody corridors and levee stability. A systematic sample of levee failures revealed primary levees that did not fail had a significantly wider woody corridor than failed levees. An analysis of the total inventory of failed levees revealed that as the width of the woody corridor decreased, the length of the levee failure increased. The conclusion reached was that the number of levee failures and their severity of damage could be reduced if woody corridors were at least 300 ft wide.

*** A comprehensive study by Woodward and Rock (Woodward, S. E., and Rock, C. A., Control of residential storm water by natural buffer strips, *Lake Reserv. Manage.*, 11, No. 1, 37, 1995) evaluated the effectiveness of natural buffer strips to remove phosphorus and suspended solids from residential storm water. Their findings indicated that the most effective removal of phosphorus and suspended solids occurred where the ground was stabilized with underbrush and had a layer of decomposing forest litter.

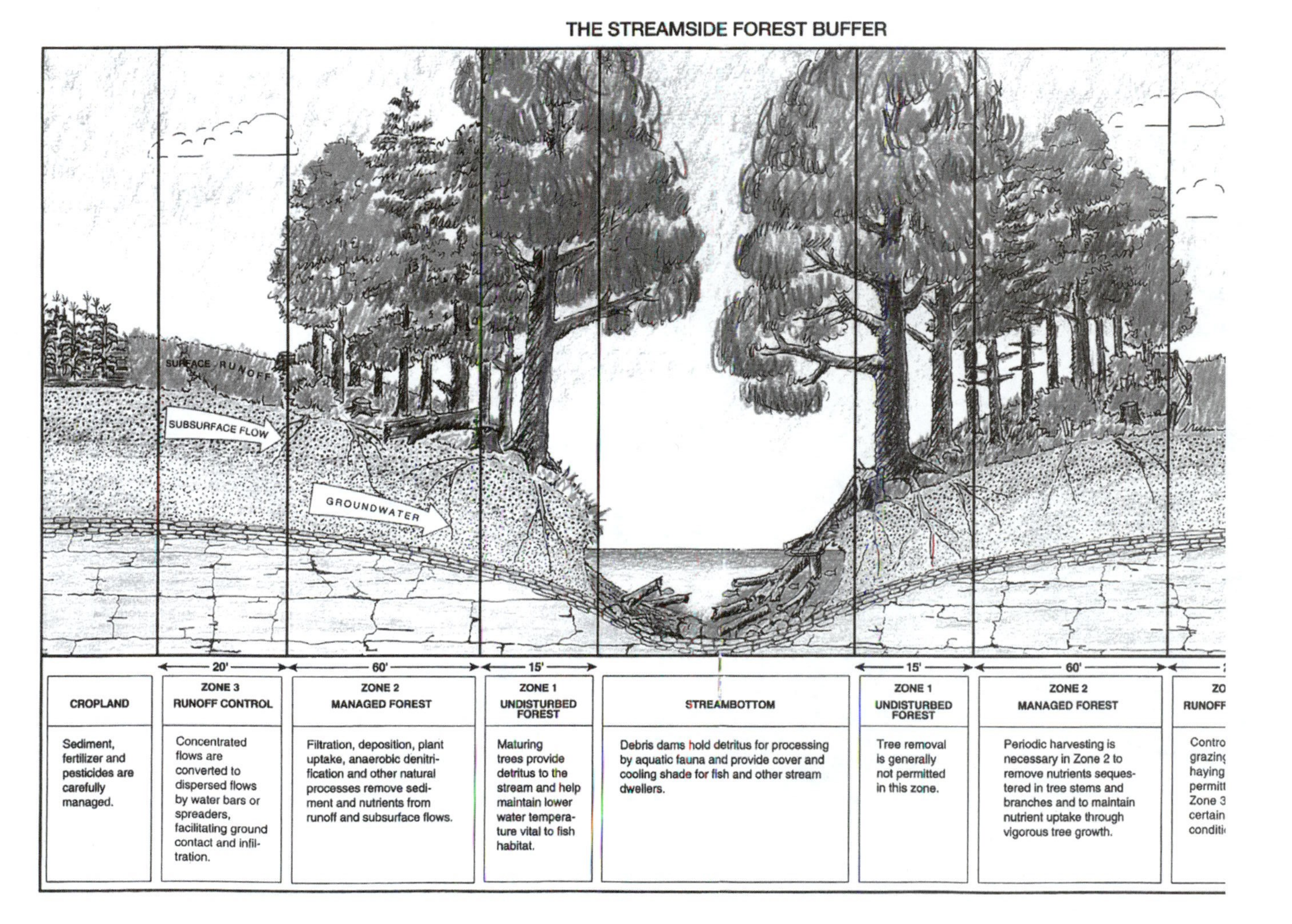

FIGURE 10.29 (From Forest Service, U.S. Department of Agriculture, Riparian Forest Buffers — Function and Design for Protection and Enhancement of Water Resources, Document No. NA-PR-07-91, 1991.)

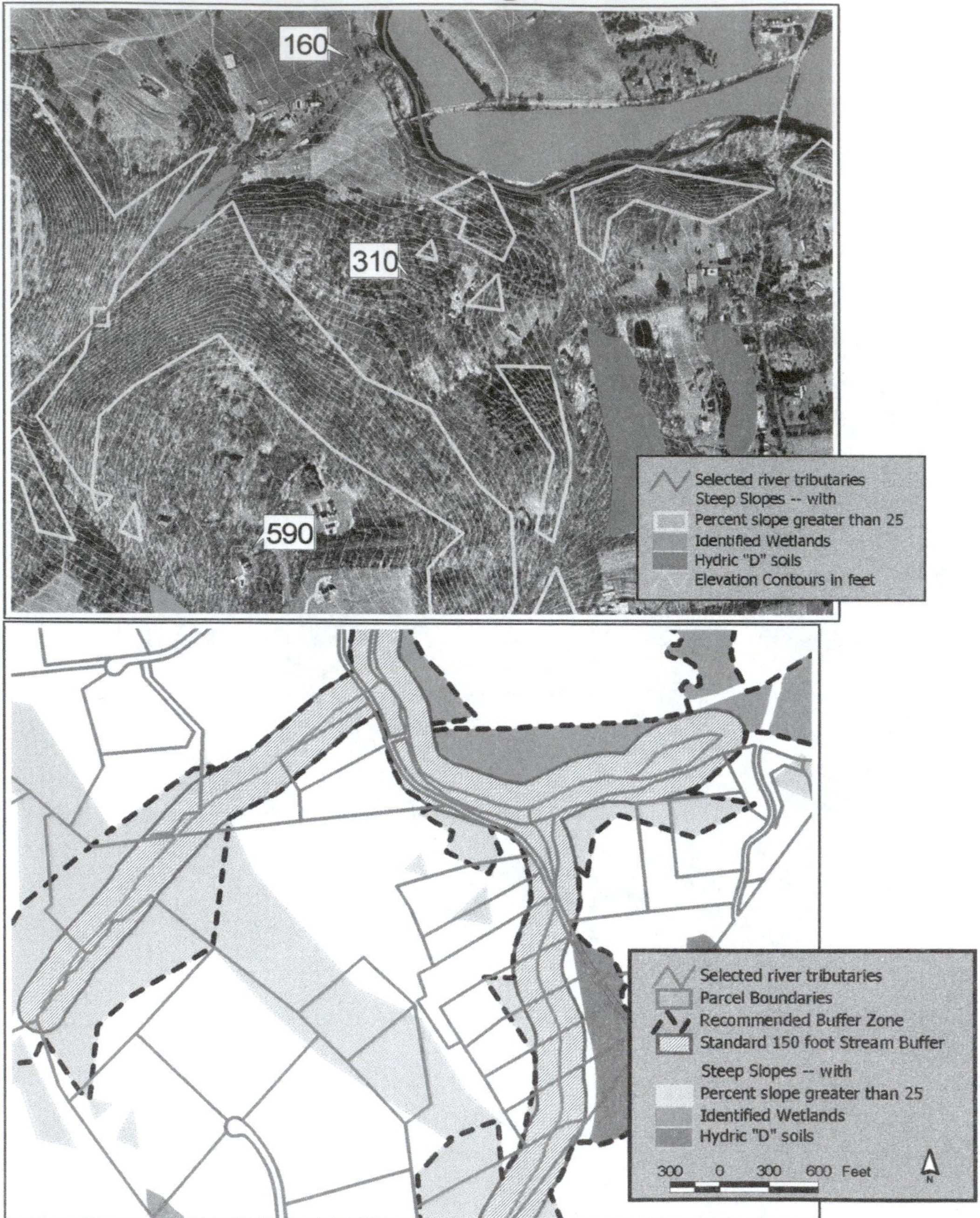

FIGURE 10.30 The top figure shows the ecological data layers collected from a typical GIS databank. The bottom figure illustrates how a stream buffer boundary line is then established based on that GIS databank. Note the variability in widths. (Graphics by Integrated Spatial Solutions, Moorestown, NJ.)

One of the greatest advantages over grassed buffers is their long-term storage of nutrients in woody material.[66] Nutrient retention by forests adjacent to agricultural land was estimated at 80% for phosphorus and 89% for nitrogen in Maryland's Rhodes River watershed.[67] Lowrance et al.[68] in a study of a Georgia coastal plain agricultural watershed also found riparian forest ecosystems

FIGURE 10.31 These close-cropped lawn areas provide little attenuation of potential pollutants and do not help slow storm water runoff.

to be excellent nutrient sinks that can effectively buffer the nutrient discharge to streams from surrounding agricultural land uses. While trees are effective filters for nutrients, their capability in this regard varies by genus and species. Generally, deciduous trees have greater nutrient demands than conifers have.[69] For example, oaks require more nutrients, especially potassium and nitrogen than spruce and pine require. Basswood, yellow poplar, dogwood, and Eastern red cedar concentrate large amounts of calcium, phosphorus, and potassium in their foliage, while beech, red spruce, and hemlock are low in their concentrations of these elements. Red and white oak, red maple, and quaking aspen do extremely well in nitrogen accumulation, but only to a point.[70] Once their requirements are met, their growth and absorption level off.[69] Additionally, Douglas fir, white ash, basswood, and yellow poplar respond well if nitrogen levels increase.

It is not unusual to find small reaches of a riparian corridor along an unimpaired waterway dominated by cultivated cropland, lawns, and pasture as well as by fallow fields, instead of woodland. Croplands are notorious sources of pollutants (including sediments, pesticides, and fertilizers), as are closely cropped lawn areas (Figure 10.31) in parks, private residences, and golf courses. In these reaches some degradation would not be unexpected; however, instream benthic macroinvertebrate communities and fish life may continue to reflect relatively unimpacted conditions. This could be the result of something as simple as dilution, expected amounts of runoff and pollutants being atypical, or full impact not being fully reflected due to a short temporal duration. It is highly unlikely that if stream degradation or pollution were particularly severe that the stream biota would continue to remain unaffected. If degradation were observable or measurable, the stream water quality rating in all likelihood would have been downgraded already to a moderately or severely impaired status — two conditions to be addressed shortly. In any event, it would be particularly prudent to recommend some retrofit activity for these areas. Because the much preferred heavily wooded vegetative cover is largely absent and the restoration of such cover is unlikely in the immediate near term, grassed buffer strips, also commonly referred to as vegetative filter strips (VFSs), are a recommended alternative. However, their effectiveness in removing waterborne pollutants is highly dependent, perhaps even more so than wooded buffers are, on their receiving

FIGURE 10.32 Although this particular level spreader has been placed along the edge of a replacement wetland, it demonstrates the technology that can be applied along the outermost periphery of a stream buffer. (From TetraTech Inc. and Delaware Technology Park. With permission.)

runoff in the form of shallow sheet flow;[71] therefore, this should be considered in their design. In addition, once these VFSs are established they generally should not be mowed to a height less than 6 in. Another alternative to protect the water quality of these waterways would be to acquire easements from adjoining landowners,* 150 to 200 ft wide on both sides of the stream, and to allow a natural succession of plant communities to repopulate the area. Ultimately, this would result in a mature woodland in as little as several decades.

After having expended considerable time on the discussion of providing *undisturbed* and appropriately sized buffer zones and maintaining the most beneficial vegetative land cover within them, we are confronted by the reality that in almost all instances the major portion of all storm water runoff from developed areas is currently bypassed around these buffers, either through constructed ditches or by underground piping. The symptoms resulting from this routine practice are all too evident: severe soil erosion at the point where the pipes or ditches enter the stream (see previous Figure 9.32), stream bank undercutting, channel widening, tree fall, and stress and death of some communities of aquatic organisms; all these symptoms occur despite the fact that such runoff may have been routed through such BMPs as detention and retention basins.** How

* As properties affected by buffer easements are sold, the deeds of conveyance need to have the buffer boundaries clearly described by metes and bounds, along with a description of permitted uses, if any, and of restricted or prohibited activities. In addition, all buffer easement boundaries should be physically marked in the field with permanent intervisible signs (see later in Figure 10.35).

** Maxted and Shaver[72] evaluated the impact on macroinvertebrate populations downstream of eight wet detention ponds. They concluded that these BMPs did not prevent the almost complete loss of sensitive taxa (e.g., mayflies, stone flies, and caddis flies) after development and did not attenuate the impact of urbanization once the watershed reached 20% impervious cover.

TABLE 10.3
Most Water-Tolerant Species

Species	Scientific Name
Ash, Carolina	*Fraxinus caroliniana*
Pumpkin	*F. profunda*
Aspen, bigtooth	*Populus grandidentata*
Baldcypress	*Taxodium distichum*
Black spruce	*Picea mariana*
Buttonbush	*Cephalanthus occidentalis*
Dogwood, red-osier	*Cornus stolonifera*
Silky	*C. amomum*
Planertree/water-elm	*Planera aquatica*
Pondcypress	*T. distichhum* var. *nutans*
Swamp-privet	*Forestiera acuminata*
Tupelo, swamp	*Nyssa sylvatica* var. *bilfora*
Water	*N. aquatica*
Willow, bankers	*Salix cottettii*
Black	*S. nigra*
Carolina	*S. caroliniana*
Purple-osier	*S. purpurea*
Sandbar	*S. interior*
Peachleaf	*S. amygdaloides*

Source: Adapted from Sykes, K. J., Perkey, A. W., and Palone, R. S., Crop Tree Management in Riparian Zones, Forest Service, USDA, Morgantown, WV. No date. With permission.

TABLE 10.4
Highly Water-Tolerant Species

Species	Scientific Name
Ash, green	*Fraxinus pennsylvanica*
American	*F. americana*
Basswood	*Tilia americana*
Balsam fir	*Abies balsamea*
Hickory, water	*Carya aquatica*
Oak, bur	*Quercus macrocarpa*
Nuttall	*Q. nuttallii*
Overcup	*Q. lyrata*
Swamp white	*Q. bicolor*
Water	*Q. nigra*
Willow	*Q. phellos*
Persimmon	*Diospyros virginiana*
Waterlocust	*Gleditsia aquatica*

Source: Adapted from Sykes, K. J., Perkey, A. W., and Palone, R. S., Crop Tree Management in Riparian Zones, Forest Service, USDA, Morgantown, WV. No date. With permission.

TABLE 10.5
Moderately Water-Tolerant Species

Species	Scientific Name
Alder	*Alnus* sp.
Ash, pumpkin	*Fraxinus profunda*
White	*F. americana*
Atlantic white-cedar	*Chamaecyparis thyoides*
Birch, river	*Betula nigra*
Yellow	*B. alleghaniensis*
Boxelder	*Acer negundo*
Catalpa	*Catalpa bignonioides*
Eastern cottonwood	*Populus deltoides*
Elm, American	*Ulmus americana*
Cedar	*U. crassifolia*
Hackberry	*Celtis occidentalis*
Hawthorn	*Crataegus* sp.
Hickory, bitternut	*Carya cordiformis*
Holly, deciduous	*Ilex decidua*
Honeylocust	*Gleditsia triacanthos*
Loblolly-bay	*Gordonia lasianthus*
Maple, red	*Acer rubrum*
Sugar	*A. saccharum*
Silver	*A. saccharinum*
Oak, Nuttall	*Quercus nuttallii*
Pin	*Q. palustris*
Southern red	*Q. falcata* var. *falcata*
Willow	*Q. phellos*
Osage-orange	*Maclura pomifera*
Persimmon	*Diospyros virginiana*
Pine, pond	*Pinus serotina*
Loblolly	*P. taeda*
Red	*P. rubens*
Shortleaf	*P. echinata*
Slash	*P. elliottii*
White	*P. strobus*
Redbay	*Persea borbonia*
Spruce, red	*Picea pungens*
White	*P. glauca*
Sweetbay	*Magnolia virginiana*
Sweetgum	*Liquidambar styraciflua*
Sycamore	*Platanus occidentalis*

Source: Adapted from Sykes, K. J., Perkey, A. W., and Palone, R. S., Crop Tree Management in Riparian Zones, Forest Service, USDA, Morgantown, WV. No date. With permission.

then does a community avoid this anachronistic engineering design and at the same time maximize the advantages of established buffer zones? Most of us intuitively already know the answer; pull the ditch and pipe outfalls back from the stream banks and reapply the collected storm water as sheet flow at the outermost fringe of the buffer.* An alternative to this damaging direct discharge

* Some caution is necessary here, particularly for woodland buffers. Trees are susceptible to waterlogging, a concern if we begin to deliver a larger proportion of storm water runoff to such areas in lieu of direct piping to the stream. Some tree species survive quite well in conditions of frequent inundation while others may die. Tables 10.3 through 10.8[70] provide more detail on species-specific tolerance to excessive wetting.

TABLE 10.6
Weakly Water-Tolerant Species

Species	Scientific Name
Birch, gray	*Betula populifolia*
Paper	*B. papyrifera*
Black locust	*Robinia pseudoacacia*
Black walnut	*Juglans nigra*
Blackgum	*Nyssa sylvatica*
Buckeye, yellow	*Aesculus octandra*
Butternut	*Juglans cinerea*
Elm, winged	*Ulmus alata*
Hazelnut	*Corylus americana*
Hickory, mockernut	*Carya tomentosa*
Pignut	*C. glabra*
Sand	*C. pallida*
Shagbark	*C. ovata*
Shellbark	*C. laciniosa*
Swamp	*C. lieodermis*
Holly, American	*Ilex opaca*
American hornbeam	*Carpinus caroliniana*
Oak, black	*Quercus velutina*
Blackjack	*Q. marilandica*
Chinkapin	*Q. muehlenbergii*
Cherrybank	*Q. falcata* var. *pagodaefolia*
Chestnut	*Q. prinus*
Laurel	*Q. laurifolia*
Live	*Q. virginiana*
Northern red	*Q. rubra*
Post	*Q. stellata*
Shingle	*Q. imbricaria*
Shumard	*Q. shumardi*
Swamp chestnut	*Q. michanxii*
Mulberry, red	*Morus rubra*
Pawpaw	*Asimina triloba*
Pecan	*Carya illinoensis*
Pine, Virginia	*Pinus virginiana*
Spruce	*P. glabra*
Redbud	*Cercis canadensis*
Sourwood	*Oxydendrum arboreum*
Southern magnolia	*Magnolia grandiflora*
Sugarberry	*Celtis laevigata*

Source: Adapted from Sykes, K. J., Perkey, A. W., and Palone, R. S., Crop Tree Management in Riparian Zones, Forest Service, USDA, Morgantown, WV. No date. With permission.

to surface water bodies is shown in Figure 10.32 where storm water outfalls are truncated at the edge of the buffer and storm water is discharged onto a shallow gabion reinforced level spreader or into an underground diffuser system that then disperses the flow along the entire periphery of the established buffer. The intent is to mimic more of a sheet flow condition. It is possible that such spreaders also could be applied in series along descending contours in areas of steeper topography (Figure 10.33) to minimize the need for detention basins that often cause considerable collateral damage to the environment both during and after their construction. This is not to imply that detention and retention basins should no longer be required. Such structures serve some useful

TABLE 10.7
Least Water Tolerant Species

Species	Scientific Name
American beech	*Fagus grandifolia*
Black cherry	*Prunus serotina*
Flowering dogwood	*Cornus florida*
Eastern red cedar	*Juniperus virginiana*
Elm, slippery	*Ulmus rubra*
Hophornbeam	*Ostrya virginiana*
Oak, white	*Quercus alba*
Pawpaw	*Asimina triloba*
Sassafras	*Sassafras albidum*
Yellow-poplar	*Liriodendron tulipifera*

Source: Adapted from Sykes, K. J., Perkey, A. W., and Palone, R. S., Crop Tree Management in Riparian Zones, Forest Service, USDA, Morgantown, WV. No date. With permission.

TABLE 10.8
Descriptions

Most water tolerant
Trees capable of living from seedling to maturity in soils that are waterlogged almost continually year after year except for short durations during drought, these species exhibit good adventitious or secondary root growth during this period
Highly water tolerant
Trees capable of living from seedling to maturity in soils that are waterlogged for 50 to 75% of the year; some new root development being expected during this period; waterlogging usually occurring during the winter, spring, and one to three months of summer
Moderately water tolerant
Trees capable of living from seedling to maturity in soils that are waterlogged about 50% of the time during the growing season; the root systems of these species producing few roots or being dormant during the waterlogged period; this period usually occurring in portions of the winter, spring, and summer
Weakly water tolerant
Trees capable of living from seedling to maturity in soils that are temporarily waterlogged for durations of 1 to 4 weeks
Least water tolerant
Trees capable of living from seedling to maturity in soils that are occasionally waterlogged for durations of a few days only

Source: Adapted from Sykes, K. J., Perkey, A. W., and Palone, R. S., Crop Tree Management in Riparian Zones, Forest Service, USDA, Morgantown, WV. No date. With permission.

purposes, despite their acknowledged limitations, and should not be discouraged. All we are proposing are changes in the end of pipe delivery.

Preserving riparian corridors by establishing appropriately sized buffer zones and devising modern ways to reapply the storm water we have collected and concentrated, provide two ways to preserve the ecological function of surface waterways found to be as yet unimpaired. Other protective measures also would include limiting the amount of impervious surface in the contributing watershed (principally for roadways and parking lots) on all new construction to 5% or less; requiring the installation of more stringent soil erosion and sediment control practices during and after the construction phase of all land development projects (this would include a daily or weekly inspection and maintenance program of all installed control practices that is much more rigorous than that presently practiced), collecting storm water through the use of biofiltration facilities

FIGURE 10.33 A series of these shallow gabion level spreaders could be placed at intervals along the contours of a steep slope to mimic a more natural overland flow and treatment of storm water runoff. (Sketch by Greg Honachefsky.)

(Figure 10.34) or sand filtration systems as close to the source as possible; and finally, the public purchase* of particularly vulnerable parcels of vacant land, such as steeply sloped areas that are covered by large unbroken tracts of woodland.

Having devised some protective strategies for those sections of streams relatively unimpacted by human development and rated as unimpaired, we can now look at those stream segments evaluated as moderately or severely impaired. These stream reaches are generally associated with a higher density of urbanization or intense agricultural operations. Space for rehabilitation or restoration practices is sparse or lacking altogether in the former, while in the latter space may not be as crucial. The question for the community here is twofold. First, what are the causes of any impairment and are they potentially correctable; second, are these stream reaches a part of the ecological infrastructure that the community can or wants to rehabilitate, or in the case of those

* States like New Jersey could have a substantial advantage in obtaining voluntary donations of buffer zones from private landowners, in particular, owners of agricultural land. New Jersey municipalities, under state law, are required to give a substantial tax reduction to farmland, until such time as it is sold for development purposes. At that time, the owner or the developer/buyer is required to pay 3 years of what are called *rollback taxes*, which are taxes based not on the value of the property as farmland but on what the value would have been had the property been assessed under its development potential based on current zoning. In lieu of collecting these rollback taxes, municipalities could instead barter for or require permanent deeded donations of riparian buffer zones; and we do not mean here the usual drainage easements for storm drains and retention or detention basins, but real honest to goodness buffers. Such donations in lieu of rollback taxes would not have to be limited to just buffers. Unique habitats or habitats of endangered or threatened species also could be acquired. In this latter regard, a natural resources inventory, as we have proposed in this book, would certainly help identify those areas worthy of preserving.

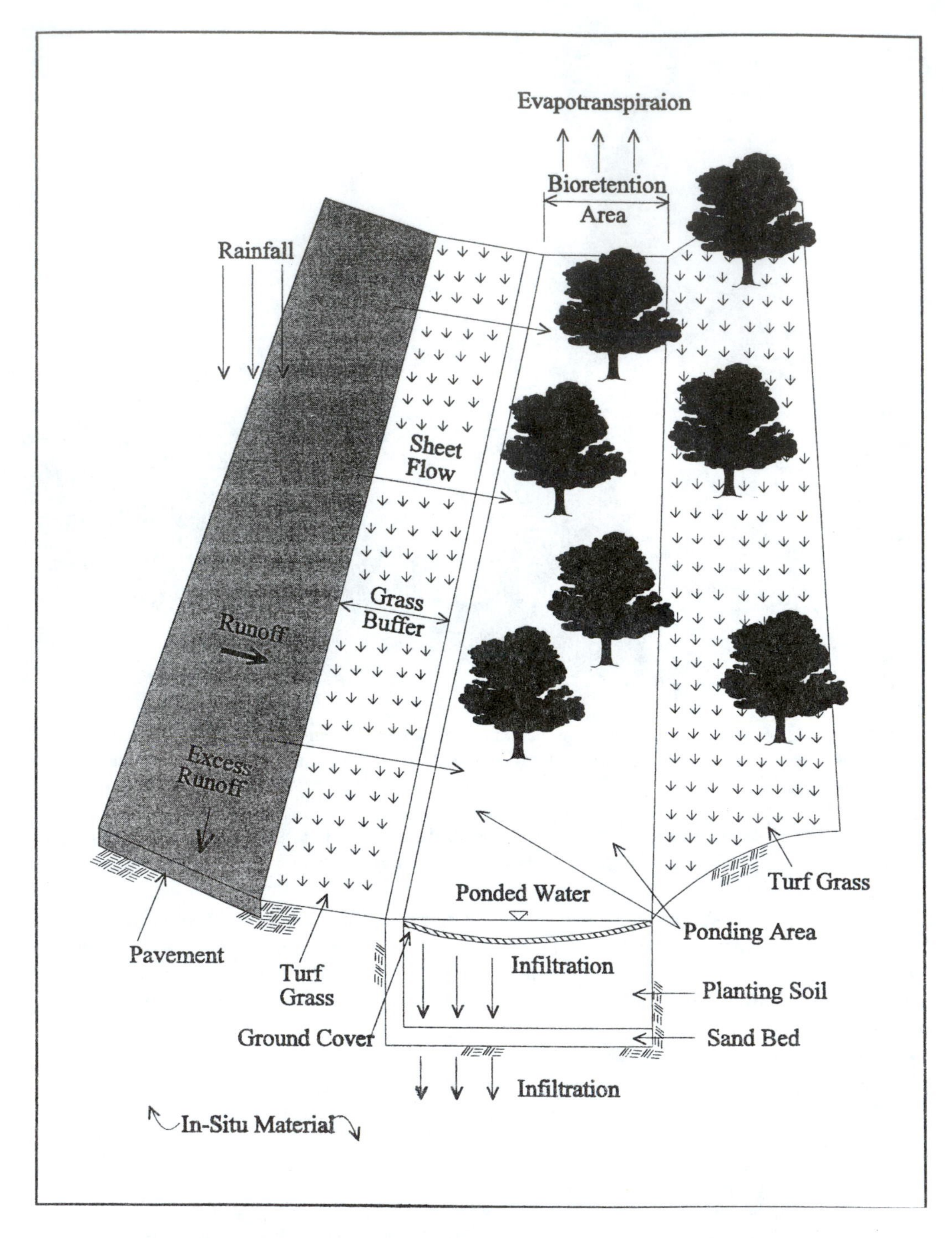

FIGURE 10.34 While this particular bioretention basin design shows direct infiltration into the ground, other modifications in less pervious soils have included the addition of underdrains following the sand bed filters to carry away the treated water. (From U.S. Department of Transportation, Evaluation and Management of Highway Runoff Water Quality, Publication No. FHWA-PD-96-032, Washington, D. C., 1996, as adapted from Prince George's County Design manual for use of Bioretention in Stormwater Management, 1993.)

TABLE 10.9
Recommended Steps for Diagnosing the Causal Agents of Degradation in Surface Waterways

1. Reviewing any state, county, or local *ambient, surface water, chemical monitoring network* data from sampling stations that may be within the affected stream reaches, for obvious anomalies in measured concentrations of parameters/pollutants
2. *Plotting all* national pollutant discharge elimination system (NPDES) permitted *point source dischargers,* contributing to the affected waterway reach, and reviewing the *discharge monitoring reports* (DMRs) for those facilities concentrating on reported violations of permit conditions and levels and number of exceedances
3. Retrieving and reviewing *land use/land cover data* of drainage area contributing to the affected stream reach; determining percentage of *impervious cover*
4. *Plotting all known hazardous waste sites including landfills and dumps*; determining if any associated groundwater pollution has been identified, if plume travel paths have been delineated, and if they are discharging to the affected waterway
5. *Assessing stream habitat quality*; if the stream bottom is being scoured, tree canopy is being removed, stream banks are undercut, channel widening is taking place, excessive tree fall and excessive sedimentation is taking place
6. *Looking for obvious land disturbance activities* occurring within and upstream of the affected stream reach, including mine and quarry operations or residential, commercial, or industrial construction
7. *Reviewing geologic and soil survey maps* for strata that may be *severely acidic* or otherwise toxic when exposed to the air
8. Checking state and local databases on historical and current application data for *lawn and garden pesticides and agricultural pesticides*; determining pounds per year applied and compounds used on specific crops grown in the contributing watershed of the affected waterway
9. Procuring and *reviewing historical aerial photography* to determine and track historical land uses in watershed contributing to the affected steam reach
10. Plotting all GPS located storm water pipe outfalls and retention and detention structures within the affected stream reach
11. *Collecting* stream *sediment* samples in affected waterway reach, *analyzing* for *possible toxic pollutants*, and compare to available standards
12. Collecting stream sediment samples and *running acute toxicity tests* on indicator organisms

waters only moderately impaired, to prevent their further deterioration? An argument can and has been made that water resource issues, generally, and impaired water quality, in particular, ought to be dealt with at the state and federal level. To be sure, that is where the primary statutory responsibility and authority exists; however, land use, primarily as dictated by local land planners, is the primary causal agent of surface water (and groundwater) quality changes over time.* Instead of passing the buck as we have conveniently done in the past, we need to acknowledge the significant role land planners play in water resource protection, never forgetting that urban stream degradation came about one relatively small project approval at a time.

Our first instinct, perhaps driven by our misgivings for past transgressions, would be to immediately designate these areas as *water quality use restoration areas.* This certainly would be a laudable action; however, some further evaluation would be in order before we could undertake this strategy. Some corroboration of the benthic macroinvertebrate assessments, if this is our initial indicator, would be helpful. This can be gleaned from the remaining items on our list of response indicators or from our list of environmental indicators. For instance, does the stream reach have a fish consumption ban or contaminated sediments, have there been repeated massive algae blooms, fish kills, or excessive weed growth, and does the waterway segment fail to meet water quality standards? Also, has a prior recreational fishery been destroyed and has other contact recreation, such as swimming been banned? Before we rush to mark up our maps in the MMP with water quality use restoration areas, we also need to see what the reasons for impairment might be. At

* Unlike easily identified point source discharges, much of the pollution and ecological degradation associated with land use and land use changes has been ascribed to *nonpoint* sources.

FIGURE 10.35 Signs like these posted along sensitive lands or dedicated buffers serve as a constant reminder to residents of the importance of these areas. (These signs are available from Site Evaluations, P.O. Box 371, Ringoes, NJ 08551.)

this point, we need to look at our inventory of *man-made infrastructure* as well as our list of stressor indicators that pertain to water quality issues. The real utility of the GIS proves its worth and investment at this stage. A simplified methodology for this latter evaluation process was proposed by Honachefsky in 1997 (Table 10.9). The spatial display capabilities of the GIS can help facilitate this evaluation process significantly. Not all the procedures and information suggested in Table 10.9 are required in all instances. For example, in some areas an impaired section of waterway may be totally surrounded by *agricultural uses*, and may be totally devoid of any NPDES permitted point source discharges or concentrated residential or commercial development. In this instance, the search for potential causes is narrowed considerably and potential solutions may be more easily derived and implemented. In this particular instance, where the problem may have been diagnosed as primarily due to agricultural operations, the contributing watershed to this reach of stream could be identified as an *agricultural conservation plan zone*. In other words, for any farm located in the drainage area contributing to the affected stream reach, an active, state-of-the-art *farm conservation plan* would be required. Such a plan would include such things as fencing to exclude livestock from close proximity to the stream corridor, conservation tillage, contour plowing, grassed waterways, manure management, strip-cropping, integrated pest and fertilizer management, dedicated stream buffers,* and so forth. In states like New Jersey, where farmland receives *preferential* and

* The state of Illinois in 1996 passed a law (Public Act 89-606) that sets the value of property containing vegetative filter strips (VFSs) at a greatly reduced rate. This new law took effect January 1, 1997 and continues in effect until December 31, 2006.

FIGURE 10.36 A biofiltration structure installed at the rear of an industrial park in the state of Delaware.

significant *reductions* in *real estate taxes*, the continuance of such preferential treatment would be conditioned on the implementation of an ongoing farm conservation plan, and this includes those farms owned by absentee landlords who own the land and lease it out for farming purposes while plans for its development are being processed.

On the other hand, for a stream reach surrounded by intense urban development whose contributing drainage area is 80 to 90% impervious surfaces and whose stream banks are lined with NPDES permitted discharge pipes and scores of storm water outfall pipes, the diagnosis of culpable contributors to the stream's degradation are more rigorous, and most likely require the full complement of suggested review procedures set forth in Table 10.9. Unfortunately, even after such an intensive investigation the actual causes of degradation may be difficult to isolate, with the possible further complication that all potential sources are equally culpable. In this case, local land planners may have *limited opportunities* to ameliorate existing conditions and community residents need to determine how much of an effort and commitment should be made to try to recover the prior purity of the stream. Sometimes, such areas can best serve as an object lesson of what our prior environmentally insensitive land use planning has wrought.

If this latter suggestion sounds too harsh or pessimistic, then by all means rehabilitation should be attempted.* In the most urbanized areas, vacant lots could be regraded and bioretention structures could be installed (Figure 10.36). Parking lots and roadway storm drains could be retrofitted with enhanced settling and sand filtration structures,** (Figures 10.37 through 10.40 show variations on this theme); and the stream itself — unless totally lined or flumed — could be rehabilitated by the (re)placement of bank vegetation, and by the installation of low-level weirs and deflector groins, providing a food source, pools, riffles, and cover for whatever stream organisms remain. Where streams have been piped completely underground, some attempts could be made to restore them

* For assistance in this regard see especially, Stream Corridor Restoration, Principles, Processes, Practices, the Federal Interagency Stream Restoration Working Group, 1998.

** An excellent treatise on this subject can be found in Bell.[73]

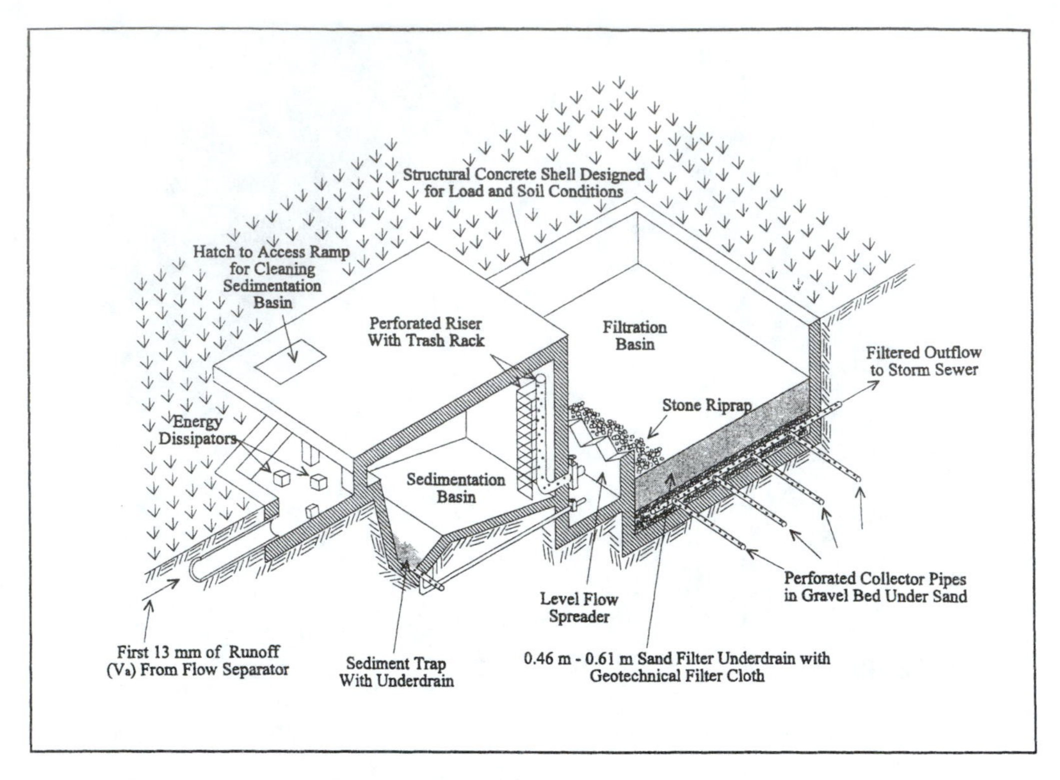

FIGURE 10.37 An Austin sand filter with full sedimentation protection. (From U.S. Department of Transportation, Evaluation and Management of Highway Runoff Water Quality, Publication No. FHWA-PD-96-032, Washington, D.C., 1996.)

once again to open channels, a process known as *daylighting*. The question is whether there is enough room provided (Figure 10.41) for these resurrected streams to carry out some of their original functions (such as floodwater retention, flow retardance, and pollutant attenuation) or they are to simply become collectors of urban debris. The cost of such retrofits in these urban areas can be particularly expensive, to say nothing of the annual expense for routine maintenance. Consequently, it is recommended that a storm water utility be established to assume responsibility for the installation and maintenance of these retrofitted structures. In the past, local communities typically financed the maintenance of storm water infrastructure through general property tax revenues. With local property owners already rebelling over continually rising property taxes, it would be unwise to suggest further increases to support the installation and maintenance of what could be a considerable number of these devices. Instead, the creation of a storm water utility, with the ability to assess user fees, would make more sense and would be more equitable and perhaps more palatable to local taxpayers. Such utilities have been successfully operated for over 25 years with the cities of Bellvue, Washington and Boulder, Colorado being some of the first cities to implement them. For the most part, user fees are based on the amount of impervious cover on each property. A common billing unit is the equivalent residential or runoff unit (ERU). An ERU represents the average impervious area associated with residential parcels (e.g., single-family homes) in the utility service area.* For example, in a community with an ERU of 2500 square ft and a billing rate of $3.00 per month per ERU, all owners of residential properties would pay a $3.00 monthly utility charge. A nonresidential property, with an impervious cover of 10,000 square ft or the equivalent of 4 ERU would pay $12.00 a month.

* See Doll, A., Lindsey, G., and Albani, R., Stormwater utilities: key components and issues, in *Adv. Urban Wet Weather Reduction Conf.*, sponsored by Water Environment Federation, Cleveland, OH, June 28–July 1, 1998.

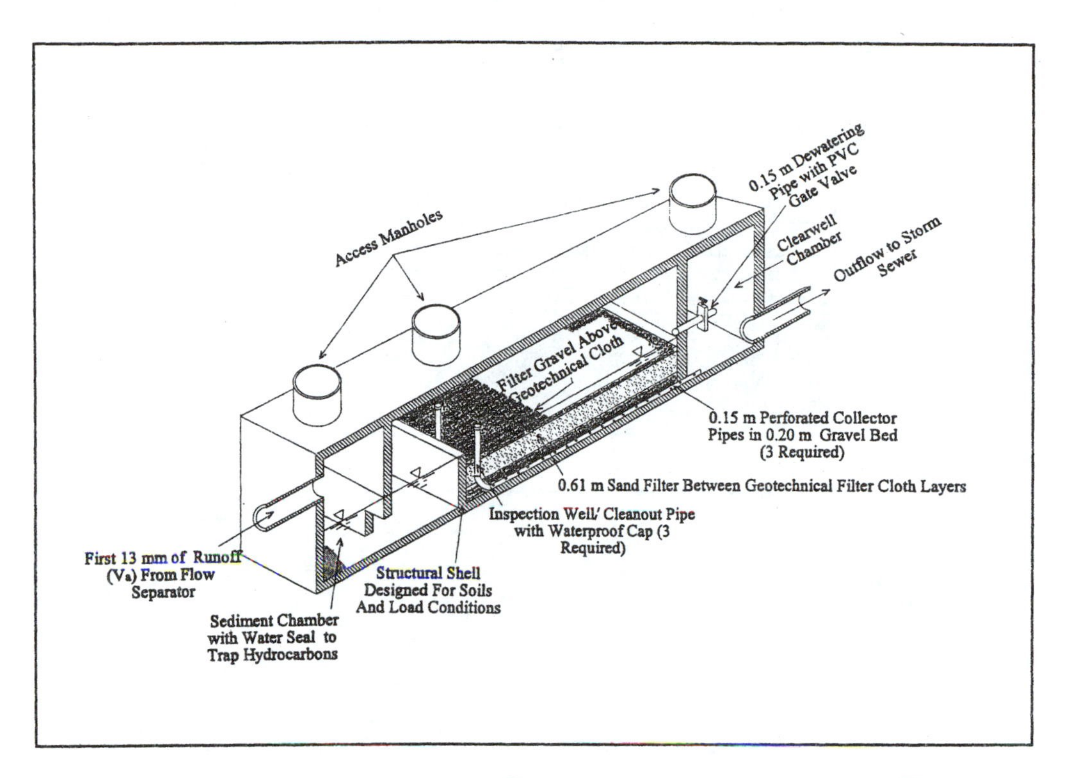

FIGURE 10.38 Original DC underground sand filter system. (From U.S. Department of Transportation, Evaluation and Management of Highway Runoff Water Quality, Publication No. FHWA-PD-96-032, Washington, D.C., 1996.)

This extended discussion demonstrates how energy, fiscal, and social costs escalate when the function or functions of natural systems are replaced by artificial methods and means. It is always more cost-effective to keep ecological systems intact and functioning and to plan any land use around them.

10.4.1.2 Groundwater

For years groundwater was an unappreciated and overlooked resource at almost all levels of government. Even as late as the early 1970s the indiscriminate and sometimes governmentally condoned dumping of hazardous and toxic chemicals into the ground surface and ultimately into underlying groundwater aquifers was a routine practice. In subsequent years, however, greater demands on groundwater resources, as an alternative to extensively contaminated surface water bodies, combined with an ever-increasing human population, quickly began to shed light on the extent to which these resources were being abused. Federal and state legislative initiatives to protect these resources eventually evolved and more of the public is aware today of the potential abuses that can occur not only from a quality point of view but also from a quantity viewpoint.

A powerful indicator of the health of a community's groundwater resources, is the number of potable water supply wells in the community that has been determined to be contaminated (see previous Figure 10.10); as a result these wells either have been permanently taken out of service or have been retrofitted with a treatment system to remove the polluting materials. Remediation in these cases can be difficult, costly, and extremely long term, hence our continuing emphasis on preventative planning. A second indicator (often regarded as both stressor and environmental) which also may be useful to local land planners in their MMP deliberations, includes the identified and mapped plumes of contaminated groundwater migrating from former spill sites,

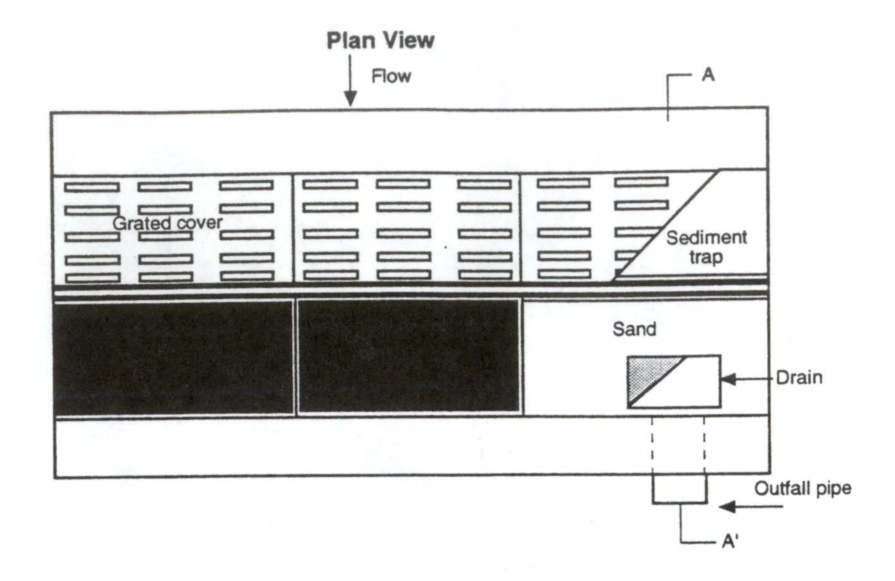

FIGURE 10.39 Delaware sand filter (see also Figure 10.40). (From U.S. Department of Transportation, Evaluation and Management of Highway Runoff Water Quality, Publication No. FHWA-PD-96-032, Washington, D.C., 1996.)

hazardous waste sites, and landfills that have not yet intercepted a sensitive receptor, such as a potable water supply well or surface watercourse (Figure 10.42). Information on such sites is readily available from state environmental protection agencies and sometimes to a lesser degree from federal environmental protection agencies. Such sites should be identified in the MMP, the areal extent of their plume contamination should be defined, and the flow path or paths should be determined so that existing potential receptors downgradient can be put on notice. In addition, no new well installation should be allowed downgradient unless provisions are incorporated into the well construction to avoid the plume or provisions are made to add an effective treatment system capable of removing the specific pollutants in question.

Some states have taken great pains to map groundwater aquifers within their jurisdiction and at the same time establish the areal extent of the surface landscape contributing to their recharge. When such information is combined with time of travel studies — derived from models and other sources — to estimate groundwater flow through subterranean geology, regulators can project how long it would take a pollutant — deposited somewhere in the contributing watershed — to reach

FIGURE 10.40 A Delaware sand filter installed in the parking lot of an elementary school in the state of Delaware.

FIGURE 10.41 This is how *not* to "daylight" a stream. It does show, however, some of thee common problems that arise from land development that does not provide appropriate buffers. Here we see leaking drums of waste solvents stored carelessly next to the stream.

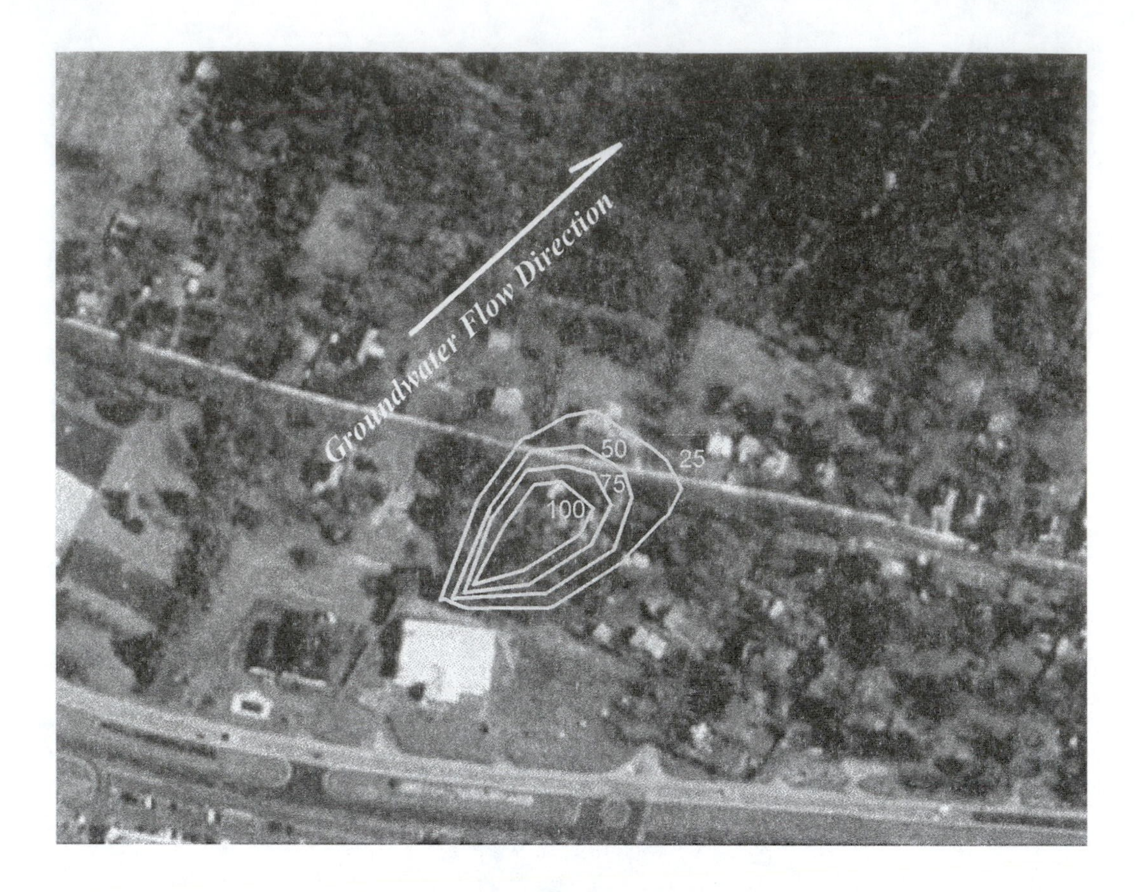

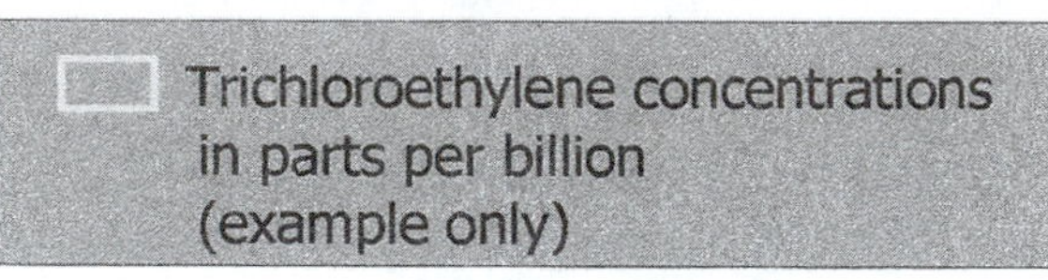

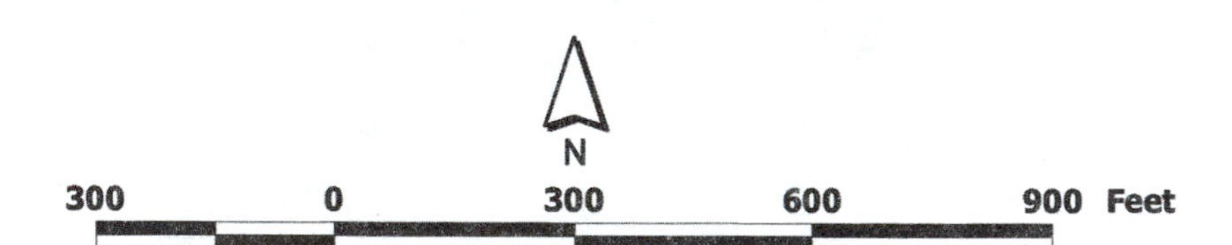

FIGURE 10.42 Identifying the areal extent and direction of flow of contaminated groundwater in the local master plan will help local planners etablish corridors of projected movement from which potable water wells or even human habitation would be excluded. (Graphics by Integrated Spatial Solutions, Moorestown, NJ.)

a susceptible receptor such as a public potable water supply well (Figure 10.43). Despite the ready availability of this information, few municipalities who rely on groundwater as their major supply of potable water have incorporated a rigorous wellhead or watershed protection strategy into their MMP.* For example, I know of one community in northern New Jersey in which the sole potable

* This occurs despite the fact that the 1986 Amendments to the Federal Safe Drinking Water Act had established a well head protection (WHP) program designed to be a proactive effort to identify potential pollutant sources near public water supply wells and to implement appropriate controls for those sources. Amendments to the same act in 1996 now provide for a source water protection (SWP) program that covers both groundwater and surface water supplies.

FIGURE 10.43 Typical wellhead protection map showing the time of travel in years of groundwater flowing from upgradient areas of the watershed. (From New Jersey Geological Survey, New Jersey Department of Environmental Protection.)

water supply comes from groundwater sources, which is indirectly purchased from a local water supply authority in an adjoining municipality. Yet the wells that produce that groundwater in the first place are located in the purchasing municipality. Both communities, however, have not thought to protect the groundwater system from which the water is derived — this despite the fact that current zoning calls for intensive commercial and residential development within the watershed both overlying and contributing recharge to the aquifer. The next logical concern is, suppose we do designate one or two or even more such areas as wellhead protection zones in the MMP. What should we allow as far as development? The simplest answer, of course, is nothing. In practice that is not generally an acceptable answer, so let us look at alternatives. The first is to limit density of development through large lot or cluster zoning and to prohibit industrial manufacturing and some commercial uses including gasoline stations and tank storage farms. A second alternative is the outright purchase of much of the contributing watershed land or at least the development rights attached to it. Many municipalities, on their own volition and with strong community support, have opted to raise real estate tax levies by several cents per hundred dollars of assessed value to buy open space and farmland. These monies often are combined with grants and loans from state and federal agencies. While many municipalities are unsure as to where these monies would best be spent, I cannot think of a better place than on parcels within the contributing watershed or watersheds of a community water supply well or wells.

Earlier we spoke of new methodologies being used to predict groundwater recharge capabilities within the municipality landscape. The areas with the highest potential may or may not coincide with or occur within the hydraulic radius of a community's own water supply well or wells, but may recharge the aquifer from which an adjoining municipality withdraws its water. Regardless of which aquifer it recharges, those areas with the *highest recharge potential* should be protected and preserved in their natural state. Again, such areas would be specifically identified in the MMP as *recharge protection areas*. Preserving the integrity of these areas could be accomplished by an outright purchase or by purchase of recharge rights, or during individual site plan approvals (Figure 10.44) the municipality could divert development away from these areas should they occur on a site and instead could redirect development to areas with a lesser potential for recharge.

10.4.1.3 Lakes and Ponds

Many of the nation's interior, freshwater lakes and ponds have been seriously neglected. Their susceptibility to extensive and accelerated deterioration (also known as advanced aging or *eutrophication*; see previous Figures 9.15 and 9.16), especially in watersheds undergoing intense development, is well documented and the symptoms of such have been discussed at some length in Chapter 9. Sections of the federal Clean Water Act, most notably Section 314, have provided funding, under the auspices of the *clean lakes program*, to local communities for rehabilitation of public lakes in their jurisdiction. Such monies have been used for *dredging* (usually the most costly form of rehabilitation due to the amount of material to be removed and the cost of disposing of what are often highly contaminated sediments), *mechanical or chemical aquatic weed removal* or suppression, *algae control*, and *installation* of stream bank *erosion control* on tributary streams — all of which could be a wasted effort if the causal agents are not forcefully and fully addressed in the MMP.

Whether a lake is already seriously impaired or relatively clean, special *lake management districts* could be established within all or part of the lake's contributing watershed. The condition of the lake would dictate the types of controls or land uses that would be permitted or required within such lake management districts. Obvious controls would include those already espoused for protecting surface waterways, such as multitiered buffer zones along the periphery of the lake or pond, more stringent storm water control measures including the establishment of permanent sediment ponds on all tributary streams, and installation of state-of-the-art technologies on all existing and any new storm water collection systems.

Groundwater Recharge Rates for a Selected Parcel

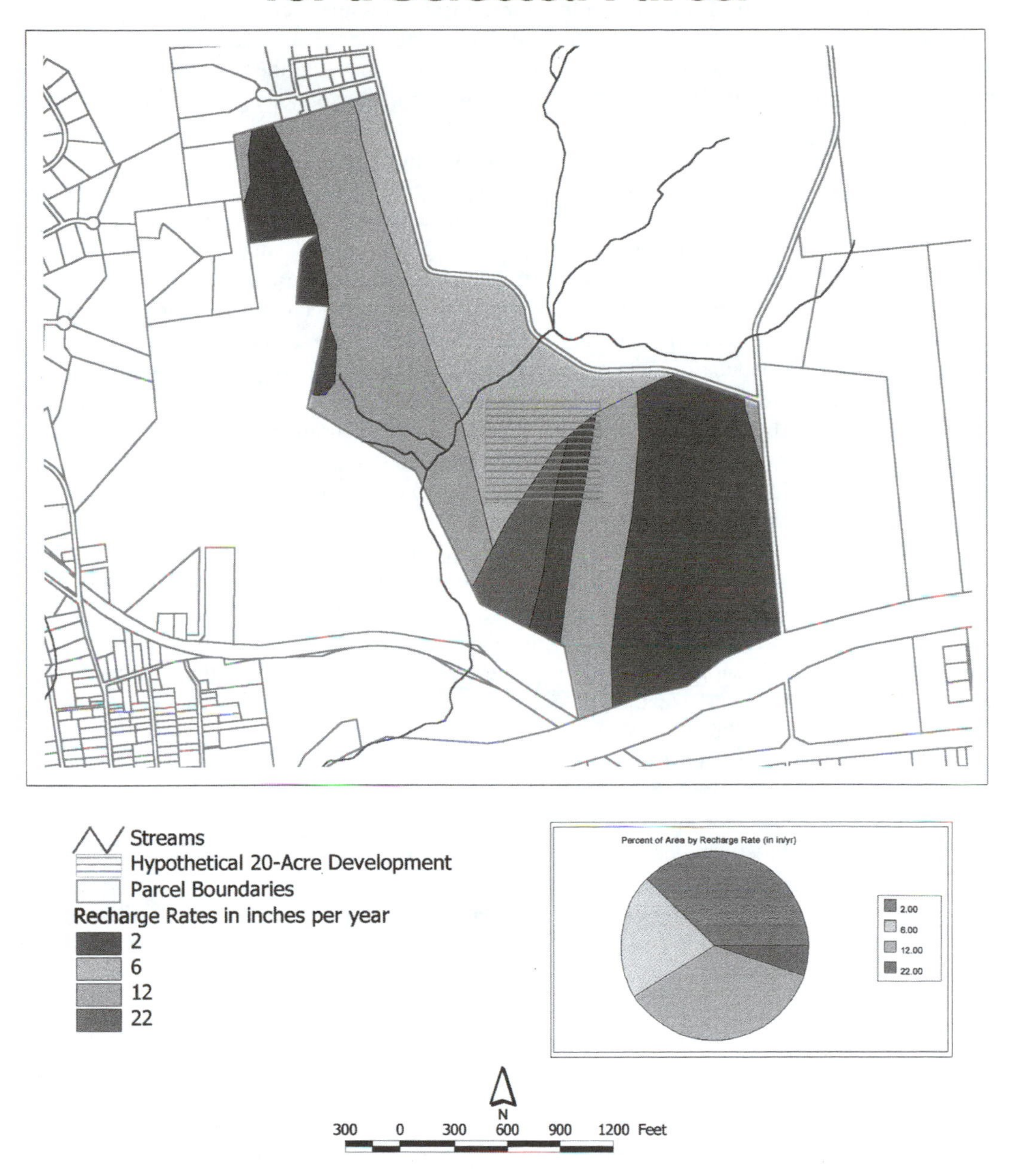

FIGURE 10.44 Areas of the community with the highest groundwater recharge potential need special protection and recognition in the municipal master plan so that individual development projects can be configured so as not to impair the critical function of these particular landscape features. (Graphics by Integrated Spatial Solutions, Moorestown, NJ.)

10.4.1.4 Wildlife

The primary objective for any municipality intent on doing more to protect its indigenous wildlife resources is, as we have repeatedly stated, to preserve and retain sufficient amounts of suitable habitat — an undertaking of no small proportions, considering the fact that wildlife and man are in constant competition to occupy the same habitat. How then do we achieve parity for our wildlife resources in this often lopsided competition? First, we must put aside the common misconception that the welfare of our wildlife resources is the *sole responsibility* of state government, despite the fact that this is what most state legislation provides. Yes, the states theoretically *own* the wildlife but the local communities really control where and how these resources survive. There obviously must be a much stronger collaboration between both groups. Second, the community must consider wildlife as a truly beneficial resource, having a tangible economic, aesthetic, and ecological value that is worthy of protection and preservation. Third, the community must, through rigorous public discussion and debate (as provided in the MMP adoption process), resolve itself to make preservation of these resources a true priority in the MMP. Fourth, it must accumulate as much information as it can about its indigenous populations of wildlife including, for example, the life cycles of each species; their current population densities; specific habitat needs; foods required; nesting preferences; and documented ability of each species to adapt to a landscape abruptly fragmented and eventually truncated by roadways, drainage ditches, electrical transmission lines, parking lots, innumerable numbers of dwellings, and commercial and industrial buildings. A great many of these latter data may be found in the records and databases of state fish and wildlife agencies, in archived university research, and among various federal wildlife agencies.* Private nonprofit agencies also have a good deal of information on these resources. All this information is an essential ingredient in the public participation component of the MMP adoption process.

There are also some common myths pertaining to habitat that we need to discuss and correct. The first is that suitable and sufficient wildlife habitat and any dedicated open space shown on a site plan or major subdivision map are synonymous. While in rare instances they can be, in most instances they are not. Second is the myth that the dedicated riparian buffer zones (aka greenways) along surface waterways often required by local planning agencies provide essentially all the habitat necessary for sustaining the well-being of a municipality's wildlife resources. While current research certainly indicates that riparian corridor habitat is highly preferred by many wildlife species,** the widths of those dedicated corridors and their connection to larger, unbroken tracts of landscape are essential to their successful function. We have discussed previously the potential widths of such buffer zones as they pertain to the protection of the homeostasis of the stream channel, generally, and to the prevention of damaging pollutants from entering the waterway, specifically. The widths, as you recall, varied from as little as 25 ft to as much as 150 to 200 ft on either side of the stream. Further research on the widths necessary to accommodate the needs of some species, particularly birds, both as a travel corridor and as a breeding habitat ranged as high as several hundred feet wide[74,75] — a significant increase and one that may raise the ire of those landowners in the community whose lands are traversed by surface water courses. Landowners, who may have been tolerant of much narrower corridors for the protection of human interests, may not be as amenable to substantial increases in corridor widths to accommodate the needs of wildlife. If past history is any indication, we know that there surely are to be concerns expressed concerning the taking of lands without just compensation.

The direct human health, safety, and welfare issues that can be successfully argued for preserving and protecting many other ecological features of a community's landscape may not be as successfully argued here, because except for such things as deer-related automobile accidents most

* An excellent reference in this regard is the USEPA *Wildlife Exposure Factors Handbook.*[77] Excerpts from this book can be found in Appendix B.

** Tilghman[76] found that woods with a stream flowing through them or adjacent to lakes had greater bird species diversity and total bird abundance.

impacts on humans are indirect, with Lyme disease caused by the deer tick being a classic example. Some empathy, however, may be induced for those wildlife species determined to be threatened or endangered. Paradoxically, the very reason many of these species are threatened in the first place is because sufficient amounts of their preferred habitat were not set aside in the past. A 25-year-old federal law has proved to be a valuable "safety net" for those species, both plant and animal, that have tottered on the brink of extinction. We are speaking of the 1973 Endangered Species Act (ESA) in which Congress found and declared, "...that various species of fish, wildlife, and plants in the United States have been rendered extinct as a consequence of economic growth and development untempered by adequate concern and conservation." Thanks to this act over 1000 endangered and threatened species in the United States now receive federal protection.

Section 7 of the ESA originally provided the means for federal agencies to authorize, fund, or carry out development projects while ensuring that such projects would not jeopardize the continued existence of threatened or endangered species. Private, nonfederal land developers, on the other hand, had no similar process and could be prevented from undertaking a project if it would result in the killing, harming, or harassment (known in legal terms as a "taking") of a threatened or endangered species. To resolve this dilemma, in 1982 Congress amended Section 10 of the ESA to allow creation of habitat conservation plans (HCPs), which were intended to allow private land development projects to move forward while protecting any threatened and endangered species that may have been identified on the site. An HCP allows the U.S. Fish and Wildlife Service to permit the "taking" of threatened or endangered species incidental to lawful activities when the taking is mitigated by conservation measures outlined in the HCP. Mitigation measures used thus far have included preservation of existing habitat, enhancement or restoration of degraded or former habitat, creation of new habitats, establishment of buffer areas around existing habitats, modifications of land use practices, and restrictions on access.

Some environmentalists, however, believe the HCP has been overly generous and that too many HCPs are developed in behind-the-scenes negotiations between the U.S. Fish and Wildlife personnel and the regulated parties without adequate public input. There is also some concern over the present "no surprises" policy that gives landowners assurances for up to 100 years that no additional conservation efforts will be required beyond those specifically agreed to in the HCP. Of particular concern is whether there will be sufficient long-term funding to pay for the implementation and monitoring of the HCP for such extended periods of time. A number of states have opted to adopt their own legislation to protect threatened and endangered species and have compiled state specific lists to complement those of the U.S. Fish and Wildlife Service. The extent of those data varies from state to state, with some states such as New Jersey having a fairly extensive electronic database that can easily retrieve data by political subdivision (municipality or county) or watershed management area (see Appendix C for examples).

Although the HCP was designed specifically to aid in the protection of threatened and endangered species, that concept also may prove to be a practical tool for the protection of wildlife in general at the local planning level. This is not to imply, however, that local land planners also could not utilize it for its originally intended purpose, that of protecting threatened and endangered species. Let us for the moment, however, examine its generalized application for any species. For this we would need to go to our previously compiled lists of response indicators already included in our MMP. Because the white-tailed deer is an acknowledged problem in many parts of the United States, it is an excellent species to use in our example. We therefore, go to our previous deer density map (see Figure 10.21). We immediately see that there are sections of the municipality where the population densities are inordinately high. Such extraordinarily high densities generally produce some easily identifiable negative ecological impacts. We can corroborate that fact by examining other response indicators including number and location of deer-related automobile accidents, number and location of areas of heavy crop damage or damage to landscape shrubbery or vegetation specifically attributable to deer, and location of sites where deer show signs of malnourishment or starvation. Undoubtedly there should be corroboration from some or all the other response indicators

TABLE 10.10
Wildlife Habitat Criteria

Recommended habitat size ranges[a,b]		Vegetative criteria[c]
Small and medium land vertebrates (e.g., mice, moles)	1.5–12.5 acres	Tree canopy height greater than 9 m (30.5 ft)
Large land vertebrates (e.g., deer)	49.5–74 acres	Tree canopy closure greater than 70%
Small woodlot birds (e.g., American robin, mockingbird, sparrow)	2.5–12.4 acres	At least three trees greater than 62 cm (2.1 ft) in diameter at breast height
Forest interior birds	≥12.4 acres	More than 0.1 snags greater than 62 cm (2.1 ft) in diameter at breast height
Forest edge species (both bird and animal species)	≥25 acres	Average shrub height greater than 0.6 m (2 ft) but less than 2.5 m (8.5 ft); shrub canopy closure greater than 6%

[a] These recommendations are broad generalizations and may very with individual species. Many species may additionally require specific edge-to-center ratios.
[b] From Vizyova, 1986; Dickman, 1987; Tilghman, 1987; cited in Adams and Dove, 1989.
[c] From Groffman *et al.*, 1989.

Source: From Metropolitan Washington Council of Governments (MWCG), Riparian Buffer Strategies for Urban Watersheds, Publication No. 95703, Prepared by Lorraine M. Herson Jones, Maureen Heraty, and Brian Jordan, Washington, D.C., 1995. With permission.

evaluated. What then is the next step? Past land planning practices would have given negligible, if any, attention to these wildlife-related problems, as site plans were reviewed and given approval, thus merely exacerbating the problem. That would not be the case in an ecologically based MMP. As applicants and/or landowners submit site plans to develop properties in these areas of acknowledged high white-tailed densities, they then would be required to prepare and submit an HCP, in which the preferred option would be the preservation of adequate habitat, primarily as a core reserve. Some idea of the extent of those core reserves can be found in Table 10.10. When evaluating core reserve size, one must also consider the long-term management of this species. If sport hunting, either by gun or by bow, is to be one of the major management tools, then the core reserve must be sized to provide adequate safety zones from occupied dwellings and other buildings, generally 450 ft. Subdivision applicants for the smallest of sites may not have sufficient area within their parcel for the preservation of an adequately sized core reserve. For them, there are other options including a partial dedication of area to enlarge an adjoining core reserve or riparian buffer, or the payment of a fee into a habitat preservation or open space fund to be used for purchasing property off-site and elsewhere in the municipality to enlarge a core reserve or riparian buffer there. Obviously, HCPs need not be confined to sites occupied by acknowledged problem species. In fact, an HCP should be required for all species and for each and every site plan submittal. In this instance, the establishment of core reserves and/or payment of a fee into a habitat preservation or open space fund also would remain a priority recommendation.

Earlier in this chapter we spoke briefly of the potential use of riparian corridors by wildlife as both habitat and passageways. Much like the riparian buffers or stream management zones that have been recommended to protect instream biota and water quality, similar buffer zones can be established for the enhancement and protection of wildlife. However, the research data are not nearly as prolific as those for water quality concerns about the appropriate widths for such buffers, which often vary by species. Despite this, there does appear to be enough research data available so that local land planners, if they wish, can establish buffer zones specifically geared to the needs of wildlife. For example, Rudolph and Dickson[78] in a study of amphibians and reptiles in eastern Texas recommended retaining streamside zones of mature trees at least 30 m (98.4 ft) wide and preferably wider. Burk et al. in a study* of wild turkeys in Mississippi recommended establishing SMZs between 84 and 104 m (275 to 341 ft) and 170 to 179 m (557 to 587 ft) wide. Croonquist and Brooks[79] in a central Pennsylvania study observed that sensitive bird species do not occur unless an undisturbed corridor >25 m (82 ft) wide on each bank is present. Keller et al.[80] in their research on bird communities on the eastern shore of Maryland and Delaware recommended that riparian forests have a total width of at least 100 m (328 ft). Finally, Dickson and Huntley,[81] in their study of squirrel populations in eastern Texas, recommended a minimum total width for the riparian zone of about 55 m (180 ft). These are only a very tiny cross section of the research done so far on wildlife and its relationship to riparian corridors. Local land planners are encouraged to gather and evaluate further the results and recommendations of many other research projects currently available.

Thus far our discussion has focused on terrestrial species; however, we must not forget their aquatic counterparts. In this case we consider our wild trout populations. For those sections of streams shown previously in Figure 10.26, we would most likely already have established protective buffers. Although we have discussed this somewhat, we need to reinforce it. Further protection must be provided in the MMP for these streams by designating them as no direct storm water discharge zones. In other words, no new storm drains would be allowed to discharge directly to any portion of these protected waterways. Instead, storm water would be discharged as sheet flow at the outermost edge of the established buffer and then allowed to flow overland from there to the stream. Along the way, it would receive the benefit of some treatment by the soil, ground litter, and vegetation.

10.4.1.5 Wetlands

It would be rare to find a local land planning board today that is not familiar with the large-scale national wetlands inventory (NWI) maps prepared by the U.S. Fish and Wildlife Service. They are one of the most common maps, besides NRCS soil maps, to be included in a municipality natural resource inventory. As a general inventory of such ecologically important landscape features, they have served their purpose quite well. Now it appears that their utility may be enhanced even further, by their combination with what are called hydrogeomorphic (HGM) features that include such characteristics as landscape position, landform, and water flow paths (i.e., inflow, outflow, and throughflow). All these can provide us with more details on the actual functions of individual wetlands — certainly valuable information for land planners intent on compiling an accurate ecologically based MMP and on assigning the appropriate protective status or buffer width, if any, to such critical landscape features. Having more detailed knowledge of the actual functions of a particular wetland also is invaluable to local land planners confronted by proposals to create replacement wetlands when construction threatens the destruction of existing ones. All wetlands are not created equal and in the past this often has been ignored, with the result that outflow wetlands — critical to maintaining low flows in adjacent waterways — have been replaced by little more than flat graded depressions that scarcely do more than trap surface water runoff. Figures 10.45 through 10.48 provide examples of the various, often significant, differences in wetlands features and functions.

On a pilot basis the U.S. Fish and Wildlife Service is enhancing conventional NWI map data by adding HGM-type attributes to the existing NWI digital database.[83] A series of function-specific maps and statistical acreage summaries are to be produced, highlighting wetlands that are likely

* This study is cited in Reference 91.

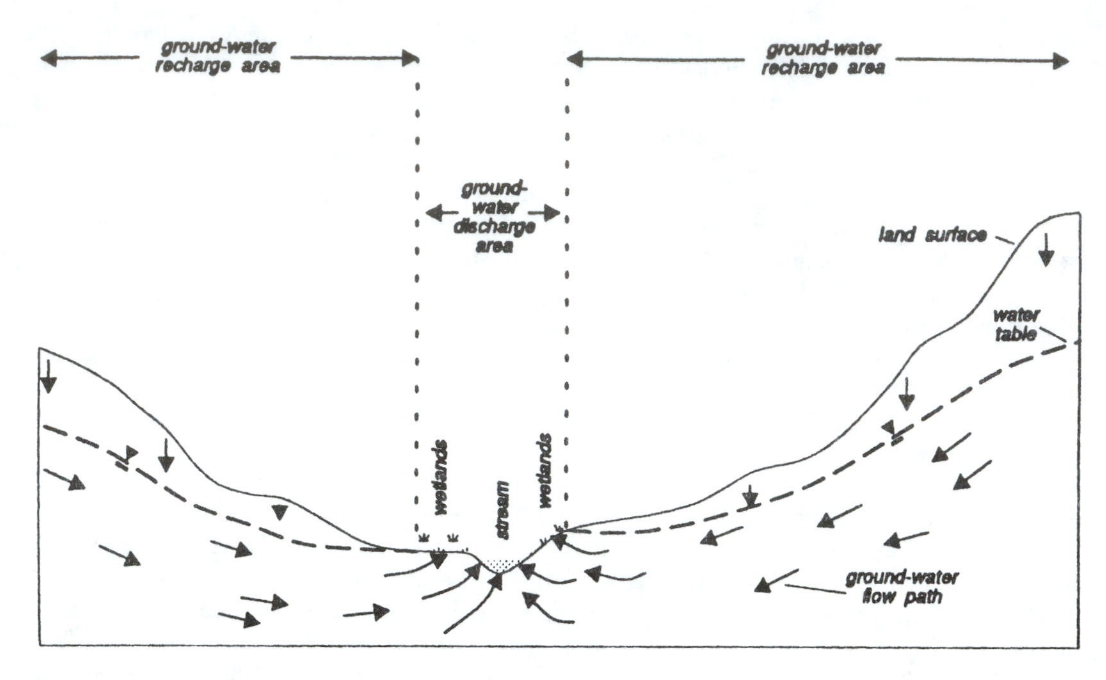

FIGURE 10.45 Example of a groundwater discharge to a stream and wetlands. (From New Jersey Department of Environmental Protection, A Method for Evaluating Ground Water Recharge in New Jersey, New Jersey Geological Survey, 1993.)

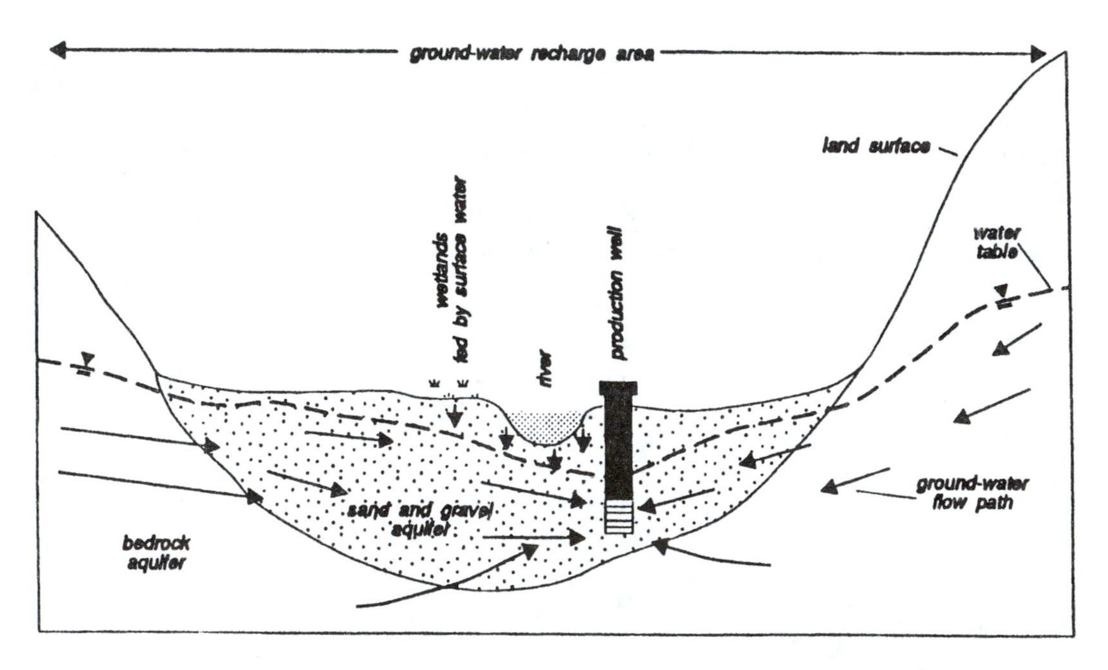

FIGURE 10.46 Wetlands fed by surface water. (From New Jersey Department of Environmental Protection, A Method for Evaluating Ground Water Recharge in New Jersey, New Jersey Geological Survey, 1993.)

to perform certain functions at significant levels (e.g., surface water detention, stream flow maintenance, nutrient cycling, sediment and other particulate retention, shoreline stabilization, fish habitat, waterfowl and waterbird habitat, other wildlife habitat, and biodiversity protection). These newest maps allow decision makers and land planners to better assess the true worth of individual

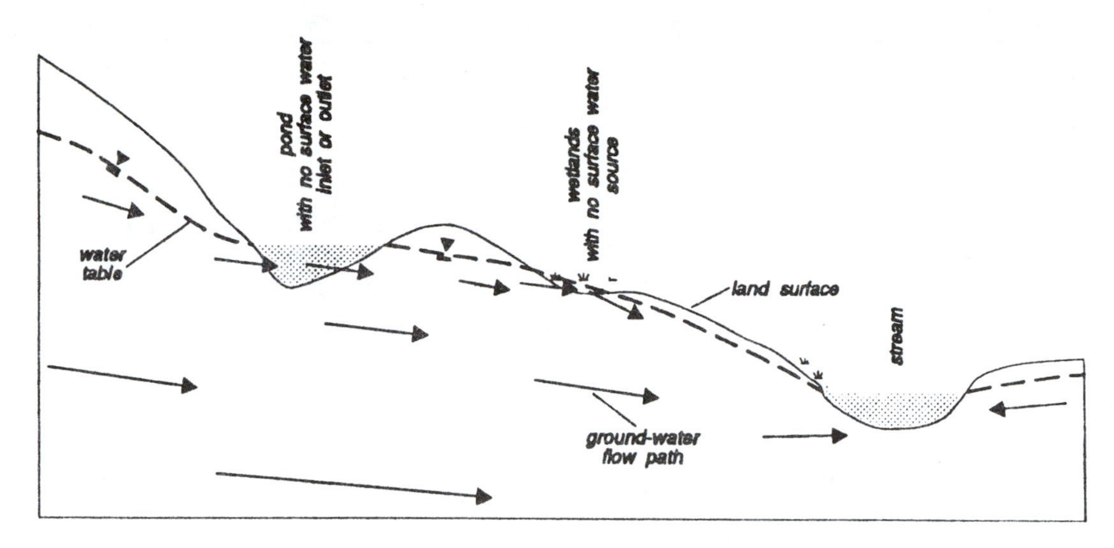

FIGURE 10.47 Examples of groundwater recharging to and discharging from a wetland. (From New Jersey Department of Environmental Protection, A Method for Evaluating Ground Water Recharge in New Jersey, New Jersey Geological Survey, 1993.)

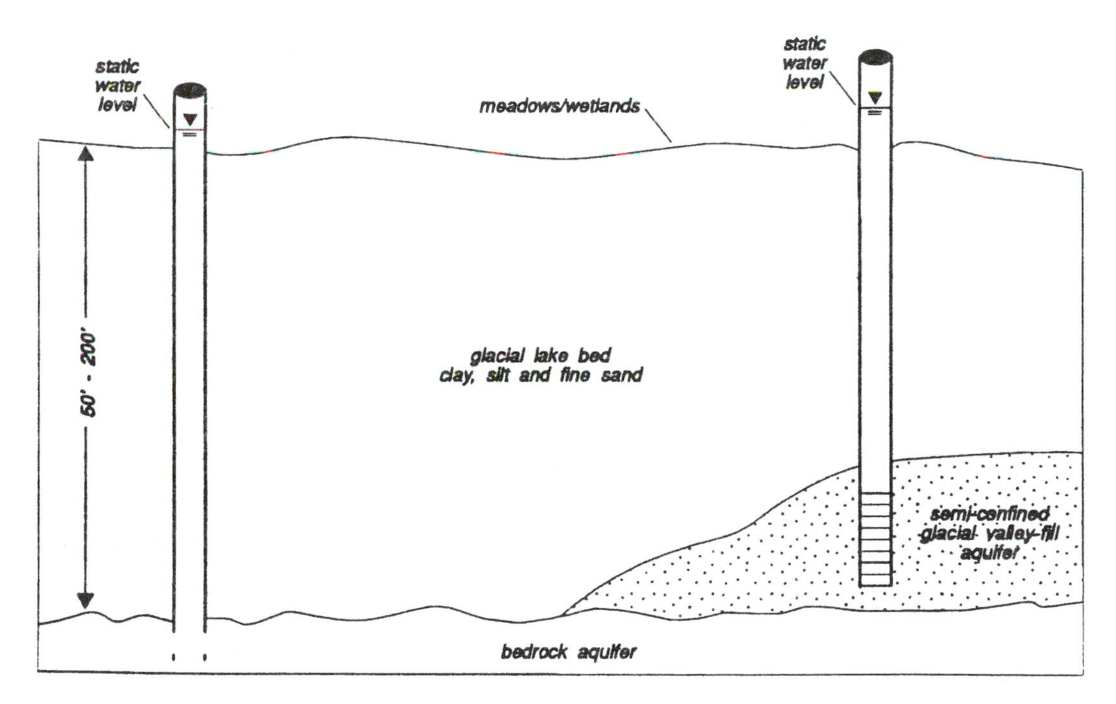

FIGURE 10.48 Example of a wetlands that is neither a significant groundwater recharge nor a discharge area. (From New Jersey Department of Environmental Protection, A Method for Evaluating Ground Water Recharge in New Jersey, New Jersey Geological Survey, 1993.)

wetlands. Figures 10.49 and 10.50 demonstrate some of the initial fieldwork accomplished thus far.[84]* As more data become available, local land planners should begin incorporating the information into their MMP.

* For more information on this process contact Ralph Tiner, Regional Wetlands Coordinator at U.S. Fish and Wildlife Service, 300 Westgate Center Drive, Hadley, MA, 01035 or by electronic mail (e-mail) Ralph_Tiner@mail.fws.gov.

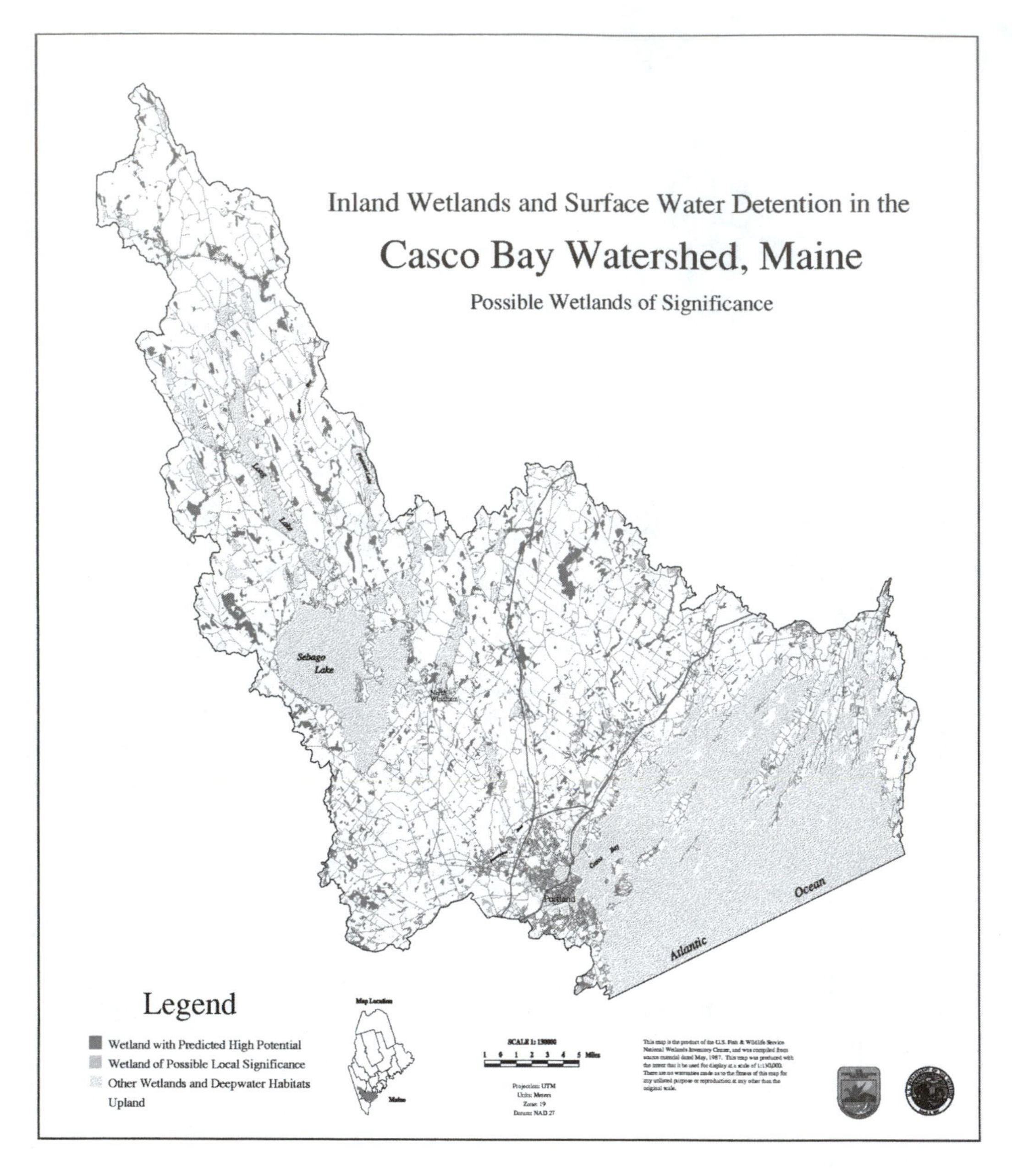

FIGURE 10.49 (From Tiner, R., Schaller, R. S., Peterson, D., Snider, K., Ruhlman, K., and Swords, J., Wetland Characterization Study and Preliminary Assessment of Wetland Functions for the Casco Bay Watershed, Southern Maine, U. S. Fish and Wildlife Service, Northeast Region, National Wetland Inventory, Ecological Services, Hadley, MA, NWI Report, 1999.)

10.4.1.6 Soils

In our discussion of soil resources in Chapter 9 we spoke of the need to preserve our prime agricultural soils. We suggested that an inventory be done and that this information be included in the MMP. Assuming that you have done that, as individual site plans are submitted for review, the soil data (now hopefully stored in your GIS database) can be retrieved and the extent of any prime agricultural soils (if any) on individual sites can be displayed and evaluated (Figure 10.51).

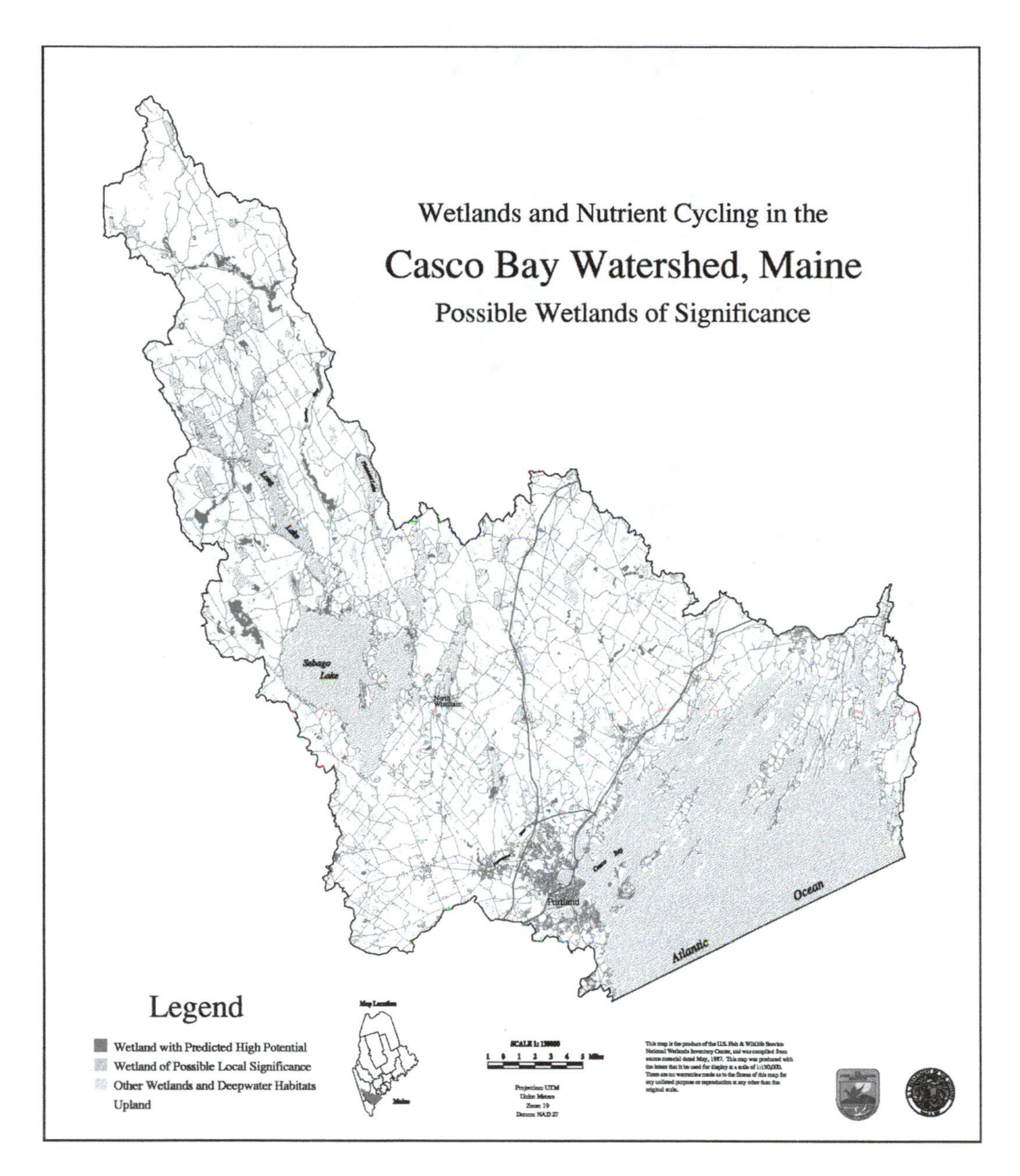

FIGURE 10.50 (From Tiner, R., Schaller, R. S., Peterson, D., Snider, K., Ruhlman, K., and Swords, J., Wetland Characterization Study and Preliminary Assessment of Wetland Functions for the Casco Bay Watershed, Southern Maine, U. S. Fish and Wildlife Service, Northeast Region, National Wetlands Inventory, Ecological Services, Hadley, MA, NWI Report, 1999.)

Hopefully, the applicant has reviewed the community MMP well before site plan submittal to ascertain the municipality's ecological priorities, not only for this natural resource but also for other ecological resources, and has incorporated them into his or her site plan design. Ideally, such soils should be left as common open space, perhaps as community gardens. There is a chance, however, that the location of these particular soils may not fit well with the applicant's proposed site design and in fact may be slated for excavation or the application of an impervious surface. If that is the situation and a design reconfiguration is out of the question, then the

Soil Suitability for Agricultural Operations

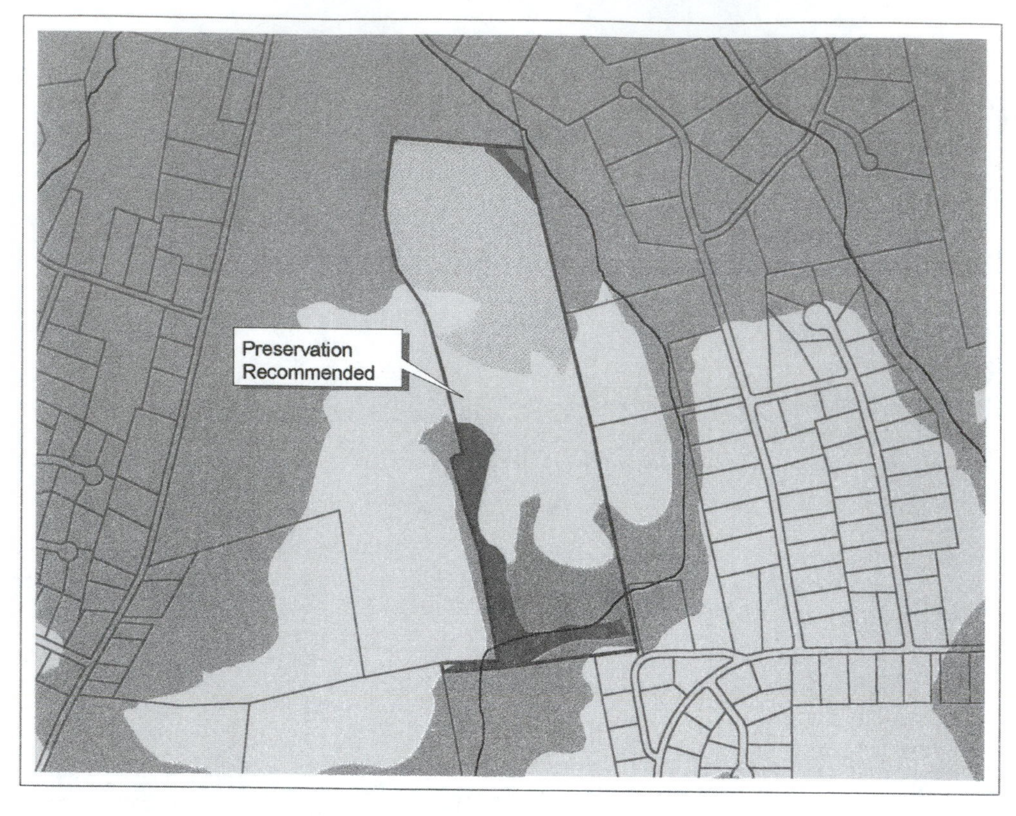

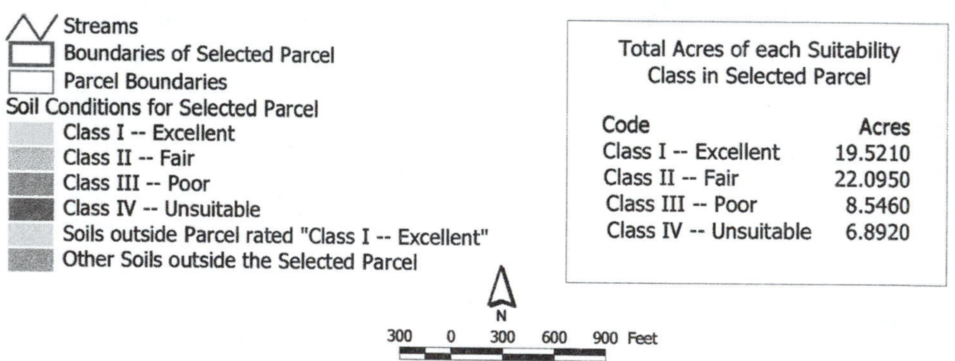

Total Acres of each Suitability Class in Selected Parcel

Code	Acres
Class I -- Excellent	19.5210
Class II -- Fair	22.0950
Class III -- Poor	8.5460
Class IV -- Unsuitable	6.8920

FIGURE 10.51 The developer of this particular parcel has been put on notice that the community will no longer accept the routine burial of its best agricultural soils under what will probably be tons of asphalt and concrete. (Graphics by Integrated Spatial Solutions, Moorestown, NJ.)

municipality has two options: (1) require the applicant to buy another tract of land equivalent in size and quality to that covered by the soils on-site — preferably in an active agricultural area — and donate it to the town; or (2) require the applicant to pay a public service replacement or impact fee into the municipality's open space, agricultural preservation, or habitat preservation fund. This would be a one time fee.

Similarly, see our previous discussion in Chapter 9 on the protection of soils with the highest recharge rates and the estimated values calculated for the water captured on a yearly basis. In this case, the applicant also would be given the same option of buying another equivalently sized tract of identified high recharge soils and donating it to the municipality or alternatively of paying an impact or public service replacement fee.

10.4.1.7 Other applications

Assessments and development of strategies need not always begin with an analysis of inventoried ecological infrastructure. Man-made infrastructure often can serve as a focal point for strategy development in an ecologically based MMP as well. For instance, the percentage of impervious cover in a watershed or subwatershed has proved to be an excellent barometer of the general health of watershed surface water quality (see previous discussion in Chapter 9). A substantial amount of scientific research indicates that watersheds with a total impervious cover of 10% or greater are likely to already have suffered some serious water quality damage. Such watersheds would, therefore, warrant special identification in the MMP as a *watershed at risk* (Figure 10.52). Strategies to prevent further deterioration would include the requirement to lessen the amount of any planned impervious cover on individual sites. This combined with other strategies, such as adequately sized stream buffer zones and prohibition on further direct storm water discharges to streams, could curtail the further deterioration in water quality.

Response indicators that are predominantly human oriented also should not be overlooked as a focal point for strategies. For example, chronic noise complaints from residents inhabiting housing located adjacent to interstate highways is a response indicator worthy of some consideration in the preparation of future development patterns of the community. It is one of the reasons that hundreds of miles of our interstate highways have been retrofitted with monstrous, million dollar a mile, 15 to 25 ft high concrete walls and other similar noise barriers. Most generally the problem originates with the initial construction of these limited access highways themselves, where vacant or agricultural land was often the predominant land use at the time of their construction. Interchanges quickly drew hotels, motels, restaurants, and gas stations, which were eventually followed by outlet stores and office buildings. The once vacant surrounding countryside soon began to sprout residential subdivisions to accommodate both employees of these new establishments and city dwellers looking to escape increasingly hostile urban environments. The vacant land immediately along the periphery of these highway corridors likewise began to fill with residential subdivisions. Under past land planning practices, little forethought was given by local land planners to establishing a substantial buffer zone along the periphery of these acknowledged noise generators for minimizing the stressful impacts such loud sound levels can produce. It is not uncommon, even today to see new residences still being constructed 100 ft or sometimes closer to the edge of such roadways. Land planners would do a great service to occupants of future subdivisions built in close proximity to such major highways if they would designate a buffer zone at least 200 ft wide along the periphery of these roadways where construction of residential dwellings would be specifically prohibited. Such buffers should have a dense vegetative cover that could be supplemented with judicially placed earthen berms. These buffers would be left as undisturbed open space (Figure 10.53) or alternatively commercial buildings could be placed between the highway and any proposed residences (Figure 10.54).

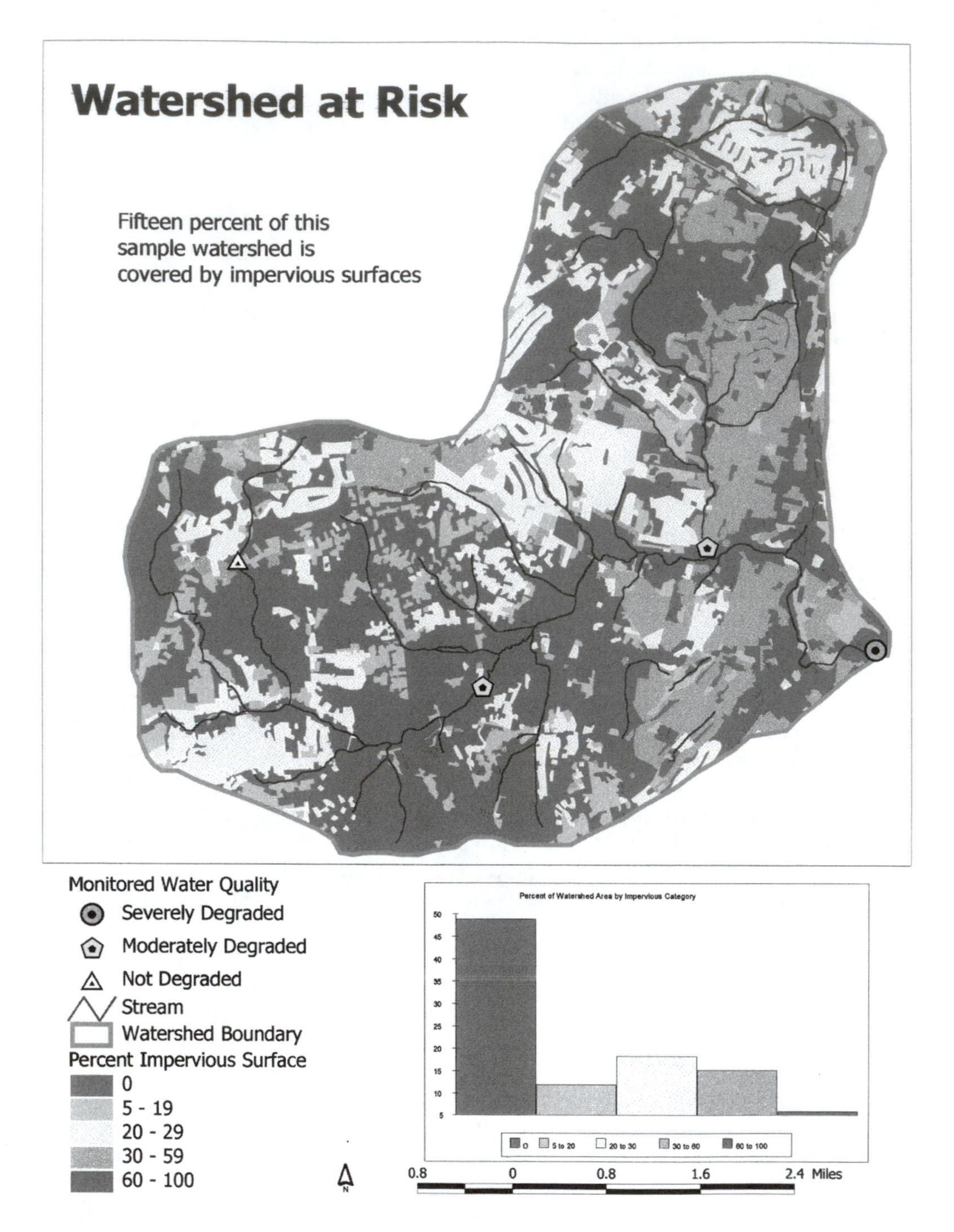

FIGURE 10.52 This watershed may very well be on the brink of irreversible damage to the quality of its water resources. With this type of information now included in an ecologically based master plan, the community has the opportunity to reverse this all too common and disastrous trend. (Graphics by Integrated Spatial Solutions, Moorestown, NJ.)

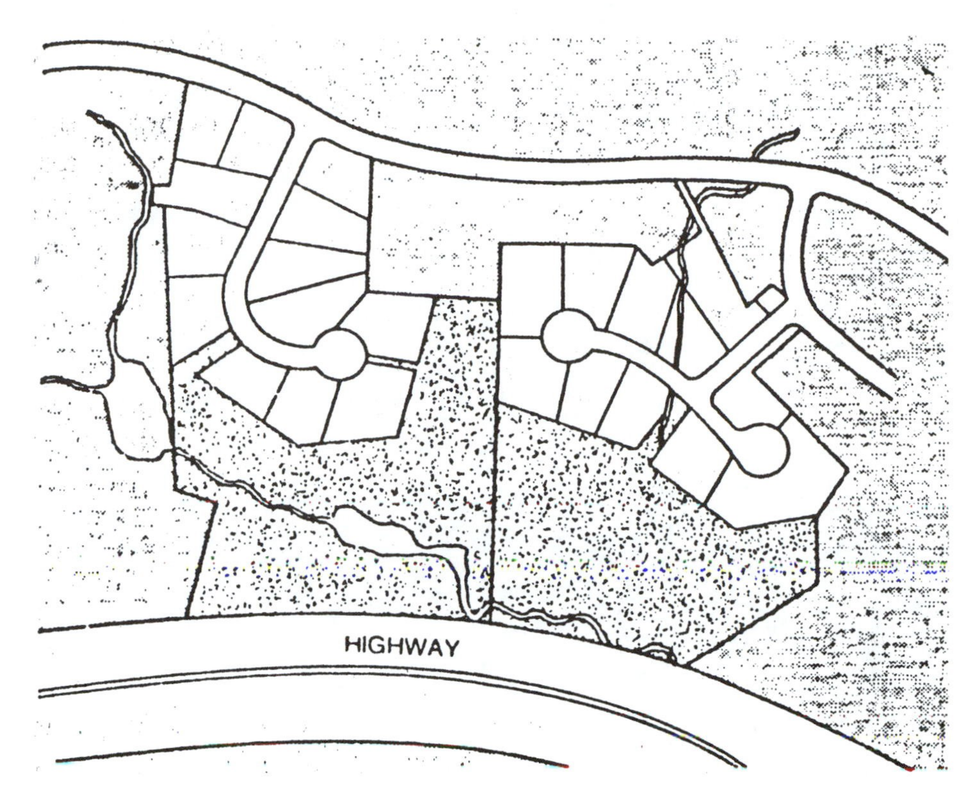

FIGURE 10.53 Open space can be left as a buffer zone between residences and a highway. (From U.S. Department of Transportation, FHA, Highway Traffic Noise, September 1980.)

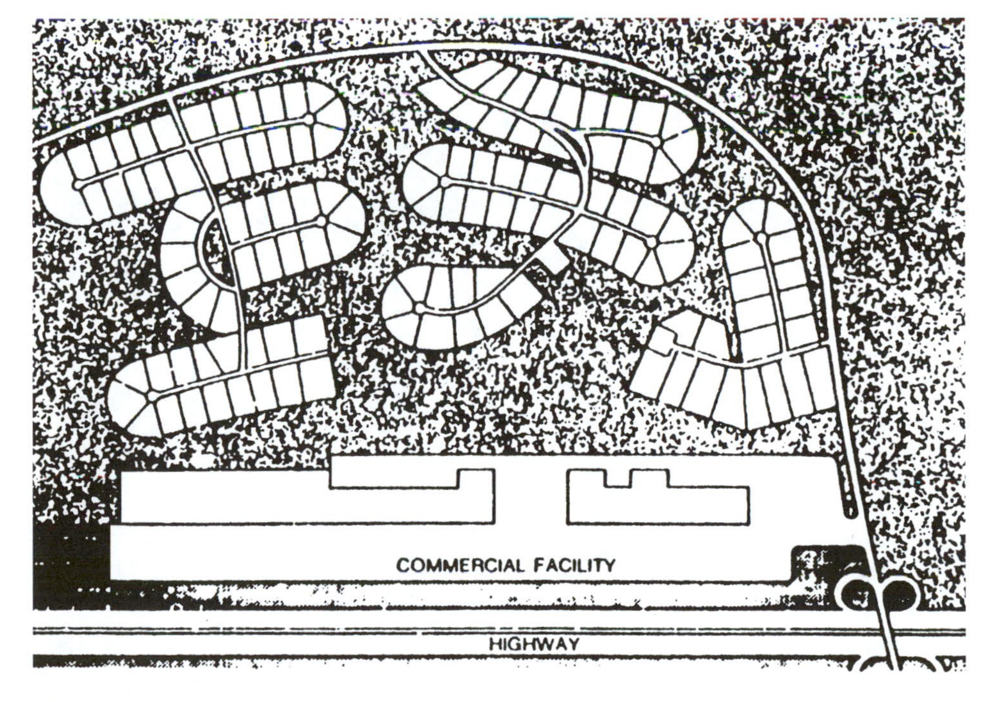

FIGURE 10.54 Using nonresidential buildings as noise buffers. (From U.S. Department of Transportation, FHA, Highway Traffic Noise, September 1980.)

11 New Ideas for a New Millennium

11.1 INTRODUCTION

One of the advantages of writing your own book is that it provides a forum for ideas and concepts that might not otherwise make the light of day. To be sure, there are those who may consider some of these ideas somewhat radical, but I would prefer to think of them as cutting edge; I would remind you that the human population once scoffed heartily at the idea that man could fly, that he could talk through a thin wire, and that pictures and images could be projected through the air only to reappear again on a screen in someone's home hundreds or thousands of miles away. Therefore, with no apologies to anyone, I offer you the following.

11.2 ONE LOT: ONE STRUCTURE

The transition from land use prescribed largely by zoning ordinances and use districts to a land use scheme that would now be guided by a holistic and ecologically based community master plan does not come easy to some — due in part to unfamiliarity with this new way of doing business, but more so because zoning has historically bestowed salubrious development values on nearly all private property. The value, I might add, did not come, for the most part, from the hard work and sacrifice of the owner, but simply from the fortuitous administrative parcelization of the landscape.

Nevertheless, an owner of 50 acres of land lying within a 1-acre minimum lot size residential zone has historically come to expect that he or she is to be allowed to divide this property into at least 50 lots, and to fight extraordinarily hard if denied that density, as any local planning board who has been sued has quickly found out. My question, therefore, is why create these artificial windfall values in the first place, and thereby raise the expectations of community landowners to the point where they now regard these development potentials as cast in stone and constitutionally guaranteed entitlements?

One possible solution to this dilemma, at least concerning vacant land parcels (and remember, this is the neo-ecoplanning era, where land use districts and zones play a more subordinate role as overlays on an underlying community-wide ecological framework), would be to simply designate all vacant parcels of land in the community, regardless of size, as entitled to a development potential/value of a single-family dwelling unit. In doing so the landowners would not be denied all use of their property and at the same time they would be guaranteed the potential for one of the long-standing, highest, and best uses of any property — that of shelter for themselves and their family. Any additional density (value) greater than that could only come about through a thorough analysis of all compiled scientific data; an evaluation of community sustainability and resources (ecological footprint?); and finally, negotiations and discussions between the landowners and the community concerning the juxtaposition of an individual site to community ecological infrastructure — all of which are most appropriately discussed at the time of the compilation of the community master plan.

The traditional zoning use districts (i.e., residential, commercial, and industrial) would now be labeled existing use zoning and would be reserved for all other parcels with preexisting land uses (that is to say, already with structures of some kind on them; and in use as a residence, commercial establishment, or industrial site) in the hope of encouraging infill within these districts, but not to the extent that it would compromise ecological infrastructure or threaten the capacities of community infrastructure.

11.3 MOUNTAINTOPS AND TOWERS

If Americans insist on having instant access to a mobile phone network, no matter where they are or where they go, then the proliferation of aesthetically unappealing transmission towers to support that frenzied desire may reach unprecedented proportions over the next decade. In some states, like New Jersey, if the present rate of phone tower installation continues unabated, it is conceivable that within 10 years or less nearly every resident of the state will be able to view at least one of these towers directly from their own backyard — a very disconcerting prospect.

Current judicial decisions have determined that such towers are a necessary public use and cannot be banned outright by municipal ordinances. Thus, municipalities have had minimal success in thwarting their installation. They have, however, been fairly successful in keeping them out of established and exclusively residential neighborhoods. Concerns about radiation exposure have been raised, essentially brushed aside, and discounted by industry representatives, and at this point in time, such an argument appears to be a futile effort to block tower installation. The real objection, as I see it, is primarily one of aesthetics. There are now areas of the northeastern United States where nearly every visible prominence sprouts a transmission tower. Once serene mountaintops, often the last bastions of undisturbed landscape, are now permanently marred by these obtrusive structures, destroying much of the aesthetic charm that persistently draws us to them in the first place. They offer us serenity and reprieve from an increasingly hostile landscape in the flatlands below. When the mountaintops are all gone, what retreats are left?

What can municipalities do? A suggested solution is the creation of ridgeline/panorama protection districts that do not necessarily prevent tower installation per se, but keep them off the topmost prominences. This could apply to other structures as well. Our suggestion, therefore, would be as follows: on slopes of 25% or greater, no building, tower, or other structure can be constructed above a contour that is 100 ft in elevation lower than the average elevation at the top or highest point on the ridgeline.

11.4 AFFORDABLE HOUSING: PROACTIVITY PAYS MUCH HIGHER DIVIDENDS

It is gratifying to see so many large housing developers and wealthy entrepreneurs stepping forward to champion the cause of the nation's low-income population, particularly the nation's urban poor, who have complained, and perhaps with some justification, that zoning ordinances, particularly large lot zoning in developing rural areas, have kept them from savoring the good life enjoyed by their higher income neighbors. I would be less suspect of this corporate altruism if it were not for the fact that these developers and entrepreneurs continue to profit handsomely from the greater housing densities they are able to coax from the courts, following lawsuits filed against municipalities alleging discrimination against the poor. In any event, if the concern is genuine, so be it. For the municipality, however, such lawsuits have been costly, not only monetarily but also physically. By physical I mean that many local long-term land use strategies have been rendered impotent by the sudden imposition of 1000 or more dwelling units on a site that may never have been envisioned for such concentrated densities or for which ecological carrying capacity is woefully inadequate to support it.

Why play such reactive roulette, when a more proactive posture could easily put affordable housing into the hands of low-income families more quickly and at locations planned by the municipality instead of dictated by the courts? It would most certainly result in dwelling densities much less than those imposed by so-called "builders' remedies." I would suggest, therefore, that each municipality preselect sites for low-cost housing (by low-cost housing, we mean single-family, detached dwellings with total floor area ranging from 880 to 1200 square ft maximum), incorporate them into the MMP, and then under its auspices or with its own revenues construct such housing itself. One-quarter acre lots would not be out of the question if city sewer and water were available.

Where such utilities are not available, lot sizes would have to be increased to 1 acre to adequately accommodate an individual septic disposal system. Furthermore, if the suburban environs are where low-income families, particularly urban low-income families, truly desire to live, would it not also make sense to redirect federal and state housing grants, loans, and subsidies to these suburban or rural sites as an aid in the construction of affordable housing in these more rural environs? Finally, aside from those low-income families who desire to migrate from heavily urbanized areas to more rural climes, each municipality needs to be able to provide affordable housing for those already in residence, particularly elderly persons on fixed incomes who desire to "downsize" and young married couples who are just starting out.

12 General Commentary on Best Management Practices

After having foolishly ignored our obligation to avoid, wherever possible, the disruption of the ecological infrastructure that helps sustain us, we further embarrass ourselves by trying to repair any damage we might have done by employing technological retrofits, commonly referred to as best management practices (BMPs). The New Jersey Department of Environmental Protection (DEP) has called particular attention to the questionable success of BMPs in its 1994 report entitled "Stormwater and Nonpoint Source Pollution Control Best Management Practices." The report concludes:

> Historically, planners have not adequately utilized the Municipal land use law as a tool for addressing stormwater and nonpoint source pollution management. This has resulted in the need to impose BMPs after the land uses have already become established because the initial planning failed to consider the environmental impacts of development. Such "after the fact" installation of BMPs has limited value on a project site because the BMPs are often unable to effectively manage the significant levels of pollution that land development tends to generate. A second limitation of "after the fact" BMP installations is that they often fail to account for the cumulative impacts of individual land uses that generate relatively few pollutants by themselves, but collectively can have a significant adverse impact on water quality in the region.

We could not have said it much better ourselves, and New Jersey is to be complimented for its frankness.

Design engineers have developed a real love for BMPs because of their good sound bite, feel good quality that offers the illusion, at least, that the solution to any problem is but a BMP away. However, is every BMP a good one and is it being appropriately applied? Sad to say, all too often the answer is no, especially in the latter instance. Local land planners, as expected, rarely challenge the appropriateness and projected effectiveness (largely due to inexperience) of proposed BMPs despite a supposedly exhaustive site plan review. Why their hired experts do not raise such issues for them, however, is not as understandable.

Our focus for this discussion is solely on the interim protective BMPs installed and operated during the construction phase of a project to control soil erosion and prevent soil migration off-site and into nearby waterways, wetlands, and other sensitive landscapes, where its impacts may be felt for decades after construction is complete. We would do well to remember that protecting our ecological infrastructure and other natural resources does not end with the adoption of an ecologically based municipal master plan (MMP). Instead, it is a continuum, carrying over to and through the construction stage and well beyond. During the construction phase our ecological resources are especially vulnerable, as vegetation is stripped from the soil surface and the earth is ripped, gouged, and excavated to install the human version of "improvements" to the land. While ecologically sensitive areas may have been designated off limits, the carryover from adjoining construction sites is a real threat, having the potential for both short-term and long-term damages. The shortcoming of soil erosion and sediment control BMPs are many. First, there is the installation of the practices themselves. For example, the commonly used fabric filter fences are often haphazardly installed, with one of the biggest flaws being the failure to properly heel in the bottom of the filter and thereby allowing sediment laden storm water to bypass the filter completely. Even where the filter fence is properly installed, the trapped sediment is not removed between storms or on any regular basis whatever to renew the filter fence trapping and filtering efficiency. Similarly, hay bale filters,

FIGURE 12.1 This hay bale and stone filter combination is not the most effective siltation control device. Newer technologies could aid the performance of these crude devices considerably. See Appendix E.

the poor man's substitute for the fabric filter fence, are often left in place for months, sometimes even a year or longer, without refurbishment with new bales and removal of trapped sediment. Where sediment basins have been installed, which by the way are one of the better erosion control BMPs during the construction phase, they also require regular maintenance and cleaning to ensure an adequate capacity during the life of the project. Some of the newer state-of-the-art techniques and structures designed to more effectively trap sediment and other associated pollutants are only infrequently applied. Fabric filter bags sold under the trade name Siltsack ®* can be placed inside newly constructed catch basins during the construction phase and could complement the typical surficial gravel and hay bale filter (Figure 12.1) usually used at such locations. Similarly, sediment-laden water, excavated from water-filled trenches and often simply discharged overland would benefit considerably from a new product called the Dirt Bag.®** See Appendix E for more details on these products.

The maintenance of erosion and sediment control BMPs is by far the biggest shortcoming during the construction phase of a project. Each site needs to have an on-site erosion control coordinator whose sole responsibility is just that. While local regulatory agencies, such as local soil conservation districts often do a commendable job of overseeing the implementation of erosion control BMPs, because of workloads they usually cannot be there everyday or even after every rainfall to ensure that such practices are fully maintained.

* This is a registered trademark of ACF Environmental, Richmond, VA 23237.

** This is a registered trademark of ACF Environmental, Richmond, VA 23237.

It should be noted that thanks to the efforts of the Soil and Water Conservation Society and several other organizations, an experienced cadre of credentialed experts in soil erosion and sediment control are now available to assist in the day-to-day oversight of soil erosion and sediment control practices on construction projects. They are certified professionals in erosion and sediment control, (CPESC), who to be certified had to pass a three-part written qualifying exam, must have attained a bachelor's degree, and must have had 6 years of professional experience in the field. Municipal land planners would do well, as part of their site plan approval conditions, to require that a CPESC be on site daily during the entire construction phase to ensure that soil erosion and sediment control practices there are adequately maintained and operated.

Appendices

APPENDIX A: WEB SITES CONTAINING ENVIRONMENTAL AND/OR ECOLOGICAL INFORMATION

	Organization	Web Site
1.	American Society of Civil Engineers	http://www.asce.org/peta/tech/nsdb01.html
2.	The National Institutes For Water Resources	wrri.nmsu.edu/niwr/index.html
3.	National Park Service	www.nps.gov
4.	U.S. Geological Survey	www.fws.gov
5.	USDA Natural Resource Conservation Service	www.nrcs.usda.gov
6.	U.S. Environmental Protection Agency	www.epa.gov/cincl
		earthl.epa.gov/awberc/awberc.htm
		www.epa.gov/ogwdw
		www.epa.gov/ord
		www.epa.gov/ost
		www.epa.gov
		www.epa.gov/surf*
		www.epa.gov/oswer
		www.epa.gov/owm
		www.epa.gov/owow
7.	The Wetlands Regulation Center	www.wetlands.com
8.	The National Wildlife Federation	www.nwf.nwf.org/nwf/
9.	The Nature Conservancy	www.tnc.org
10.	The Sierra Club	www.sierraclub.org
11.	Soil and Water Conservation Society	www.swcs.org
12.	The Terrene Institute	www.terrene.org
13.	Trout Unlimited	www.tu.org.
14.	The Water Environment Federation	www.wcf.org
15.	National Water Resources Assoc.	www.nwra.org
16.	National Environmental Training Center For Small Communities	www.estd.wvu.edu/netc
17.	WATERSHEDS	h2osparc.wq.nscu.edu/wresourc.html
18.	Nonpoint Education for Municipal Officials (NEMO)	www.canr.uconn.edu/ces/nemo
19.	Great Lakes Information Network	www.great.lakes.net
20.	National Farm A Syst Program	www.wisc.edu/farmasyst/
21.	U.S. Fish and Wildlife Service	www.fws.gov

*Highly recommended

See also numerous web sites operated by state environmental protection agencies. Query internet by state.

United States Environmental Protection Agency

Office of Research and Development Washington DC 20460

EPA/600/R-93/187a
December 1993

EPA

Wildlife Exposure Factors Handbook

Volume I of II

2.1.10. American Woodcock (woodcock and snipe)

Order *Charadriformes*, Family *Scolopacidae*. These inland members of the sandpiper family have a stocky build, long bill, and short legs. However, their habitats and diet are distinct. Woodcock inhabit primarily woodlands and abandoned fields, whereas snipe are found in association with bogs and freshwater wetlands. Both species use their long bills to probe the substrate for invertebrates. The woodcock and snipe are similar in length, although the female woodcock weighs almost twice as much as the female snipe.

Selected species

The American woodcock (*Scolopax minor*) breeds from southern Canada to Louisiana throughout forested regions of the eastern half of North America. The highest breeding densities are found in the northern portion of this range, especially in the Great Lakes area of the United States, northern New England, and southern Canada (Gregg, 1984; Owen et al., 1977). Woodcock winter primarily in the southeastern United States and are year-round residents in some of these areas. Woodcock are important game animals over much of their range (Owen et al., 1977).

Body size. Woodcock are large for sandpipers (28 cm bill tip to tail tip), and females weigh more than males (Keppie and Redmond, 1988). Most young are full grown by 5 to 6 weeks after hatching (Gregg, 1984).

Habitat. Woodcock inhabit both woodlands and abandoned fields, particularly those with rich and moderately to poorly drained loamy soils, which tend to support abundant earthworm populations (Cade, 1985; Owen and Galbraith, 1989; Rabe et al., 1983a). In the spring, males use early successional open areas and woods openings, interspersed with low brush and grassy vegetation, for singing displays at dawn and dusk (Cade, 1985; Keppie and Redmond, 1985). Females nest in brushy areas of secondary growth woodlands near their feeding areas, often near the edge of the woodland or near a break in the forest canopy (Gregg, 1984). During the summer, both sexes use second growth hardwood or early successional mixed hardwood and conifer woodlands for diurnal cover (Cade, 1985). At night, they move into open pastures and early successional abandoned agricultural fields, including former male singing grounds, to roost (Cade, 1985; Dunford and Owen, 1973; Krohn, 1970). During the winter, woodcock use bottomland hardwood forests, hardwood thickets, and upland mixed hardwood and conifer forests during the day. At night, they use open areas to some degree, but also forested habitats (Cade, 1985). Diurnal habitat and nocturnal roosting fields need to be in close proximity to be useful for woodcock (Owen et al., 1977).

Food habits. Woodcocks feed primarily on invertebrates found in moist upland soils by probing the soil with their long prehensile-tipped bill (Owen et al., 1977; Sperry, 1940). Earthworms are the preferred diet, but when earthworms are not available, other soil invertebrates are consumed (Miller and Causey, 1985; Sperry, 1940; Stribling and Doerr, 1985). Some seeds and other plant matter may also be consumed (Sperry, 1940). Krohn (1970) found that during summer most feeding was done in wooded areas prior to entering fields at night, but other studies have indicated that a significant amount of food is acquired during nocturnal activities (Britt, 1971, as cited in Dunford and Owen, 1973). Dyer and Hamilton (1974) found that during the winter in southern Louisiana, woodcock exhibited three feeding periods: early morning (0100 to 0500 hours) in the nocturnal habitat, midday (1000 to 1300 hours) in the diurnal habitat, and at dusk (1700 to 2100 hours) again in the nocturnal fields; earthworms and millipedes were consumed in both habitat types. Most of the woodcocks' metabolic water needs are met by their food (Mendall and Aldous, 1943, as cited in Cade, 1985), but captive birds have been observed to drink (Sheldon, 1967). The chicks leave the nest soon after hatching, but are dependent on the female for food for the first week after hatching (Gregg, 1984).

Molt. Woodcock molt twice annually. The prenuptial molt involves body plumage, some wing coverts, scapulars, and tertials and occurs in late winter or early spring; the complete postnuptial molt takes place in July or August (Bent, 1927).

Migration. Fall migration begins in late September and continues through December, often following the first heavy frost (Sheldon, 1967). The migration may take 4 to 6 weeks (Sheldon, 1967). Some woodcock winter in the south Atlantic region, while those that breed west of the Appalachian Mountains winter in Louisiana and other Gulf States (Martin et al., 1969, as cited in Owen et al., 1977). Woodcock are early spring migrants, leaving their wintering grounds in February and arriving on their northern breeding grounds in lâte March to early April (Gregg, 1984; Sheldon, 1967; Owen et al., 1977). Dates of woodcock arrival at their breeding grounds vary from year to year depending on the timing of snowmelt (Gregg, 1984). Sheldon (1967) summarizes spring and fall migration dates by States from numerous studies.

Breeding activities and social organization. From their arrival in the spring, male woodcock perform daily courtship flights at dawn and at dusk, defending a site on the singing grounds in order to attract females for mating (Owen et al., 1977; Gregg, 1984). Often several males display on a single singing ground, with each defending his own section of the area. Females construct their nests on the ground, usually at the base of a tree or shrub located in a brushy area adjacent to an opening or male singing ground (Gregg and Hale, 1977; McAuley et al., 1990; Owen et al., 1977). Females are responsible for all of the incubation and care of their brood (Trippensee, 1948). The young leave the nest soon after hatching and can sustain flight by approximately 18 days of age (Gregg, 1984).

Home range and resources. The home range of woodcocks encompasses both diurnal cover areas and nocturnal roosting areas and varies in size depending on season and the distribution of feeding sites and suitable cover. During the day, movements are usually limited until dusk, when woodcock fly to nocturnal roost sites. Hudgins et al. (1985) and Gregg (1984) found spring and summer diurnal ranges to be only 1 to 10 percent of the total home range. Movement on the nocturnal roost sites also is limited; however, during winter, woodcock are more likely to feed and move around at night (Bortner, pers. comm.). Singing males generally restrict their movements more than non-singing males, juveniles, and females (Owen et al., 1977).

Population density. The annual singing-ground survey conducted by the United States and Canada provides information on the population trends of woodcock in the northern states and Canada during the breeding season (note from B. Bortner, U.S. Fish and Wildlife Service, Office of Migrating Bird Management, to Susan Norton, January 9, 1992). Gregg (1984) summarized results of several published singing-ground surveys and found estimates to vary from 1.7 male singing grounds per 100 ha in Minnesota (Godfrey, 1974, cited in Gregg, 1984) to 10.4 male singing grounds per 100 ha in Maine (Mendall and Aldous, 1943, cited in Gregg, 1984). Although this method is appropriate for assessing population trends, flushing surveys, telemetry, and mark-recapture are better methods for estimating woodcock densities because there are variable numbers of females and nonsinging males associated with active singing grounds (Dilworth, Krohn, Riffenberger, and Whitcomb pers. comm., cited by Owen et al., 1977). For example, Dwyer et al. (1988) found 2.2 singing males per 100 ha in a wildlife refuge in Maine, but with mark-recapture techniques, they found yearly summer densities of 19 to 25 birds per 100 ha in the same area.

Population dynamics. Woodcocks attempt to raise only a single brood in a given year but may renest if the initial clutch is destroyed (McAuley et al., 1990; Sheldon, 1967). In 12 years of study in Wisconsin, Gregg (1984) found 42 percent of all nests to be lost to predators and another 11 percent lost to other causes. Survival of juveniles in their first year ranges from 20 to 40 percent, and survival of adults ranges from 35 to 40 percent for males to approximately 40 to 50 percent for females (Dwyer and Nichols, 1982; Krohn et al., 1974). Derleth and Sepik (1990) found high adult survival rates (0.88 to 0.90 for both sexes) between June and October in Maine, indicating that adult mortality may occur primarily in the winter and early spring. They found lower summer survival rates for young woodcock between fledging and migration than for adults during the same months, with most losses of young attributed to predation.

Similar species *(from general references)*

- The **common snipe** (*Gallinago gallinago*) is similar in length (27 cm) to the woodcock, although lighter in weight. Snipe are primarily found in association with bogs and freshwater wetlands and feed on the various invertebrates associated with wetland soils. Snipe breed primarily in boreal forest regions and thus are found slightly north of the woodcock breeding range, with some areas of overlap in the eastern half of the continent. The breeding range of the snipe, however, extends westward to the Pacific coast and throughout most of Alaska, thus occupying a more extensive east-west range than the woodcock.

General references

Cade (1985); Dwyer et al. (1979); Dwyer and Storm (1982); Gregg (1984); National Geographic Society (1987); Owen et al. (1977); Sheldon (1967); Trippensee (1948).

American Woodcock (*Scolopax minor*)

Factors	Age/Sex/ Cond./Seas.	Mean	Range or (95% CI of mean)	Location	Reference	Note No.
Body Weight (g)	A M	176		throughout range	Nelson & Martin, 1953	
	A F	218				
	A M April	134.6 ± 2.9 SE		Maine	Dwyer et al., 1988	
	A M May	133.8 ± 5.8 SE				
	A M June	151.2 ± 9.5 SE				
	A M summer	145.9	127 - 165	central Massachusetts	Sheldon, 1967	
	J M summer	140.4	117 - 152			
	A F summer	182.9	162 - 216			
	J F summer	168.8	151 - 192			
	A M fall	169		Minnesota	Marshall (unpubl.)	1
	J M fall	164				
	A F fall	213				
	J F fall	212				
	at hatching	13.0	9 - 16	Wisconsin	Gregg, 1984	
Egg Weight (g)	at laying	18 - 19		Wisconsin	Gregg, 1984	
	near hatching	14 - 16				
Chick Growth Rate (g/day)	M	5.1		Maine	Dwyer et al., 1982	
	F	6.2				
Metabolic Rate (kcal/kg-day)	A F basal	115		s Michigan	Rabe et al., 1983b	2
	A M basal	126			estimated	3
	A F basal	118				
	A F free-living	315		s Michigan	Rabe et al., 1983b	4
	A F nesting	553				
	A M free-living	313	(148 - 662)		estimated	5
	A F free-living	296	(140 - 627)			

American Woodcock (*Scolopax minor*)

Factors	Age/Sex/ Cond./Seas.	Mean	Range or (95% CI of mean)	Location	Reference	Note No.
Food Ingestion Rate (g/g-day)	A B winter (earthworm diet)	0.77	0.11 - 1.43	Louisiana (captive)	Stickel et al., 1965	
Water Ingestion Rate (g/g-day)	A M A F	0.10 0.10			estimated	6
Inhalation Rate (m^3/day)	A M A F	0.11 0.13			estimated	7
Surface Area (cm^2)	A M A F	314 362			estimated	8

Dietary Composition	Spring	Summer	Fall	Winter	Location/Habitat (measure)	Reference	Note No.
earthworms Diptera Coleoptera Lepidoptera other animals plants		67.8 6.9 6.2 3.3 5.3 10.5			North America/NS (% volume; stomach contents)	Sperry, 1940	
earthworms beetle larvae grit (inorganic) other organic		58 10 31 1			Maine/fields (% wet weight; mouth esophagus, stomach, & proventriculus contents)	Krohn, 1970	9
earthworms other invertebrates				99 + <1	N Carolina/soybean fields (% wet weight; digestive tract)	Stribling & Doerr, 1985	
earthworms Coleoptera Hymenoptera				87 11 2	Alabama/NS (% volume; esophagus contents)	Miller & Causey, 1985	10

American Woodcock (*Scolopax minor*)

Population Dynamics	Age/Sex Cond./Seas.	Mean	Range	Location/Habitat	Reference	Note No.
Home Range Size (ha)	A M inactive A M active A M singing	3.1 (median) 73.6 (median) 10.5 (median)	0.3 - 6.0 38.2 - 171.2 4.6 - 24.1	Pennsylvania/mixed forests with shrubs and fields	Hudgins et al., 1985	
	B B summer A F with brood	32.4 ± 27.6 SD 4.5	7 - 98	Wisconsin/woods, open areas, brush	Gregg, 1984	
Population Density (birds/ha)	 B B winter B B winter B B winter	 3.38 0.20 0.034		North Carolina/agricultural: untilled soy stubble untilled corn stubble rebedded corn fields	Connors & Doerr, 1982	
	nests in spring	0.21 (nests/ha)		Pennsylvania/mixed pine and hardwoods, open fields	Coon et al., 1982	
	A M summer A F summer J B summer B B summer	0.035 0.056 0.125 0.223	0.026 - 0.046 0.037 - 0.074 0.108 - 0.143 0.190 - 0.250	Maine/second growth forest, meadows, and ponds	Dwyer et al., 1988	
Clutch Size		4	3 - 5	throughout range and habitats	Bent, 1927	
	1st clutch 2nd clutch	3.8 ± 0.42 SD 3.0 ± 0.67 SD		Maine/mixed forests, agricultural fields	McAuley et al., 1990	
Clutches/ Year		1 but renest if 1st lost		throughout range and habitats	McAuley et al., 1990	
Percent Nests Hatching		about 50		Maine/mixed forests, fields	McAuley et al., 1990	
Days Incubation		19 - 21		NS/NS	Mendall & Aldous, 1943; Pettingill, 1936	11
Age at Fledging		18 - 19 days		Wisconsin/woods, open areas, brush	Gregg, 1984	

American Woodcock (*Scolopax minor*)

Population Dynamics	Age/Sex Cond./Seas.	Mean	Range	Location/Habitat	Reference	Note No.
Age at Sexual Maturity	M F	< 1 year 1 year		throughout range and habitats	Sheldon, 1967	
Annual Mortality Rates	A M east A M central J M east J M central A F east A F central J F east J F central	65 ± 5.2 SD 60 ± 15 SD 80 ± 4.8 SD 64 ± 12 SD 51 ± 7.3 SD 47 ± 9.6 SD 64 ± 7.7 SD 69 ± 9.4 SD		eastern and central United States/NS	Dwyer & Nichols, 1982	

Seasonal Activity	Begin	Peak	End	Location	Reference	Note No.
Mating/Laying	early February early April		mid-March	Texas Maine	Whiting & Boggus, 1982 Dwyer et al., 1982	
Hatching	early February late February late March mid-April	 early May mid-May	 early June	Louisiana Virginia Connecticut Massachusetts Maine	Pettingill, 1936 Pettingill, 1936 Pettingill, 1936 Sheldon, 1967 Dwyer et al., 1982	1 1 1
Molt		August to early September		NS/NS	Owen & Krohn, 1973	12
Migration spring	mid-February March	 April	early March	leaving North Carolina arriving in northern range	Connors & Doerr, 1982 Gregg, 1984	
fall	October late September		December mid-December	arriving North Carolina leaving Canada	Sheldon, 1967 Owen et al., 1977	

American Woodcock (*Scolopax minor*)

1 As cited in Sheldon (1967).
2 Metabolic rate estimated by authors from equation of Aschoff and Pohl (1970).
3 Estimated using equation 3-28 (Lasiewski and Dawson, 1967) and summer body weights from Nelson and Martin (1953).
4 Estimate of free-living metabolism based on energy budget model. Metabolism during nesting estimated for peak needs during egg-laying.
5 Estimated using equation 3-37 (Nagy, 1987) and summer body weights from Nelson and Martin (1953).
6 Estimated using equation 3-15 (Calder and Braun, 1983) and summer body weights from Nelson and Martin (1953).
7 Estimated using equation 3-19 (Lasiewski and Calder, 1971) and summer body weights from Nelson and Martin (1953).
8 Estimated using equation 3-21 (Meeh, 1879 and Rubner, 1883, as cited in Walsberg and King, 1978) and summer body weights from Nelson and Martin (1953).
9 Grit comprised only 14 percent of total digestive tract contents volume.
10 Should provide a more accurate estimate of proportion of soft-bodied earthworms consumed than would including other portions of the digestive tract.
11 Cited in Trippensee (1948).
12 Cited in Owen et al. (1977).

2.2.2. Red Fox (foxes and coyotes)

Order *Carnivora*, Family *Canidae*. Unlike the more social wolves, foxes and coyotes tend to hunt alone, although coyotes may hunt larger prey in pairs. Foxes and coyotes are primarily carnivorous, preying predominantly on small mammals, but they also may eat insects, fruits, berries, seeds, and nuts. Foxes are found throughout most of the United States and Canada, including the arctic, as are coyotes with the exception of the southeastern United States. Foxes and coyotes are active primarily at night.

Selected species

Red foxes (*Vulpes vulpes*) are present throughout the United States and Canada except in the southeast, extreme southwest, and parts of the central states. Red fox prey extensively on mice and voles but also feed on other small mammals, insects, hares, game birds, poultry, and occasionally seeds, berries, and fruits (Palmer and Fowler, 1975). Twelve subspecies are recognized in North America (Ables, 1974).

Body size. The dog-sized red fox has a body about 56 to 63 cm in length, with a 35 to 41 cm tail (Burt and Grossenheider, 1980). They weigh from 3 to 7 kg, with the males usually outweighing the females by about 1 kg (Voigt, 1987; see table).

Habitat. As the most widely distributed carnivore in the world, the red fox can live in habitats ranging from arctic areas to temperate deserts (Voigt, 1987). Red foxes utilize many types of habitat--cropland, rolling farmland, brush, pastures, hardwood stands, and coniferous forests (MacGregor, 1942; Eadie, 1943; Cook and Hamilton, 1944; Ables, 1974). They prefer areas with broken and diverse upland habitats such as occur in most agricultural areas (Ables, 1974; Samuel and Nelson, 1982; Voigt, 1987). They are rare or absent from continuous stands of pine forests in the southeast, moist conifer forests along the Pacific coast, and semiarid grasslands and deserts (Ables, 1974).

Food habits. The red fox feeds on both animal and plant material, mostly small mammals, birds, insects, and fruit (Korschgen, 1959; Samuel and Nelson, 1982). Meadow voles are a major food in most areas of North America; other common prey include mice and rabbits (Korschgen, 1959; Voigt, 1987). Game birds (e.g., ring-necked pheasant and rufted grouse) and waterfowl are seasonally important prey in some areas (Pils and Martin, 1978; Sargeant, 1972; Voigt and Broadfoot, 1983). Plant material is most common in red fox diets in summer and fall when fruits, berries, and nuts become available (Johnson, 1970; Major and Sherburne, 1987). Red foxes often cache food in a hole for future use (Samuel and Nelson, 1982). They also are noted scavengers on carcasses or other refuse (Voigt, 1987). Most activity is nocturnal and at twilight (Nowak and Paradiso, 1983).

Temperature regulation and molt. In winter, foxes do not undergo hibernation or torpor; instead, they are active year-round. They undergo one molt per year, which usually begins in April and is finished by June. The winter coat is regrown by October or November in northern latitudes (Voigt, 1987).

Breeding activities and social organization. Breeding occurs earlier in the south than in the red fox's northern ranges (Samuel and Nelson, 1982) (see table). A mated pair maintains a territory throughout the year, with the male contributing more to its defense than the female (Preston, 1975). Pups are born and reared in an underground den, and the male assists the female in rearing young, bringing food to the den for the pups (Samuel and Nelson, 1982). Pups first emerge from the den when 4 to 5 weeks old (Samuel and Nelson, 1982). Once considered solitary, red foxes now are reported to exhibit more complex social habits (MacDonald and Voigt, 1985). A fox family, the basic social unit, generally consists of a mated pair or one male and several related females (MacDonald, 1980; Voigt, 1987). The additional females are usually nonbreeders that often help the breeding female (Voigt, 1987).

Home range and resources. The home ranges of individuals from the same family overlap considerably, constituting a family territory (Sargeant, 1972; Voigt and MacDonald, 1984). Territories of neighboring red fox families are largely nonoverlapping and contiguous, usually resulting in all parts of a landscape being occupied by foxes. Territory sizes range from less than 50 to over 3,000 ha (see table). Territories in urban areas tend to be smaller than those in rural areas (Ables, 1969). Adults visit most parts of their territory on a regular basis; however, they tend to concentrate their activities near to their dens, preferred hunting areas, abundant food supplies, and resting areas (Ables, 1974; Keenan, 1981). Territory boundaries often conform to physical landscape features such as well-traveled roads and streams (Ables, 1974). Territory defense is primarily by nonaggressive mechanisms involving urine scent-marking and avoidance behaviors. Scent marking occurs throughout the territory; there is little patrolling of territory boundaries. Each fox or family usually has a main underground den and one or more other burrows within the home range (Nowak and Paradiso, 1983). Most dens are abandoned burrows of other species (e.g., woodchucks, badgers) (Samuel and Nelson, 1982). Tunnels are up to 10 m in length and lead to a chamber 1 to 3 m below the surface (Nowak and Paradiso, 1983). Pup-rearing dens are the focal point of fox activity during spring and early summer. Foxes have some rest sites and usually forage away from the den (Voigt, 1987).

Population density. One red fox family per 100 to 1,000 ha is typical (Voigt, 1987; see table). Red foxes have larger home ranges where population densities are low and in poorer habitats (Voigt, 1987). Most young foxes, especially males, disperse before the age of 1 (Voigt, 1987), usually during September to March, with peaks in dispersal in October and November (Phillips et al., 1972; Storm et al., 1976).

Population dynamics. Foxes usually produce pups their first year, except in extremely high density areas and in some years in northern portions of their range where they may delay breeding until the next season (Allen, 1984; Harris, 1979; Storm et al., 1976; Voigt and MacDonald, 1984). Litter size generally averages four to six pups (see table). The pups leave the den about 1 month after birth, and they are weaned by about 8 to 10 weeks of age (Ables, 1974). Red foxes incur high mortality rates as a result of shooting, trapping, disease, and accidents (e.g., roadkills) (Storm et al., 1976). Two factors that tend to limit red fox abundance are competition with other canids, especially coyotes, and seasonal limits on food availability (Voigt, 1987). Fecundity is higher in areas of high mortality and low population densities (Voigt, 1987).

Similar species (from general references)

- The **arctic fox** (*Alopex lagopus*) is smaller than the red fox (body length approximately 51 cm; weight 3.2 to 6.7 kg) and is restricted in its distribution to the arctic, found in the United States only in Alaska. This species primarily scavenges for food but also eats lemmings, hares, birds, and eggs as well as berries in season.

- The **swift fox** (*Vulpes velox*) is smaller than the red fox (body length 38 to 51 cm; weight 1.8 to 2.7 kg) and inhabits the deserts and plains of the southwest and central United States. It dens in ground burrows and feeds on small mammals and insects.

- The **kit fox** (*Vulpes macrotis*) is similar in size to the swift fox and is considered by some to be the same species, although it has noticeably larger ears. It inhabits the southwestern United States and prefers open, level, sandy areas and low desert vegetation. It feeds on small mammals and insects.

- The **gray fox** (*Urocyon cinereoargenteus*) is similar in size (body length 53 to 74 cm; weight 3.2 to 5.8 kg) to the red fox and ranges over most of the United States except the northwest and northern prairies, inhabiting chaparral, open forests, and rimrock regions. Secretive and nocturnal, gray foxes will climb trees to evade enemies. They feed primarily on small mammals but also eat insects, fruits, acorns, birds, and eggs.

- The **coyote** (*Canis latrans*) is much larger (body length 81 to 94 cm; weight 9 to 22 kg) than the red fox and is found throughout most of the United States (except possibly eastern), western Canada, and Alaska. It inhabits prairies, open woodlands, brushy and boulder-strewn areas, and dens in the ground. Coyotes share some feeding habits with the red fox but also scavenge and hunt larger prey in pairs.

General references

Ables (1974); Burt and Grossenheider (1980); Palmer and Fowler (1975); Voigt (1987).

Red Fox (*Vulpes vulpes*)

Factors	Age/Sex/ Cond./Seas.	Mean	Range or (95% CI of mean)	Location	Reference	Note No.
Body Weight (kg)	A M spring	5.25 ± 0.18 SE	4.54 - 7.04	Illinois	Storm et al., 1976	
	A F spring	4.13 ± 0.11 SE	3.27 - 4.72			
	A M fall	4.82 ± 0.081 SE	4.13 - 5.68	Iowa	Storm et al., 1976	
	A F fall	3.94 ± 0.079 SE	2.95 - 4.59			
	neonate B	0.102 ± 0.12 SD	0.071 - 0.109	Wisconsin	Storm & Ables, 1966	
	at weaning B	0.70		North Dakota	Sargeant, 1978	
Pup Growth Rate (g/day)	birth to weaning	15.9		North Dakota/lab	Sargeant, 1978	
Metabolic Rate (kcal/kg-day)	J summer	193 ± 56 SD		Ohio/lab	Vogtsberger & Barrett, 1973	
	A M basal	47.9			estimated	1
	A F basal	51.1				
	A M free-living	161	(68 - 383)		estimated	2
	A F free-living	168	(71 - 400)			
Food Ingestion Rate (g/g-day)	J 5-8 wks	0.16		North Dakota/lab	Sargeant, 1978	
	J 9-12 wks	0.12				
	J 13-24 wks	0.11				
	A before whelp	0.075		North Dakota/captive	Sargeant, 1978	3
	F after whelp	0.14				
	A nonbreeding	0.069		North Dakota/captive	Sargeant, 1978	
Water Ingestion Rate (g/g-day)	A M	0.084			estimated	4
	A F	0.086				
Inhalation Rate (m^3/day)	A M	2.0			estimated	5
	A F	1.7				

Red Fox (*Vulpes vulpes*)

Factors	Age/Sex/ Cond./Seas.	Mean	Range or (95% CI of mean)	Location	Reference	Note No.
Surface Area (cm^2)	A M A F	3,220 2,760			estimated	6

Dietary Composition	Spring	Summer	Fall	Winter	Location/Habitat (measure)	Reference	Note No.
rabbits				44.4	Nebraska/statewide	Powell & Case, 1982	
small mammals				33.0			
pheasant				8.4	(% wet volume; stomach contents)		
other birds				11.2			
misc.				2.0			
not accounted for				1.0			
mammals	92.2	37.1	61.7	65.0	Illinois/farm and woods	Knable, 1974	
birds	2.4	43.2	0.2	8.6			
arthropods	0.2	11.6	4.2	<0.1			
plants	4.6	6.3	31.1	26.1	(% wet weight; stomach contents)		
unspecified/other	0.6	1.8	2.8	0.3			
rabbits	24.8	10.7	36.5	38.7	Missouri	Korschgen, 1959	
mice/rats	24.2	6.2	21.3	22.5			
other mammals	4.0	1.4	8.1	8.2	(% wet volume; stomach contents)		
poultry	21.0	45.0	16.3	11.6			
carrion	12.9	13.0	6.5	7.4			
livestock	9.8	0.3	2.0	5.4			
birds	0.6	1.2	1.1	3.8			
invertebrates	trace	15.3	1.6	trace			
plant foods	2.7	6.9	6.6	2.1			
mammals				81.4	Maryland/Appalachian Province (fall & winter)	Hockman & Chapman, 1983	
birds				4.8			
arthropods				2.8			
plants				7.0	(% wet weight; stomach contents)		
unspecified/other				4.0			

Red Fox (*Vulpes vulpes*)

Population Dynamics	Age/Sex/ Cond./Seas.	Mean	Range	Location/Habitat	Reference	Note No.
Territory size (ha)	A B summer	1,611	277 - 3,420	nw British Columbia/ alpine and subalpine	Jones & Theberge, 1982	
	A M summer	1,967	514 - 3,420			
	A F summer	1,137	277 - 1,870			
	A F spring	699 ± 137 SD	596 - 855	ec Minnesota/woods, fields, swamp	Sargeant, 1972	
	A M all year	717		Wisconsin/diverse	Ables, 1969	
	A F all year	96	57 - 170			
Population Density (N/ha)	B B spring	0.001		Canada/northern boreal forests/arctic tundra	Voigt, 1987	
	B B spring	0.01		s Ontario, Canada/southern habitats	Voigt, 1987	
	B B		0.046 - 0.077	"good fox range" in North America	Ables, 1974	
Litter Size		5.5		s Wisconsin/farm, marsh, pasture	Pils & Martin, 1978	7
		6.8	2 - 9	Illinois/farm and woods	Storm et al., 1976	8
		6.7	3 - 12	Iowa/farm and woods	Storm et al., 1976	7
		4.2		upper Michigan/NS	Switzenberg, 1950	8
		4.1		North Dakota/prairie potholes	Allen, 1984	7
Litters/Year		1		NS/NS	Samuel & Nelson, 1982	
Days Gestation		51 - 54		New York/NS	Sheldon, 1949	9
Age at Weaning		8 - 10 weeks		NS/NS	Ables, 1974	
Age at Sexual Maturity	F	10 months		Illinois, Iowa/farm woods	Storm et al., 1976	

Red Fox (*Vulpes vulpes*)

Population Dynamics	Age/Sex/ Cond./Seas.	Mean	Range	Location/Habitat	Reference	Note No.
Annual Mortality Rates (percent)	B B	79.4		s Wisconsin/various	Pils & Martin, 1978	
	J M J F A F A B	83 81 74 77		Illinois/Iowa/ farms and woods	Storm et al., 1976	
Longevity		< 1.5 yrs	up to 6 yrs	NS/NS	Storm et al., 1976	

Seasonal Activity	Begin	Peak	End	Location	Reference	Note No.
Mating	early Dec. late December late January February	late January Jan. - Feb.	late February March early February March	Iowa New York southern Ontario, Canada northern Ontario, Canada	Storm et al., 1976 Layne & McKeon, 1956; Sheldon, 1949 Voigt, 1987 Voigt, 1987	 9
Parturition		March late March, April		southern CAN e North Dakota	Voigt, 1987 Sargeant, 1972	
Molt	April		June	NS/NS	Voigt, 1987	
Disperal	late September		March	Illinois, Iowa	Storm et al., 1976	

1 Estimated using extrapolation equation 3-45 (Boddington, 1978) and body weights from Storm et al. (1976) (Illinois).
2 Estimated using extrapolation equation 3-47 (Nagy, 1987) and body weights from Storm et al. (1976) (Illinois).
3 Food consumption of an adult pair for 11 days prior to whelping (i.e., parturition) and of the adult female for the first 4 weeks after whelping.
4 Estimated using extrapolation equation 3-17 (Calder and Braun, 1983) and body weights from Storm et al. (1976) (Illinois).
5 Estimated using extrapolation equation 3-20 (Stahl, 1967) and body weights from Storm et al. (1976) (Illinois).
6 Estimated using extrapolation equation 3-22 (Stahl, 1967) and body weights from Storm et al. (1976) (Illinois).
7 Litter size determined from embryo count. Using placental scars generally overestimates litter size, and counting live pups often underestimates litter size (Allen, 1983; Lindstrom, 1981).
8 Method of determining litter size not specified.
9 Cited in Samuel and Nelson (1982).

APPENDIX C

NATURAL LANDS MANAGEMENT

CAUTIONS AND RESTRICTIONS ON NATURAL HERITAGE DATA

The quantity and quality of data collected by the Natural Heritage Program is dependent on the research and observations of many individuals and organizations. Not all of this information is the result of comprehensive or site-specific field surveys. Some natural areas in New Jersey have never been thoroughly surveyed. As a result, new locations for plant and animal species are continuously added to the data base. Since data acquisition is a dynamic, ongoing process, the Natural Heritage Program cannot provide a definitive statement on the presence, absence, or condition of biological elements in any part of New Jersey. Information supplied by the Natural Heritage Program summarizes existing data known to the program at the time of the request regarding the biological elements or locations in question. They should never be regarded as final statements on the elements or areas being considered, nor should they be substituted for on-site surveys required for environmental assessments. The attached data is provided as one source of information to assist others in the preservation of natural diversity.

This office cannot provide a letter of interpretation or a statement addressing the classification of wetlands as defined by the Freshwater Wetlands Act. Requests for such determination should be sent to the DEP Land Use Regulation Program, CN 401, Trenton, NJ 08625-0401.

This cautions and restrictions notice must be included whenever information provided by the Natural Heritage Database is published.

N.J. Department of Environmental Protection Division of Parks & Forestry

EXPLANATIONS OF CODES USED IN NATURAL HERITAGE REPORTS

FEDERAL STATUS CODES

The following U.S. Fish and Wildlife Service categories and their definitions of endangered and threatened plants and animals have been modified from the U.S. Fish and Wildlife Service (F.R. Vol. 50 No. 188; Vol. 55, No. 35; F.R. 50 CFR 17.11 and 17.12). Federal Status codes reported for species follow the most recent listing.

LE Taxa formally listed as endangered.

LT Taxa formally listed as threatened.

PE Taxa already proposed to be formally listed as endangered.

PT Taxa already proposed to be formally listed as threatened.

C1 Taxa for which the Service currently has on file substantial information on biological vulnerability and threat(s) to support the appropriateness of proposing to list them as endangered or threatened species.

C1* Taxa which may be possibly extinct (although persuasive documentation of extinction has not been made--compare to 3A status).

C2 Taxa for which information now in possession of the Service indicates that proposing to list them as endangered or threatened species is possibly appropriate, but for which substantial data on biological vulnerability and threat(s) are not currently known or on file to support the immediate preparation of rules.

C3 Taxa that are no longer being considered for listing as threatened or endangered species. Such taxa are further coded to indicate three subcategories, depending on the reason(s) for removal from consideration.

3A Taxa for which the Service has persuasive evidence of extinction.

3B Names that, on the basis of current taxonomic understanding, do not represent taxa meeting the Act's definition of "species".

3C Taxa that have proven to be more abundant or widespread than was previously believed and/or those that are not subject to any identifiable threat.

S/A Similarity of appearance species.

STATE STATUS CODES

Two animal lists provide state status codes after the Endangered and Nongame Species Conservation Act of 1973 (NSSA 23:2A-13 et. seq.): the list of endangered species (N.J.A.C. 7:25-4.13) and the list defining status of indigenous, nongame wildlife species of New Jersey (N.J.A.C. 7:25-4.17(a)). The status of animal species is determined by the Nongame and Endangered Species Program (ENSP). The state status codes and definitions provided reflect the most recent lists that were revised in the New Jersey Register, Monday, June 3, 1991.

D Declining species-a species which has exhibited a continued decline in population numbers over the years.

E Endangered species-an endangered species is one whose prospects for survival within the state are in immediate danger due to one or many factors - a loss of habitat, over exploitation, predation, competition, disease. An endangered species requires immediate assistance or extinction will probably follow.

EX Extirpated species-a species that formerly occurred in New Jersey, but is not now known to exist within the state.

I Introduced species-a species not native to New Jersey that could not have established itself here without the assistance of man.

INC Increasing species-a species whose population has exhibited a significant increase, beyond the normal range of its life cycle, over a long term period.

T Threatened species-a species that may become endangered if conditions surrounding the species begin to or continue to deteriorate.

P Peripheral species-a species whose occurrence in New Jersey is at the extreme edge of its present natural range.

S Stable species-a species whose population is not undergoing any long-term increase/decrease within its natural cycle.

U Undetermined species-a species about which there is not enough information available to determine the status.

Status for animals separated by a slash(/) indicate a duel status. First status refers to the state breeding population, and the second status refers to the migratory or winter population.

Plant taxa listed as endangered are from New Jersey's official Endangered Plant Species List N.J.S.A. 131B-15.151 et seq.

E Native New Jersey plant species whose survival in the State or nation is in jeopardy.

REGIONAL STATUS CODES FOR PLANTS

LP Indicates taxa listed by the Pinelands Commission as endangered or threatened within their legal jurisdiction. Not all species currently tracked by the Pinelands Commission are tracked by the Natural Heritage Program. A complete list of endangered and threatened Pineland species is included in the New Jersey Pinelands Comprehensive Management Plan.

EXPLANATION OF GLOBAL AND STATE ELEMENT RANKS

The Nature Conservancy has developed a ranking system for use in identifying elements (rare species and natural communities) of natural diversity most endangered with extinction. Each element is ranked according to its global, national, and state (or subnational in other countries) rarity. These ranks are used to prioritize conservation work so that the most endangered elements receive attention first. Definitions for element ranks are after The Nature Conservancy (1982: Chapter 4, 4.1-1 through 4.4.1.3-3).

GLOBAL ELEMENT RANKS

G1 Critically imperiled globally because of extreme rarity (5 or fewer occurrences or very few remaining individuals or acres) or because of some factor(s) making it especially vulnerable to extinction.

G2 Imperiled globally because of rarity (6 to 20 occurrences or few remaining individuals or acres) or because of some factor(s) making it very vulnerable to extinction throughout its range.

G3 Either very rare and local throughout its range or found locally (even abundantly at some of its locations) in a restricted range (e.g., a single western state, a physiographic region in the East) or because of other factors making it vulnerable to extinction throughout it's range; with the number of occurrences in the range of 21 to 100.

G4 Apparently secure globally; although it may be quite rare in parts of its range, especially at the periphery.

G5 Demonstrably secure globally; although it may be quite rare in parts of its range, especially at the periphery.

GH Of historical occurrence throughout its range i.e., formerly part of the established biota, with the expectation that it may be rediscovered.

GU Possibly in peril range-wide but status uncertain; more information needed.

GX Believed to be extinct throughout range (e.g., passenger pigeon) with virtually no likelihood that it will be rediscovered.

G? Species has not yet been ranked.

STATE ELEMENT RANKS

S1 Critically imperiled in New Jersey because of extreme rarity (5 or fewer occurrences or very few remaining individuals or acres). Elements so ranked are often restricted to very specialized conditions or habitats and/or restricted to an extremely small geographical area of the state. Also included are elements which were formerly more abundant, but because of habitat destruction or some other critical factor of its biology, they have been demonstrably reduced in abundance. In essence, these are elements for which, even with intensive searching, sizable additional occurrences are unlikely to be discovered.

S2 Imperiled in New Jersey because of rarity (6 to 20 occurrences). Historically many of these elements may have been more frequent but are now known from very few extant occurrences, primarily because of habitat destruction. Diligent searching may yield additional occurrences.

S3 Rare in state with 21 to 100 occurrences (plant species in this category have only 21 to 50 occurrences). Includes elements which are widely distributed in the state but with small populations/acreage or elements with restricted distribution, but locally abundant. Not yet imperiled in state but may soon be if current trends continue. Searching often yields additional occurrences.

S4 Apparently secure in state, with many occurrences.

S5 Demonstrably secure in state and essentially ineradicable under present conditions.

SA Accidental in state, including species (usually birds or butterflies) recorded once or twice or only at very great intervals, hundreds or even thousands of miles outside their usual range; a few of these species may even have bred on the one or two occasions they were recorded; examples include european strays or western birds on the East Coast and visa-versa.

SE Elements that are clearly exotic in New Jersey including those taxa not native to North America (introduced taxa) or taxa deliberately or accidentally introduced into the State from other parts of North America (adventive taxa). Taxa ranked SE are not a conservation priority (viable introduced occurrences of G1 or G2 elements may be exceptions).

SH Elements of historical occurrence in New Jersey. Despite some searching of historical occurrences and/or potential habitat, no extant occurrences are known. Since not all of the historical occurrences have been field surveyed, and unsearched potential habitat remains, historically ranked taxa are considered possibly extant, and remain a conservation priority for continued field work.

SN Regularly occurring, usually migratory and typically nonbreeding species for which no significant or effective habitat conservation measures can be taken in the state; this category includes migratory birds, bats, sea turtles, and cetaceans which do not breed in the state but pass through twice a year or may remain in the winter (or, in a few cases, the summer); included also are certain lepidoptera which regularly migrate to a state where they reproduce, but then completely die out every year with no return migration. Species in this category are so widely and unreliably distributed during migration or in winter that no small set of sites could be set aside with the hope of significantly furthering their conservation. Other nonbreeding, high globally-ranked species (such as the bald eagle, whooping crane or some seal species) which regularly spend some portion of the year at definite localities (and therefore have a valid conservation need in the state) are not ranked SN but rather S1, S2, etc.

SR Elements reported from New Jersey, but without persuasive documentation which would provide a basis for either accepting or rejecting the report. In some instances documentation may exist, but as of yet, its source or location has not been determined.

SRF Elements erroneously reported from New Jersey, but this error persists in the literature.

SU Elements believed to be in peril but the degree of rarity uncertain. Also included are rare taxa of uncertain taxonomical standing. More information is needed to resolve rank.

SX Elements that have been determined or are presumed to be extirpated from New Jersey. All historical occurrences have been searched and a reasonable search of potential habitat has been completed. Extirpated taxa are not a current conservation priority.

SXC Elements presumed extirpated from New Jersey, but native populations collected from the wild exist in cultivation.

T Element ranks containing a "T" indicate that the infraspecific taxon is being ranked differently than the full species. For example *Stachys palustris* var. *homotricha* is ranked "G5T? SH" meaning the full species is globally secure but the global rarity of the var. *homotricha* has not been determined; in New Jersey the variety is ranked historic.

Q Elements containing a "Q" in the global portion of its rank indicates that the taxon is of questionable, or uncertain taxonomical standing, e.g., some authors regard it as a full species, while others treat it at the subspecific level.

.1 Elements documented from a single location.

Note: To express uncertainty, the most likely rank is assigned and a question mark added (e.g., G2?). A range is indicated by combining two ranks (e.g., G1G2, S1S3).

IDENTIFICATION CODES

These codes refer to whether the identification of the species or community has been checked by a reliable individual and is indicative of significant habitat.

Y Identification has been verified and is indicative of significant habitat.

BLANK Identification has not been verified but there is no reason to believe it is not indicative of significant habitat.

? Either it has not been determined if the record is indicative of significant habitat or the identification of the species or community may be confusing or disputed.

Revised September 1991

01 AUG 1996

WATERSHED MANAGEMENT AREA 8
RARE SPECIES AND NATURAL COMMUNITIES PRESENTLY RECORDED IN
THE NEW JERSEY NATURAL HERITAGE DATABASE

NAME	COMMON NAME	FEDERAL STATUS	STATE STATUS	REGIONAL STATUS	GRANK	SRANK	DATE OBSERVED	IDENT.
*** ALEXANDRIA TWP								
PASSERCULUS SANDWICHENSIS	SAVANNAH SPARROW		T/T		G5	S2	1984-05-??	Y
PASSERCULUS SANDWICHENSIS	SAVANNAH SPARROW		T/T		G5	S2	1971-??-??	
*** BEDMINSTER TWP								
ACCIPITER COOPERII	COOPER'S HAWK		E		G4	S2	1993-05-03	
ACCIPITER COOPERII	COOPER'S HAWK		E		G4	S2	1990-07-06	Y
ALASMIDONTA UNDULATA	TRIANGLE FLOATER				G4	S3	1994-05-27	Y
AMMODRAMUS SAVANNARUM	GRASSHOPPER SPARROW		T/T		G5	S2	1988-07-??	Y
AMMODRAMUS SAVANNARUM	GRASSHOPPER SPARROW		T/T		G5	S2	1991-07-04	Y
AMMODRAMUS SAVANNARUM	GRASSHOPPER SPARROW		T/T		G5	S2	1987-06-30	Y
AMMODRAMUS SAVANNARUM	GRASSHOPPER SPARROW		T/T		G5	S2	1992-07-03	Y
AMMODRAMUS SAVANNARUM	GRASSHOPPER SPARROW		T/T		G5	S2	1981-SUMMR	Y
AMMODRAMUS SAVANNARUM	GRASSHOPPER SPARROW		T/T		G5	S2	1988-07-??	Y
AMMODRAMUS SAVANNARUM	GRASSHOPPER SPARROW		T/T		G5	S2	1988-08-??	Y
AMMODRAMUS SAVANNARUM	GRASSHOPPER SPARROW		T/T		G5	S2	1981-SUMMR	Y
AMMODRAMUS SAVANNARUM	GRASSHOPPER SPARROW		T/T		G5	S2	1988-07-??	Y
AMMODRAMUS SAVANNARUM	GRASSHOPPER SPARROW		T/T		G5	S2	1992-06-21	Y
AMMODRAMUS SAVANNARUM	GRASSHOPPER SPARROW		T/T		G5	S2	1988-06-??	Y
ARDEA HERODIAS	GREAT BLUE HERON		T/S		G5	S2	1992-04-27	Y
BARTRAMIA LONGICAUDA	UPLAND SANDPIPER		E		G5	S1	1980-??-??	Y
CAREX FRANKII	FRANK'S SEDGE				G5	S3	1954-08-04	Y
CAREX FRANKII	FRANK'S SEDGE				G5	S3	1976-06-14	Y
CLEMMYS INSCULPTA	WOOD TURTLE		T		G4	S3	1987-03-26	Y
CLEMMYS INSCULPTA	WOOD TURTLE		T		G4	S3	1994-09-10	Y
DOLICHONYX ORYZIVORUS	BOBOLINK		T/T		G5	S2	1993-07-06	Y
DOLICHONYX ORYZIVORUS	BOBOLINK		T/T		G5	S2	1988-06-??	Y
DOLICHONYX ORYZIVORUS	BOBOLINK		T/T		G5	S2	1988-06-??	Y
DOLICHONYX ORYZIVORUS	BOBOLINK		T/T		G5	S2	1991-07-04	Y
DOLICHONYX ORYZIVORUS	BOBOLINK		T/T		G5	S2	1987-SUMMR	Y

01 AUG 1996

WATERSHED MANAGEMENT AREA 8
RARE SPECIES AND NATURAL COMMUNITIES PRESENTLY RECORDED IN
THE NEW JERSEY NATURAL HERITAGE DATABASE

NAME	COMMON NAME	FEDERAL STATUS	STATE STATUS	REGIONAL STATUS	GRANK	SRANK	DATE OBSERVED	IDENT.
DOLICHONYX ORYZIVORUS	BOBOLINK		T/T		G5	S2	1987-SUMMR	Y
DOLICHONYX ORYZIVORUS	BOBOLINK		T/T		G5	S2	1988-06-??	Y
DOLICHONYX ORYZIVORUS	BOBOLINK		T/T		G5	S2	1988-06-??	Y
DOLICHONYX ORYZIVORUS	BOBOLINK		T/T		G5	S2	1988-??-??	Y
DOLICHONYX ORYZIVORUS	BOBOLINK		T/T		G5	S2	1988-??-??	Y
DOLICHONYX ORYZIVORUS	BOBOLINK		T/T		G5	S2	1988-07-??	Y
DOLICHONYX ORYZIVORUS	BOBOLINK		T/T		G5	S2	1988-06-??	Y
DOLICHONYX ORYZIVORUS	BOBOLINK		T/T		G5	S2	1994-05-11	Y
DOLICHONYX ORYZIVORUS	BOBOLINK		T/T		G5	S2	1987-06-??	Y
DOLICHONYX ORYZIVORUS	BOBOLINK		T/T		G5	S2	1987-SUMMR	Y
DOLICHONYX ORYZIVORUS	BOBOLINK		T/T		G5	S2	1987-SUMMR	Y
DOLICHONYX ORYZIVORUS	BOBOLINK		T/T		G5	S2	1987-SUMMR	Y
DOLICHONYX ORYZIVORUS	BOBOLINK		T/T		G5	S2	1987-SUMMR	Y
DOLICHONYX ORYZIVORUS	BOBOLINK		T/T		G5	S2	1988-??-??	Y
DOLICHONYX ORYZIVORUS	BOBOLINK		T/T		G5	S2	1990-06-05	Y
FLOODPLAIN FOREST	FLOODPLAIN FOREST				G4	S3?	1988-04-13	Y
GOMPHUS ABBREVIATUS	SPINE-CROWNED CLUBTAIL				G3G4	S?	1984-06-12	Y
PASSERCULUS SANDWICHENSIS	SAVANNAH SPARROW		T/T		G5	S2	1991-05-09	
PASSERCULUS SANDWICHENSIS	SAVANNAH SPARROW		T/T		G5	S2	1980-??-??	Y
POOECETES GRAMINEUS	VESPER SPARROW		E		G5	S2	1970-??-??	Y
POOECETES GRAMINEUS	VESPER SPARROW		E		G5	S2	1987-07-??	Y
SELAGINELLA RUPESTRIS	LEDGE SPIKE-MOSS				G5	S2	198?-??-??	Y
STRIX VARIA	BARRED OWL		T/T		G5	S3	1993-04-05	Y
*** BERNARDS TWP								
BUTEO LINEATUS	RED-SHOULDERED HAWK		E/T		G5	S2	1995-06-19	Y
*** BETHLEHEM TWP								
BAT HIBERNACULUM	BAT HIBERNACULUM				G?	S?	1994-09-13	Y
*** BRANCHBURG TWP								

NATURAL LANDS MANAGEMENT

NATURAL HERITAGE PRIORITY SITE MAPS

The Priority Site Maps identify boundaries of some of the most important sites in the state for endangered and threatened plants, animals and ecosystems. These maps do not contain all of the important areas in the state for endangered biological diversity. They only depict the boundaries of priority sites which have been delineated by the Office of Natural Lands Management to date. These areas should be considered to be top priorities for the preservation of biological diversity. If these areas are allowed to be degraded or destroyed, we may lose some of the unique components of our natural heritage.

N.J. Department of Environmental Protection Division of Parks & Forestry

STANDARD SITE BOUNDARY LINE
(sites smaller than 3,200 acres)

MACROSITE BOUNDARY LINE
(sites larger than 3,200 acres)

SITE LOCATOR DOT

LOCATOR DOT FOR SITES SMALLER THAN DOT

01 AUG 1996

NATURAL HERITAGE PRIORITY SITES WITHIN
WATERSHED MANAGEMENT AREA 8

SITE NAME	BIODIV. SIGNIF.	BIODIVERSITY COMMENTS	USGS QUADRANGLE	COUNTY	MUNICIPALITY
ABRAITYS PINE STAND SITE	B4	Only occurrence of a state listed endangered plant.	STOCKTON	NJHUNT	DELAWARE TWP
BLACK RIVER MACROSITE	B5	Contains a large contiguous forested patch which provides habitat for a State Threatened forest interior nesting raptor. Also provides additional buffer for two standard sites with high significance.	CHESTER MENDHAM	NJMORR	CHESTER TWP RANDOLPH TWP ROXBURY TWP
BUDD LAKE BOG	B3	This site contains the northern most stand of a federally threatened plant species, a state imperiled natural community (black spruce swamp), and a number of state rare and endangered plant species.	STANHOPE TRANQUILITY HACKETTSTOWN	NJMORR	MOUNT OLIVE TWP
BUDD LAKE OUTLET	B5	One extant state listed endangered plant plus additional historical records.	HACKETTSTOWN	NJMORR	MOUNT OLIVE TWP
BURNT MILLS	B5	The site contains a good quality mature floodplain forest community. Perhaps the best remaining floodplain forest stand on the Raritan River.	RARITAN GLADSTONE	NJSOME	BEDMINSTER TWP BRANCHBURG TWP
CHESTER RAILROAD SITE	B3	One State listed plant species.	CHESTER	NJMORR	CHESTER TWP
EAST AMWELL GRASSLANDS MACROSITE	B4	The site contains good quality occurrences of at least five state endangered or threatened grassland bird species.	HOPEWELL STOCKTON	NJHUNT NJMERC NJSOME	EAST AMWELL TWP HOPEWELL TWP RARITAN TWP HILLSBOROUGH TWP DELAWARE TWP
HELL MOUNTAIN	B5	Historical location for federally listed plant and extant site for state listed endangered plant species.	CALIFON	NJHUNT	TEWKSBURY TWP
IRONIA	B3	Contains two globally imperiled plant species and a state significant collection of plant species.	CHESTER MENDHAM	NJMORR	RANDOLPH TWP ROXBURY TWP

01 AUG 1996

NATURAL HERITAGE PRIORITY SITES WITHIN
WATERSHED MANAGEMENT AREA 8

SITE NAME	BIODIV. SIGNIF.	BIODIVERSITY COMMENTS	USGS QUADRANGLE	COUNTY	MUNICIPALITY
ORCHARD DRIVE GRASSLANDS	B4	Contains 2 State endangered bird species and 3 State threatened bird species.	RARITAN	NJSOME	HILLSBOROUGH TWP
SAND BROOK	B5	Single known New Jersey occurrence for state listed plant species.	STOCKTON	NJHUNT	DELAWARE TWP
SOLBERG AIRPORT	B5	Site contains one State endangered and three State threatened bird species.	RARITAN FLEMINGTON	NJHUNT	READINGTON TWP

12 Records Processed

14 JUL 1992

NATURAL HERITAGE PRIORITY SITES
QUADRANGLE MAP KEY

QUADNAME*** CHESTER

SITE LOCATOR NUMBER	SITECODE	SITENAME
3	S.USNJHP1*58	BLACK RIVER MACROSITE
2	S.USNJHP1*167	CHESTER RAILROAD SITE
1	S.USNJHP*39	IRONIA

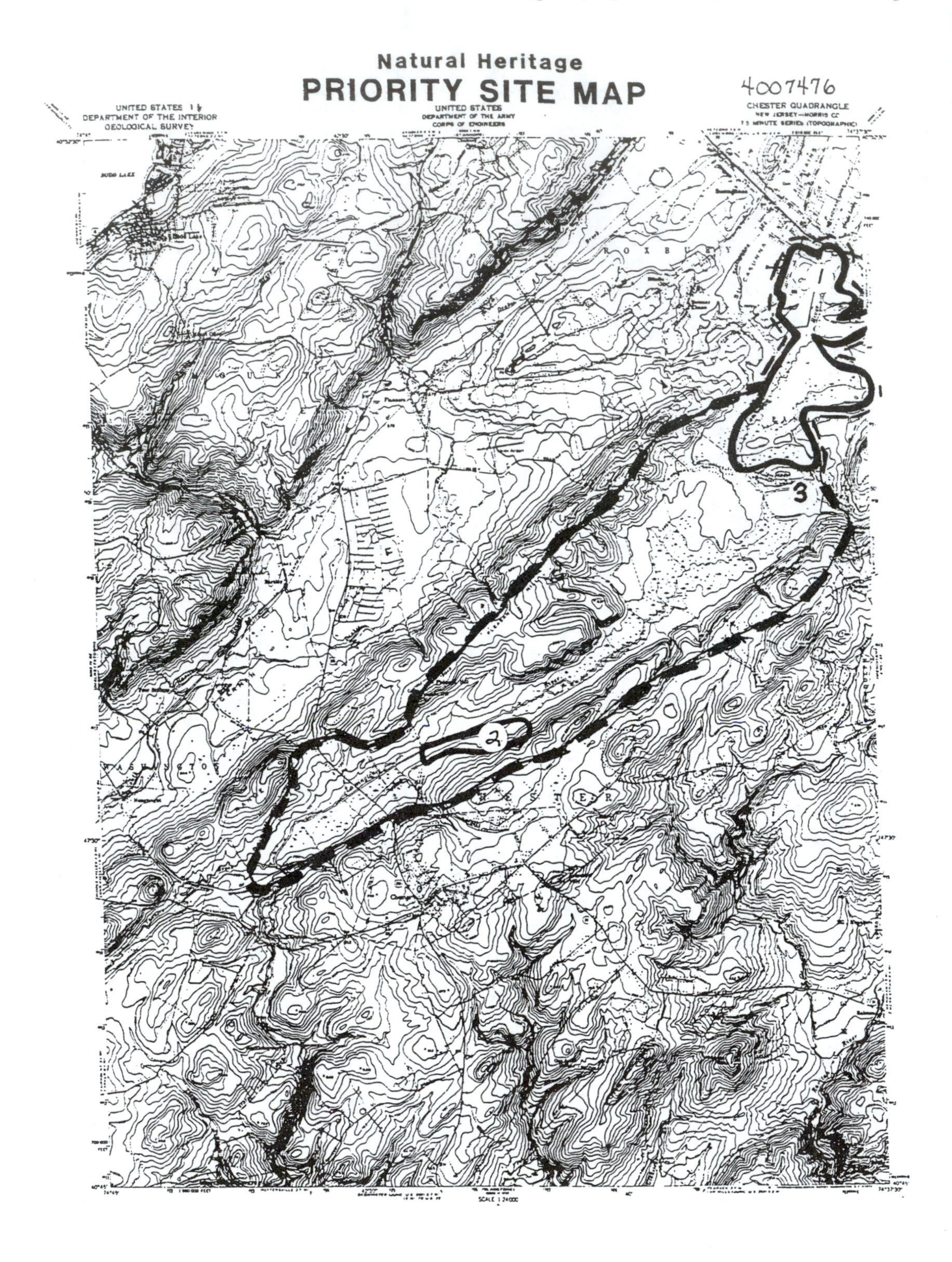
Natural Heritage
PRIORITY SITE MAP
4007476
UNITED STATES
DEPARTMENT OF THE INTERIOR
GEOLOGICAL SURVEY
UNITED STATES
DEPARTMENT OF THE ARMY
CORPS OF ENGINEERS
CHESTER QUADRANGLE
NEW JERSEY—MORRIS CO.
7.5 MINUTE SERIES (TOPOGRAPHIC)
ROXBURY
CHESTER
1
2
3
SCALE 1:24000

APPENDIX D

New Jersey Geological Survey
Geological Survey Report GSR-32

A METHOD FOR EVALUATING GROUND-WATER-RECHARGE AREAS IN NEW JERSEY

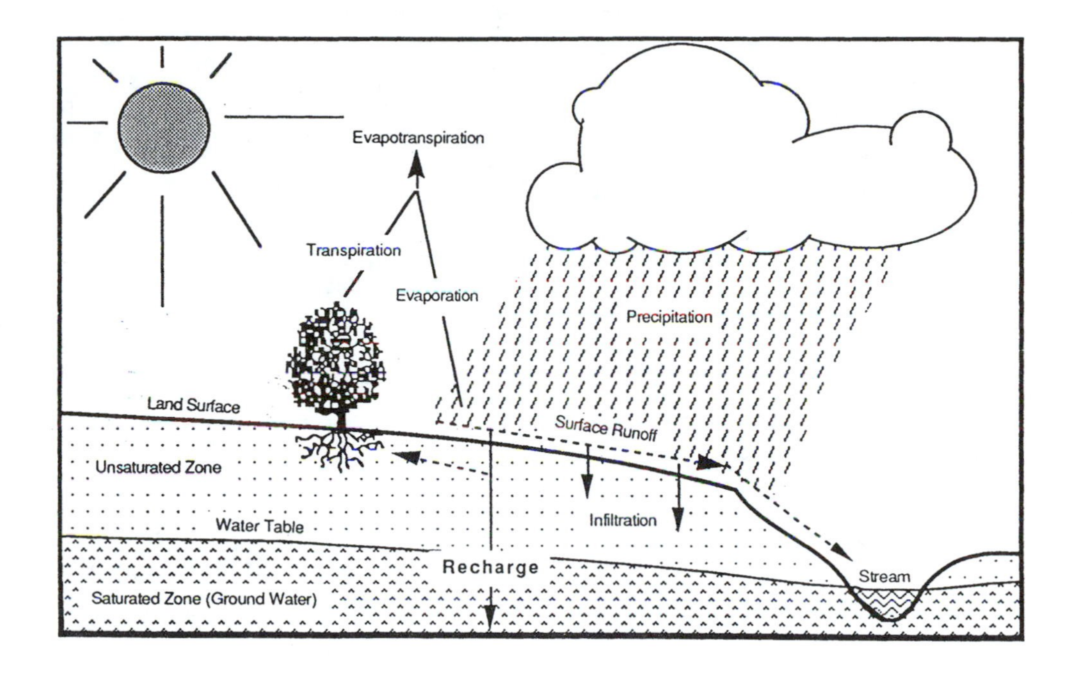

N.J. Department of Environmental Protection and Energy - Division of Science and Research

New Jersey Geological Survey
Geological Survey Report GSR-32

A Method for Evaluating Ground-Water-Recharge Areas in New Jersey

by
Emmanuel G. Charles, Cyrus Behroozi,
Jack Schooley, and Jeffrey L. Hoffman

New Jersey Department of Environmental Protection and Energy
Division of Science and Research
Geological Survey
CN 427
Trenton, NJ 08625
1993

CONVERSION FACTORS

For the convenience of the reader, units used in this report may be obtained or converted using the following factors:

Multiply	by	to obtain
inch (in.)	0.08333	foot (ft.)
inch (in.)	2.540	centimeter (cm)
acre	43,560	square feet ($ft.^2$)
square inch ($in.^2$) on 1:24,000 scale map	91.83	acres
square mile ($mi.^2$)	640	acres
cubic inch ($in.^3$)	0.0005787	cubic foot ($ft.^3$)
cubic foot ($ft.^3$)	7.481	gallons
area (acres) x recharge (inches/year)	27,156	gallons/year
map ($in.^2$) x recharge (inches/year)	2,493,667	gallons/year
degrees Fahrenheit (°F)	°C = (°F - 32) x 0.5555	degrees Celsius (°C)
degrees Celsius (°C)	°F = 1.8 x °C + 32	degrees Fahrenheit(°F)

New Jersey Geological Survey Reports (ISSN 0741-7357) are published by the New Jersey Geological Survey, CN 427, Trenton, NJ 08625. This report may be reproduced in whole or part provided that suitable reference to the source of the copied material is provided.

Additional copies of this and other reports may be obtained from:

Maps and Publications Sales Office
Bureau of Revenue
CN 417
Trenton, NJ 08625

A price list is available on request.

CONTENTS

Page

Abstract . . . 1

I. Background . . . 2
Introduction . . . 2
Acknowledgements . . . 3
Ground-water-recharge concepts . . . 3
Development of the method . . . 4

II. Mapping ground-water-recharge areas . . . 7
1. Acquire source data . . . 8
2. Prepare composites and mylar templates . . . 9
3a. Prepare land-use/land-cover (LULC) overlay . . . 10
3b. Prepare soil-group overlay . . . 12
4. Prepare coded LULC/soil-group combination map . . . 13
5. Prepare spreadsheet and calculate recharge . . . 14
6. Prepare recharge base map . . . 16
7. Mapping by soil unit for more accurate results . . . 16
8. Recharge estimates for specific percentages of impervious area . . . 16

III. Classifying and ranking ground-water-recharge areas . . . 29
Frequency-weighted classification . . . 29
Volumetric-recharge classification . . . 32
Comparing recharge volume on a parcel-specific basis . . . 43

IV. Limitations of the ground-water-recharge map . . . 44
Recharge-value accuracy . . . 44
Map accuracy . . . 45
Classification accuracy . . . 45
Basin-wide baseflow adjustment . . . 46

References . . . 47

Glossary . . . 48

Figures

Figure 1. Ground-water recharge in the hydrologic cycle . . . 4
2. Registration . . . 9
3. Producing a mylar template . . . 10
4. Producing a LULC overlay . . . 12
5. Producing a soil-group overlay . . . 13
6. Producing a combination map . . . 13
7. Spreadsheet format for calculating recharge . . . 14
8. Adding LULC and soil-group codes to the spreadsheet . . . 14
9. Adding R-factor and R-constant to the spreadsheet . . . 15
10. Completed spreadsheet with calculated recharge . . . 15
11. Recharge base map . . . 16
12. Steps in producing a ground-water-recharge map . . . 17
13. Step 2: USGS quadrangle composite . . . 18
14. Step 2: NWI quadrangle composite . . . 19

Figure 15. Step 2: Mylar template 20
16. Step 2: Mylar orthophotquad composite 21
17. Step 3a: Land-use/land-cover overlay 22
18. Step 3b: Soil map composite 23
19. Soil legend for example study area 24
20. Step 3b: Soil group overlay showing soil-group symbols, boundaries and shaded hydric soils 25
21. Step 4 : LULC-soil combination map 26
22. Step 5: Prepare spreadsheet and calculate recharge 27
23. Step 6: Recharge base map 28
24. Example ranking with shaded recharge map 29
25. Partial spreadsheet after steps 1, 2, and 3 of the frequency method 30
26. Partial spreadsheet after step 4 of frequency method 30
27. Full spreadsheet after step 7 of frequency method 31
28. Examples of frequency-method classifications using sample data 32
29. Partial spreadsheet after step 1 of volumetric method 33
30. Partial spreadsheet after step 2 of volumetric method 33
31. Converting 0.1-inch to 1.0-inch grouped data 33
32. Partial spreadsheet showing cumulative area (step 3 of volumetric method) for 0.1-inch recharge group 34
33. Partial spreadsheet showing cumulative percentage (step 3 of volumetric method) for 0.1-inch recharge group 35
34. Partial spreadsheet showing cumulative volume (step 3 of volumetric method) for 0.1-inch recharge group 35
35. Full spreadsheet completed (step 3 of volumetric method) for 1.0-inch recharge group 36
36. Area vs. recharge group, NJGS test case 37
37. Volume vs. recharge group, NJGS test case 37
38. Cumulative area vs. recharge group, NJGS test case 38
39. Cumulative volume vs. recharge group, NJGS test case 38
40. Cumulative volume and area vs. recharge group, NJGS test case 39
41. Full spreadsheet showing step 5 of volumetric method for 0.1-inch grouped data 40
42. Significant class intervals and percent volume for volumetric method sample data 41
43. Example classifications of volumetric method sample data 41
44. Recharge map with parcels used in volume comparison 43

Appendixes

Appendix 1. Legislation 51
2. Land-use/land-cover definitions by LULC code 52
3. Soil recharge group by soil unit 53
4. Recharge constants and factors by recharge soil group 55
5. Recharge factors and constants by soil series 56
6. Climate factors by New Jersey municipality 68
7. Development and application of the soil-water budget to the method 73
8. Wetlands and streams as zones of ground-water discharge and recharge 92
9. Soil Conservation Service field office addresses and phone numbers 95

State of New Jersey
Department of Environmental Protection and Energy
Division of Science and Research
New Jersey Geological Survey
CN-029
Trenton, NJ 08625
Tel. # 609-292-1185
Fax. #609-633-1004

Jeanne M. Fox
Commissioner

Haig F. Kasabach
State Geologist

Dear User:

The purpose of "A Method for Evaluating Ground-Water-Recharge Areas in New Jersey" is to provide municipalities in New Jersey a means to assess ground-water-recharge areas and rank their importance as required by N.J.S.A. 58:11 A, 12-16 et seq. The maps produced using the methodology show land areas of similar recharge characteristics and their relative contribution to the overall amount of recharge. The quantity and quality of ground-water recharge can be managed largely by wise use of the land through which it is replenished. Thus the effect of present or future land uses on recharge in the study area can be evaluated and considered in land-use planning.

We have attempted to make the procedures and explanations in this methodology as clear and user friendly as possible. This is part of the New Jersey Geological Survey's continuing effort to provide the best possible assistance on ground-water issues to government and the public.

Haig F. Kasabach
State Geologist

ABSTRACT

The purpose of this method is to provide municipal planners with a means to make ground-water-recharge maps that can be used in their planning decisions. Awareness of ground-water-recharge areas is important because land use and land cover have a large effect on the recharge that is necessary for most water supplies, wetlands and surface-water bodies. The recharge maps can be used to help decide where, how, and to what extent to develop land.

The method estimates ground-water recharge rather than aquifer recharge. Ground-water recharge includes, but does not distinguish between, recharge to aquifers and non-aquifers. Application of this method does not require specialized equipment or specialized training in hydrology or mapping.

Because in New Jersey recharge occurs throughout much of the land area, soil-water budgets were used to simulate recharge for all combinations of soil, land use/land cover, and climate based on the equation:

recharge = precipitation - surface runoff - evapotranspiration - soil-moisture deficit

These simulations showed that estimates of long-term recharge could be made using factors developed for climate, soil, and land-use/land-cover. The method utilizes tables of climate factors, recharge factors, basin factors, and recharge constants in a simple recharge formula that can be applied to any combination of soil, climate, and land-use/land-cover:

recharge = (recharge-factor x climate-factor x basin-factor) - recharge-constant

To prepare a recharge map, the study area is divided into parcels using county soil surveys and land-use/land-cover mapped according to categories developed for use in estimating recharge. Then the appropriate recharge-factor and recharge-constant are read from the tables and assigned to each parcel. Finally, recharge (inches/year) is calculated by using the recharge factor, recharge constant, basin factor, and a municipal climate-factor in the recharge formula. Recharge (or discharge) from surface-water bodies, wetlands and hydric soils are not evaluated using the method. These areas are eliminated from the assessment.

The basin factor is used to calibrate calculated volumetric recharge against basin-wide stream baseflow estimates. The basin factor that results in the most satisfactory calibration for the basins tested is 1.3. Further research may define separate basin factors for different watersheds.

Instructions are given for ranking the ground-water-recharge areas. The recommended procedure is based on the actual quantity of ground-water recharge within the study area. The ranking scheme (for example, high, moderate, low) is designed to adapt to any study area and any set of ground-water protection practices.

A METHOD FOR EVALUATING GROUND-WATER-RECHARGE AREAS IN NEW JERSEY

I. BACKGROUND

Introduction

State legislation (NJSA 58:11A,12-16, *et. seq.)* requires the Department of Environmental Protection and Energy (DEPE) to publish a methodology to map and rank aquifer-recharge areas. In addition, the legislation (appendix 1) requires the DEPE to publish ground-water protection practices designed to encourage ecologically sound development in aquifer-recharge areas. DEPE must also publish and periodically update aquifer-recharge maps of the state.

The New Jersey Geological Survey (NJGS) has undertaken two tasks in response to the legislation. The first, presented in this report, is to develop and publish a methodology that will enable municipalities to map and rank land areas according to their ability to transmit water to the subsurface. The second is to produce aquifer-recharge maps of the entire state.

The procedures in this report are for estimating ground-water recharge (the volume of water transmitted to the subsurface through soils) rather than aquifer recharge (recharge to geologic formations which can yield economically significant quantities of water to wells or springs). Ground-water recharge is critical to aquifers, wetlands, streams, and lakes. The method is thus useful for evaluating the effect of present and future land uses on these resources. In addition, ground-water recharge values are being used in conjunction with maps of the water-transmitting characteristics of geologic formations to prepare aquifer-recharge maps.

The procedure developed by the NJGS is designed for application by municipalities as part of their land-use planning. A primary consideration was that the method should not require advanced knowledge of hydrology or mapping, but still provide a reliable assessment of recharge. The method is designed for use by environmental planners, environmental scientists, and engineers. The ground-water-recharge maps that result are to be used at the discretion of municipalities and as one of many considerations in land-use planning. Because the quantity and quality of ground-water recharge can be managed largely by wise use of the land through which it occurs, the recharge maps should be used in conjunction with ecologically sound land-use regulations or ground-water protection practices.

This background section presents an overview of ground-water-recharge concepts, reviews the requirements for the method, and then describes the method chosen. Instructions for estimating recharge and producing ground-water-recharge-area maps follow in Section II. Guidance for classifying and ranking the recharge areas on the map is given in Section III. Section IV discusses the limitations of the methodology that should be considered when using the recharge maps. Understanding and applying sections I through IV do not require an advanced knowledge of hydrology and mapping. A glossary is included to explain the necessary terminology. The appendixes contain data required in the mapping procedure and additional technical documentation.

Acknowledgements

We gratefully acknowledge the exceptional assistance that Daryl Lund, Jack Tibbetts, and Paul Welle of the Soil Conservation Service provided throughout the development of this method. Rich Volkert of NJGS also provided valuable technical assistance. This document benefited from reviews by Suzanne Hess and Caroline Swartz of the Hunterdon County Planning Board, Richard Guilick of the Randolph Township (Morris County) Planning Office, Ken Lechner of Franklin Township (Gloucester County) Department of Planning and Zoning, Glenn Carter of the Plainsboro Township (Middlesex County) Department of Community Development, and Otto Zapecza and Gregory McCabe, Jr., of the United States Geological Survey.

We acknowledge valuable conceptual discussions with William Alley, Scott Andres, Jim Boyle, Robert Canace, Fred Charles, Rick Clawges, Wayne Hutchinson, Laura Nicholson, Ron Taylor and Ron Witte. Valuable assistance with developing the application methodology came from Gail Carter, Mark French, Bill Graff, Evelyn Hall, Ron Pristas and Terry Romagna. Special thanks to Dan van Abs, Jim Gaffney, and the Cook College Office of Continuing Education for oportunity to test and refine the application through a short course. Useful literature for the project was suggested or supplied by Maria Baratta, Dorothy McLaughlin Alibrando, George Blyskun and Don Cramer. Editing by Butch Grossman greatly clarified the document.

Ground-water-recharge concepts

The principal processes that affect ground-water recharge can be summarized by tracing the path that water from precipitation would take (fig. 1). The potential for natural ground-water recharge begins with precipitation (rain, snow, hail, sleet). Some of the precipitation never seeps into the soil, but instead leaves the system as surface runoff. The water that seeps into the soil is infiltration. Part of the water that does infiltrate is returned to the atmosphere through evapotranspiration. Evapotranspiration refers to water that is returned to the atmosphere from vegetated areas by evaporation from the soil and plant surfaces (dew and rain) and soil water that is taken up by plant roots and transpired through leaves or needles. Infiltrated water that is not returned to the atmosphere by evapotranspiration moves vertically downward and, upon reaching the saturated zone, becomes ground water. This ground water could be in geologic material that is either an aquifer or non-aquifer, depending on whether it can yield satisfactory quantities of water to wells.

Many climate, soil, and vegetation factors influence the processes that control ground-water recharge. The most important climatic factors are the amount, intensity, and form of precipitation. Climate also influences recharge through the effect of wind, humidity, and air temperature on evapotranspiration. Soil properties are decisive factors in the recharge process. These properties exert strong control on permeability, water-holding capacity, water content prior to a precipitation event, and depth of plant roots (Balek, 1988). In addition, land use and land cover affect the surface condition of the soil, which can enhance or reduce infiltration. Under conditions prevalent in New Jersey, slope of the land surface does not have a significant affect on the total volume of surface runoff and infiltration, but does affect the rate of surface runoff and peak discharge (Paul Welle, U.S. Department of Agriculture, Soil Conservation Service, oral communication, July 26, 1990). The type of vegetation influences recharge through its effects on evapotranspiration, interception of precipitation, and surface runoff.

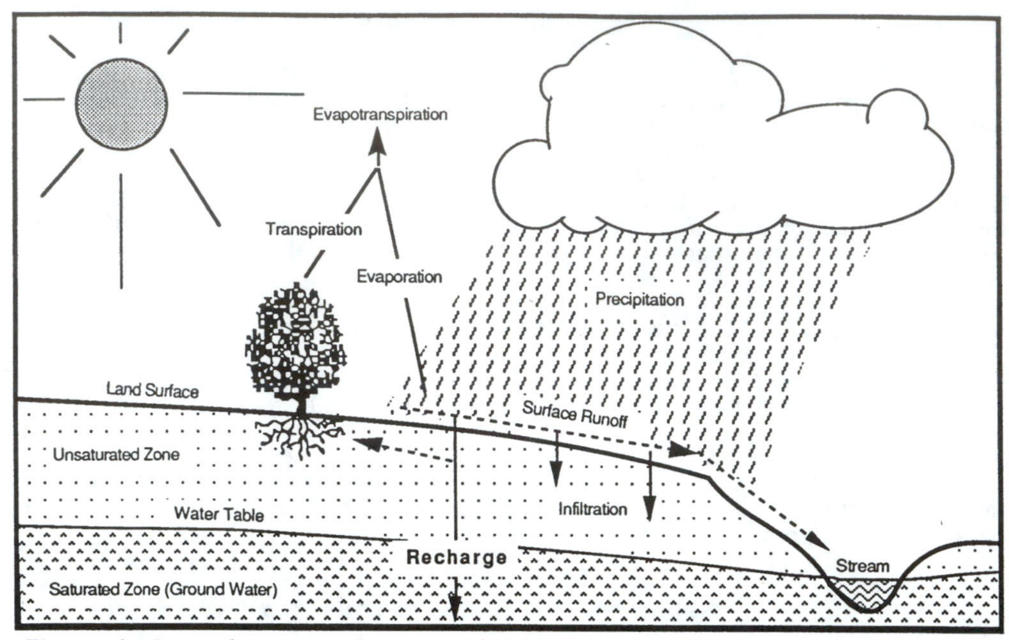

Figure 1. Ground-water recharge in the hydrologic cycle.

Development of the method

A primary consideration in developing the method was that it provide a reliable estimate of ground-water recharge without requiring an advanced knowledge of hydrology or mapping. In addition, the approach was to use readily available equipment and data. Maps produced by the method should be at a large enough scale to clearly delineate a few acres, and also should indicate areas of similar recharge characteristics and their relative contribution to long-term recharge. This is because the method may be used to help decide between alternative development plans, and because land development is long-lasting and maintenance of recharge is a long-term concern.

These requirements were met using a soil-water budget as the basis for recharge calculations. A soil-water budget estimates recharge by subtracting water that is unavailable for recharge (surface runoff and evapotranspiration) from precipitation (the initial budget amount). Any deficit in water storage in the unsaturated zone (soil-moisture deficit) must be made up before ground-water recharge can occur. The resulting equation is:

$$\text{recharge} = \text{precipitation} - \text{surface runoff} - \text{evapotranspiration} - \text{soil-moisture deficit} \quad (1)$$

Although recharge to ground water is a highly variable and complex process, a soil-water budget can account for the principal mechanisms and provide reasonable recharge estimates. Appendix 7 provides a comprehensive technical explanation of the data and calculations used to develop the method, and how the results were adapted for the mapping procedure. Briefly, the method was developed as follows:

An expanded form of equation 1 was used to simulate monthly recharge for all reasonable combinations of climate, soil and land-use/land-cover found in New

Jersey. Recharge was based on statewide ranges of precipitation and the principal factors that control surface runoff and evapotranspiration. Data on five environmental factors were necessary for the simulations: precipitation, soil, land-use/land-cover, surface runoff, and evapotranspiration.

Daily precipitation data were selected from 32 of the 126 National Oceanic and Atmospheric Administration (NOAA) climate stations in New Jersey on the basis of their even geographic distribution and complete record. Thirty years of data were used in the simulations because it is the standard length of climate record for comparison purposes (Linsley, Kohler, and Paulhus, 1982).

The soil data were hydrologic-soil group, soil type, depth and type of root barriers, and available water capacities. These were developed from a database of New Jersey soils maintained by the state SCS office. These data were used in the surface runoff and evapotranspiration calculations.

Land-use/land-cover is an important consideration that was used in both surface runoff and evapotranspiration calculations. A land-use/land-cover classification of 14 categories (appendix 2) was designed specifically for this method. The classification system was derived largely from a system used in the Soil Conservation Service (SCS) curve-number method for calculating runoff (U.S. Department of Agriculture, 1986). The number of categories was reduced to reflect useful long-term land-use distinctions and limitations inherent in mapping from aerial photos.

Surface runoff was calculated using a modification of the SCS curve-number method. Because the curve-number method is designed for calculating runoff from the largest annual storms, adjustments were made so the results more accurately reflect runoff observed in New Jersey from smaller storms (appendix 7). These adjustments are applicable only to recharge calculations and are important because frequent smaller storms contribute most of the long-term recharge.

Evapotranspiration was computed for each of the 32 climate stations using a method developed by Thornthwaite and Mather (1957). Evapotranspiration calculations incorporated the effects of land-use/land-cover. Adjustments were made to the evapotranspiration results so they would more closely approximate evapotranspiration from naturally-watered, open, vegetated areas in New Jersey (see appendix 7).

The simulations showed that average annual recharge could be estimated on the basis of climate, soil characteristics and land-use/land-cover. The results were incorporated in a simple formula which allows one to calculate average annual recharge in inches per year from a climate factor (C-factor), a recharge factor (R-factor), a basin factor (B-factor) and a recharge constant (R-constant):

annual ground-water recharge = (R-factor x C-factor x B-factor) - R-constant (2)

The basin factor (B-factor), a constant of 1.3, was assigned by calibrating predicted volumetric ground-water recharge to reported basin-wide stream baseflow values.

Climate factors were developed for every municipality (appendix 6). Recharge factors and recharge constants (appendixes 4, 5 and 6) were developed for every possible combination of soil characteristics and land use/land cover found in New Jersey.

A user can conveniently carry out the procedure either manually or with a Geographic Information System (GIS). The procedures are designed to be applied at the 1:24,000 scale. First, maps are prepared showing land-use/land-cover according to the categories in appendix 2 and soil units based on SCS data. These two maps are combined to show the distinct areas for which recharge is to be calculated. R-factors and R-constants are then looked up in appendix 4 (by recharge soil group) or appendix 5 (by soil unit). Finally the climate factor for the municipality is found in appendix 6 and ground-water recharge calculated.

There are four primary qualifiers of the method. First, the method estimates ground-water recharge (recharge to both aquifers and non-aquifers) rather than aquifer recharge. Second, a fundamental assumption when using a soil-water budget to estimate ground-water recharge is that all water which migrates below the root zone recharges ground water (Rushton, 1988). Third, the method addresses only natural ground-water recharge. Intentional and unintentional artificial recharge, withdrawals of ground water, and natural discharge are not addressed. Fourth, wetlands and water bodies are eliminated from the analysis before recharge mapping is begun. This is because the direction of flow between ground-water and surface water or wetlands depends on site specific factors and can also change seasonally (appendix 8). Incorporating these complexities was beyond the resources of this study.

II. MAPPING GROUND-WATER-RECHARGE AREAS

The step-by-step directions below enable one to produce ground-water-recharge maps. This section specifies what data to acquire, how to prepare overlays and combine them in order to calculate recharge, and how to produce a ground-water-recharge map. The mapping may be done manually or with a computerized Geographic Information System (GIS). The directions that follow assume application to a municipality using the manual method, but the steps are easily adapted to GIS application.

The procedure involves mapping land-use/land-cover, combining the land-use/land-cover maps with soil maps to delineate areas with distinct recharge characteristics, and then calculating ground-water recharge using the map information and tables developed for New Jersey soils and climate. It is recommended that you read and thoroughly understand the procedure before performing any of the steps. This will enable you to consider options that yield greater accuracy and detail.

The entire procedure is outlined below and on a flow chart on page 17 (fig. 12). Following the flow chart is a series of full-page figures that are referred to throughout the mapping procedure explanation. Copy figures 15, 17, 20, and 21 onto 8 1/2 x 11 inch transparencies and simulate the method by following the procedure in the document. The actual workings of the method will seem straightforward after working through the example.

Included in the outline below is the estimated time required to complete each step using the manual method. The estimates assume that a microcomputer spreadsheet is used for the calculations and approximately 4 hours of field checking is made for land-use/land-cover verification. The low numbers of the ranges correspond to small municipalities (study areas less than 2 square miles) and the high ones correspond to large municipalities (greater than 20 square miles).

Step	Description	Approximate Staff Hours
1.	Acquire source data	2 - 6
2.	Prepare composite maps and mylar templates	6 - 8
3a.	Prepare land-use/land-cover (LULC) overlay	8 - 96
3b.	Prepare soil group overlay	4 - 40
4.	Prepare coded LULC/soil group combination map	4 - 12
5.	Prepare spreadsheet and calculate recharge	16 - 56
6.	Prepare recharge base map	6 - 20
	total staff hours:	46 - 238

Overall time requirements might be significantly reduced if a GIS were used. Staff hour estimates would be somewhat higher for steps 3a and 3b (preparing the overlays), but much lower for later steps. Steps 3a and 3b would require more time with a GIS because the overlays would need to be compiled and drafted manually and then drafted again digitally (or scanned) on the computer. Once computerized, however, steps 4, 5, and 6 would be a matter of a few keystrokes with a GIS. If using a GIS, a separate spreadsheet would not be required for step 5. Also, depending on which classification scheme were chosen, time requirements for portions of the classification (section III) would be greatly reduced by using a GIS. In summary, the initial time spent on computerizing the overlay maps would be more than offset by the time saved in subsequent tasks.

The advantage of the manual procedure is that it requires no special equipment or GIS expertise; the disadvantage is that it does not offer the time savings and flexibility of a GIS. However, when using the manual method in anything other than a small study area, using a microcomputer spreadsheet for calculations can save time. Also, a light table is strongly recommended, but not essential, for use in both manual and GIS applications.

The mapping procedure uses materials readily available statewide. Source documents and map scales are consistent throughout the state. This uniformity simplified development of the methodology and allows the results of one study to be easily compared to those of another. A user who develops an understanding of the methodology may decide to use locally available maps or information which is not discussed here. Substitutions are certainly recommended where they save time and money, as long as the overall accuracy of the maps and the final calculations are not compromised.

1. Acquire source data

Before beginning the mapping, acquire the following documents:

USGS quads - U.S. Geological Survey 1:24,000-scale topographic quadrangle maps (USGS quads) will be used to produce a base map. These maps can be purchased from the DEPE Maps and Publications Sales Office (609-777-1038), directly from the USGS (USGS Map Sales DFC; Box 25286 MS 306; Denver, CO 80225; (303-236-7477), or at many retail map stores. Purchase all of the USGS quads needed to cover the area of interest.

Photoquads

Photoquads - The DEPE has produced orthophotographic quadrangles (photoquads) from March 1986 aerial photography. These register to the 1:24,000-scale USGS topographic quadrangles. The photoquads are used as a base on which land-use/land-cover will be delineated. Photoquads are preferable to conventional aerial photography because they provide a high quality, high resolution, uniform, easy to use interpretive tool and because the distortion inherent in aerial photography has been removed. The photoquads are available from MARKHURD (the manufacturer, 800-627-4873). Alternatively, variable quality and lower resolution paper diazo prints are available from the DEPE Maps and Publications Sales Office (609-777-1038). At the time of publication, more recent photoquads (photographed in 1991) were being prepared. These more recent photoquads are preferable and should be used if available. Order the photoquads corresponding to the USGS quads for the area of interest.

NWI quads - The U.S. Fish and Wildlife Service has produced 1:24,000-scale National Wetlands Inventory quadrangles (NWI quads) for all of New Jersey. These maps show the extent and types of freshwater wetlands and surface water (together referred to as "wet areas") as of the aerial photography date (1976). NWI quads are available at the DEPE Map and Publication Sales Office and many map stores.

Soil surveys - The U.S. Department of Agriculture, Soil Conservation Service (SCS) has published a soil survey for each county in the state except Essex and Hudson. The county soil maps are compiled as soft-cover books containing soil descriptions and properties in text and tables. Also included are soil map sheets, which are aerial photographs overprinted with soil boundaries and symbols. They are used to delineate soils with distinct properties.

To acquire a soil survey for any county, contact the local SCS Field Office (addresses and phone numbers are given in appendix 9). For GIS applications, inquire into whether digitized soil coverage is available. If you are in one of the two counties, Essex and Hudson, that lack soil surveys, contact the SCS office about available information.

2. Prepare composites and mylar templates

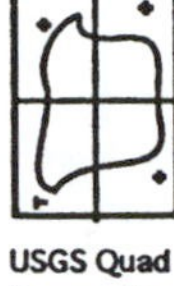

USGS quad composite - Use a blade and straightedge to remove the USGS quad borders to facilitate edgematching the sheets. Carefully edgematch and firmly join the sheets covering your study area to produce a USGS quad composite. The quads will not edgematch perfectly. Try to distribute the error evenly along the quad edges. Delineate the study-area boundary on the USGS quad composite. If the study area is a municipality, the boundary is probably already marked as a dashed line on the quadrangle. Use caution with municipal boundaries on older quads because some are reputed to be inaccurate. In any case, highlight the study area boundary and municipal boundaries (if present) within it.

Choose at least four tick mark locations near the corners of the composite. Marks should be made on features, such as road intersections, which are clearly identifiable on the USGS quads, photoquads, NWI quads, and soil survey maps. Make a well defined "+" to serve as a tick mark at each location (fig. 13). Highlight the tick marks with a marker.

The product of this step is a composite map of those USGS quads which cover the study area. The composite includes a highlighted study area boundary and tick marks.

NWI quad composite - The method omits wetland areas and surface-water bodies in calculating ground-water recharge. The NWI quad composite is prepared so that these wet areas can be eliminated from the recharge evaluation.

NWI Quad composite

Use a blade and straightedge to remove the NWI quad borders. Do not edgematch the NWI quads. Instead, carefully overlay and register each NWI quad to the underlying USGS quad composite independently. Since the NWI quad base comes directly from the USGS quad, there will be many features to use for registration (for example road intersections, topographic contours, etc.).

The NWI quads will probably not match the USGS quad composite perfectly due to printing distortions. Distribute the error evenly throughout each map sheet. The NWI quads may overlap slightly or have gaps between them in some places, but the final NWI quad composite should register well with the USGS quad composite. After the NWI quads are registered and temporarily attached to the USGS quad composite, firmly join them into one large composite. Transfer the tick marks and study-area boundary from the underlying USGS composite onto the NWI quad composite (fig. 2).

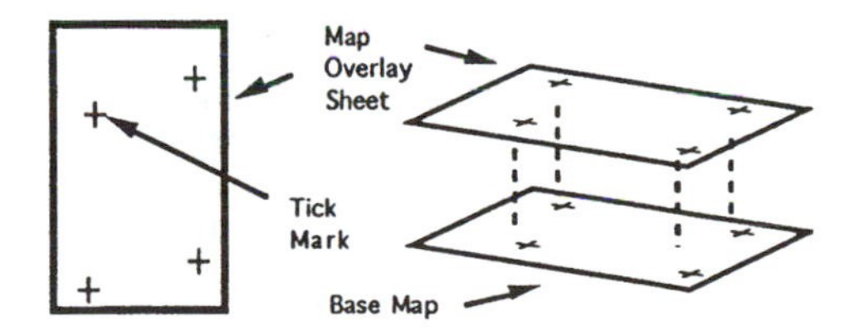

Figure 2. Registration.

The product of this step is a composite map of those NWI quads that cover the study area (fig. 14). The composite includes the highlighted study-area boundary and tick marks.

Mylar templates - Overlay a continuous blank sheet of mylar on the NWI quad composite. Transfer the tick marks and study-area boundary onto the template. Outline the boundaries of all wet areas (wetlands and surface water bodies) with a dark pencil or marker and shade everything inside the wet area boundaries (fig. 3 and 15).

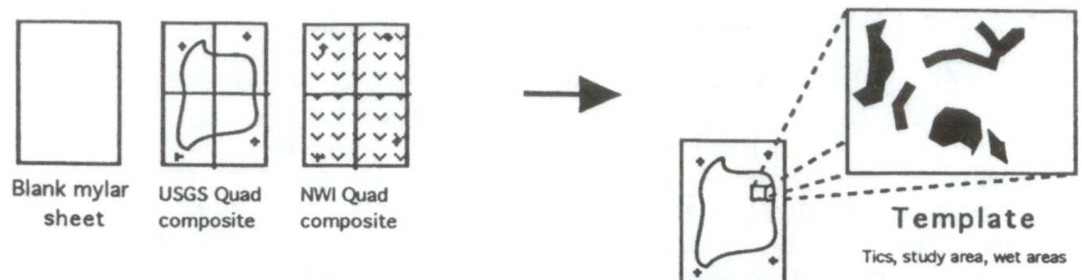

Figure 3. Producing a mylar template.

Verify surface-water bodies from the NWI quad composite using the USGS quad composite. Overlay the mylar template on the USGS quad composite. Any surface water bodies which were not mapped from the NWI composite will show in light blue (or purple, if added to the USGS quads during a recent photorevision). Be careful because not all purple photorevision will be wet areas. For example, it could be easy to mistake the shape of a disturbed area for a surface water body. Trace any remaining surface-water boundaries (excluding small streams) onto the mylar, and complete the shading of wet areas.

Produce two additional copies of the completed template. This can be done manually by tracing the original template, or photographically by a professional reproduction company. Professional reproduction is preferred because manual tracing introduces more error.

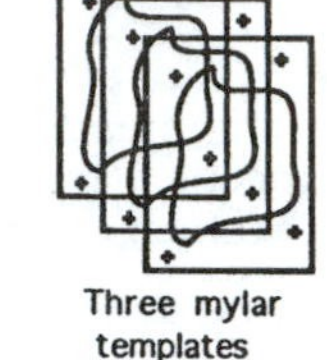

The products of this step are three duplicate mylar templates containing the ticks, study-area boundary, and shaded wet areas.

3a. Prepare land-use/land-cover (LULC) overlay

When preparing the overlays discussed below, use a different colored pencil for each separate overlay so they are distinct and recognizable. This will simplify the mapping procedure.

The land-use/land-cover map to be prepared in this step is very specific in terms of the land classification scheme and accuracy standards. It is not the same as a land-use map that might be found in a municipal master plan, and municipal land-use maps can not be substituted for the land-use/land-cover maps specified in this procedure. Municipal land-use maps can, however, be used as a reference in preparing the map.

Municipal land-use maps commonly show only land use (not land cover) and generally only on a lot-by-lot basis. For example, a school might have large tracts of lawn or forest within the property boundary. On a municipal land-use map, the entire property might be classified as "public," "quasi-public," or "institutional." On the land-use/land-cover map required for recharge mapping, the forest, lawn, and impervious areas need to be classified separately if such tracts are larger than 5 acres.

1:24,000-scale photoquads are to be used as a base for delineating land-use/land-cover (LULC). Use a blade and straightedge to remove the photoquad borders. Notice that the photo image extends beyond the line representing the USGS quad edge. Trim the borders of the photoquads to the line which represents the USGS quad edge. Do not edgematch the photoquads. Instead, carefully register each photoquad to the USGS quad composite independently.

The photoquads will not match the USGS quads perfectly. Distribute the error evenly throughout each map sheet. The photoquads may overlap slightly or have gaps between them in some places, but the final photoquad composite should register well with the USGS quad composite. After each photoquad is registered and temporarily attached to the USGS quad composite, firmly join the photoquads into one large composite. Transfer the tick marks and study-area boundary to the photoquad composite (fig. 16).

Use the LULC categories in appendix 2 to delineate LULC polygons directly on the photoquad composite (a red pencil shows up best on a grey-tone photoquad). Use the codes given in the appendix to label each distinct LULC area. If you will be estimating recharge for a specific percent impervious cover instead of the ranges shown in appendix 2, note such areas on the map with your own symbol. Tests performed by NJGS show that the easiest LULC delineation sequence from start to finish is:

land use/land cover	LULC code
agricultural	8
wooded areas	9
landscaped open space	0
landscaped commercial	5
unlandscaped commercial	6
unvegetated	7
residential 1/8 acre lots	1
residential 1/8 to 1/2 acre lots	2
residential 1/2 to 1 acre lots	3
residential 1 to 2 acre lots	4

At the 1:24,000 scale of the maps, areas of less than 5 acres should not be mapped (a 5-acre parcel is 470 by 470 feet, about the diameter of a pencil eraser at 1:24,000 scale). For example, if a 2-acre residential lot is in the middle of a wooded area, the entire area should be mapped as "wooded area." Most highways will be so narrow (slivers) on a 1:24,000 map that it is appropriate to absorb them into surrounding polygons. Omitting small parcels simplifies the mapping effort and leads to a more readable final map.

LULC delineation is designed to be a tabletop procedure done primarily from existing knowledge and photoquad interpretation. Calculation of average lot size for residential districts is not necessary, but make sure the average lot sizes are consistent with knowledge of the area. Map estimates of lot size can be easily obtained by counting the number of lots (houses) that appear through a one-quarter-inch diameter hole punched in a file card:

Houses/hole	Average lot size	LULC code
10 - 40	1/8 to 1/2 acre	2
5 - 10	1/2 to 1 acre	3
2 - 5	1 to 2 acres	4

Notice that two of the general LULC categories (agricultural and wooded areas) can be subdivided. Use of the subdivided categories (cropland, permanent pasture; brush,woods and orchards) will yield a more accurate final recharge map. If a more general recharge map is adequate, the general LULC distinctions will probably be sufficient. However if a more detailed map is desired, the subdivisions are recommended. This choice is provided because some users may not be able to justify the possible extra work required to distinguish the subdivisions of LULC. If the user is already quite familiar with the LULC in the study area, use of the subdivisions will require little additional effort. Otherwise airphotos, maps, and reports can aid in the mapping process.

Regardless of the level of detail used for LULC delineation, it is essential that a few of your interpretations be field checked to validate your photo interpretations. Questions concerning airphoto interpretation and land-use/land-cover mapping may be resolved by consulting Avery and Berlin (1985).

After the delineation process is complete, overlay, register, and temporarily attach a mylar template to the photoquad composite. Transfer all LULC boundaries and labels to the template, disregarding the shaded wet areas.

The product of this step is a mylar overlay containing ticks, study-area boundary, shaded wet areas, LULC boundaries, and LULC codes (figs. 4 and 17).

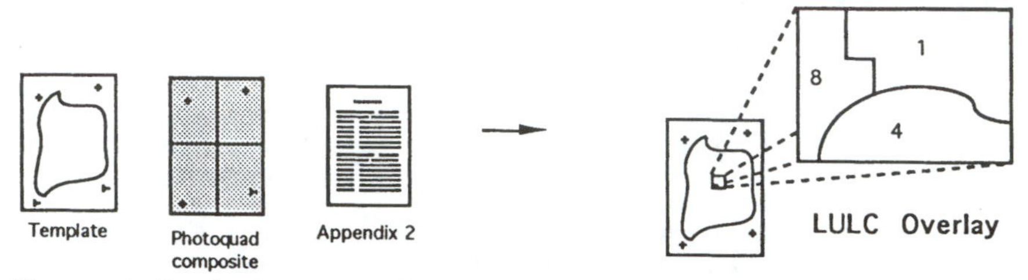

Figure 4. Producing a LULC overlay.

3b. Prepare soil-group overlay

In the SCS county soil survey, find the map sheets covering your study area. These maps are originally at either 1:15,840 or 1:20,000. They must be re-scaled to 1:24,000. County map sheets at 1:15,840 must be reduced to 66 percent of the original size. Sheets at 1:20,000 must be reduced to 83.33 percent of the original size. It is important that the soil-map sheets be accurately reduced. A photocopier introduces some distortion. A professional photographic reduction introduces much less distortion and maintains image quality. Therefore it is preferable to have the sheets professionally reduced.

Use a blade and straightedge to trim the borders from the reduced soil-map sheets. Do not edgematch the soil map sheets. Carefully register each sheet to the USGS quad composite independently. The soil-map sheets were produced using air photo bases containing some distortion, and will therefore not match the USGS quad composite in some places. Distribute the registration error throughout each soil map sheet as evenly as possible before temporarily attaching it to the USGS quad composite. There may be gaps or overlaps between soil-map sheets in the final soil-map composite. After each sheet is registered and temporarily attached to the USGS quad composite, firmly join them into one large composite. Transfer the tick marks to the soil-map composite (fig. 18).

Register and temporarily attach a mylar template to the soil-map composite. Do not trace any soil boundaries yet. Using the soil-group table in appendix 3 (fig. 19 for example), find the recharge soil-group code (A, B, C, etc.). Write this code on the mylar template over the soil symbol. Appendix 3 lists the full name of each soil unit rather than the symbol found on the soil-map sheets because the symbols vary from county to county; symbol-to-unit name translation is given in the county soil survey book.

Some symbols refer to soil complexes (areas with soils of two or more soil series) which are not listed as such in the appendix. The predominant unit in a soil complex is the first name given and should be taken as the soil type. For example, in Mercer County the symbol "SyB" refers to the soil complex "Sassafras-Woodstown," so "Sassafras" would be the unit to look up in Appendix 7.

If the map unit name includes rock outcrop in any form, the full map unit name and associated recharge soil group will be listed in the appendix.

Still other symbols refer to urban land complexes. If the soil survey lists a soil series associated with urban land or urban land complex, use that soil as the map unit to look up in the appendix. For example, in Mercer County, the symbol "Ug" refers to "Urban land, Galestown Material", so "Galestown" would be the unit to look up in the appendix. The Soil Conservation Service should be contacted for advice on urban land or urban land complexes that do not have an associated soil series.

After all the recharge soil group codes have been added, trace the boundaries separating soils of different groups on the mylar template. Eliminate any map unit smaller than 5 acres (pencil eraser size). Smooth out any boundary discontinuities between map sheets. Finally, shade all polygons that contain hydric soils (recharge soil group L).

The product of this step is a mylar overlay containing registration ticks, the study-area boundary, soil-group boundaries, and soil-group codes (fig. 5 and 20).

Figure 5. Producing a soil group overlay.

4. Prepare coded LULC/soil-group combination map

On a light table, register the LULC overlay with the soil-group overlay. Secure the two maps on the table with drafting tape, then register and tape your third clean mylar template over these. First add the shaded hydric soil areas from the soil-group overlay to the combination map. Then trace the lines over all non-shaded map areas from both underlying maps onto the mylar template to produce a combination LULC/soil-group map. In areas where the combination of soil and LULC boundaries produces slivers or polygons smaller than 5 acres, absorb and smooth the lines into the neighboring polygons, but give preference to the LULC boundaries. Finally, on the combination map, assign each polygon a unique numeric code to give each an identifier (figs. 6 and 21).

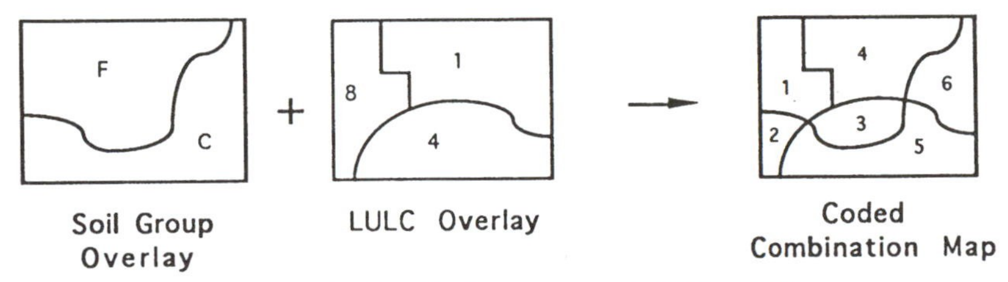

Figure 6. Producing a combination map.

5. Prepare spreadsheet and calculate recharge

Create a computer spreadsheet template (fig. 7) for data needed to calculate groundwater recharge.

Code	LULC	Soil Group	R-Factor	R-Constant	C-Factor	Recharge
1						
2						
3						
etc.						

Figure 7. Spreadsheet format for calculating recharge.

Refer to the two original maps to determine LULC and soil-group codes for each polygon on the combination map, and add these to the spreadsheet (fig. 8):

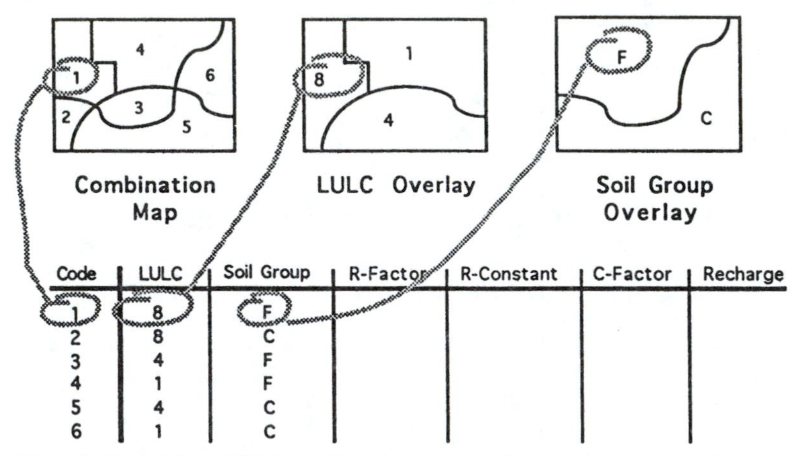

Code	LULC	Soil Group	R-Factor	R-Constant	C-Factor	Recharge
1	8	F				
2	8	C				
3	4	F				
4	1	F				
5	4	C				
6	1	C				

Figure 8. Adding LULC and soil group codes to the spreadsheet.

Using appendix 4, add recharge factors (R-factors) and recharge constants (R-constants) to the spreadsheet. The example below depicts the look-up procedure for polygon number 1 of figure 9.

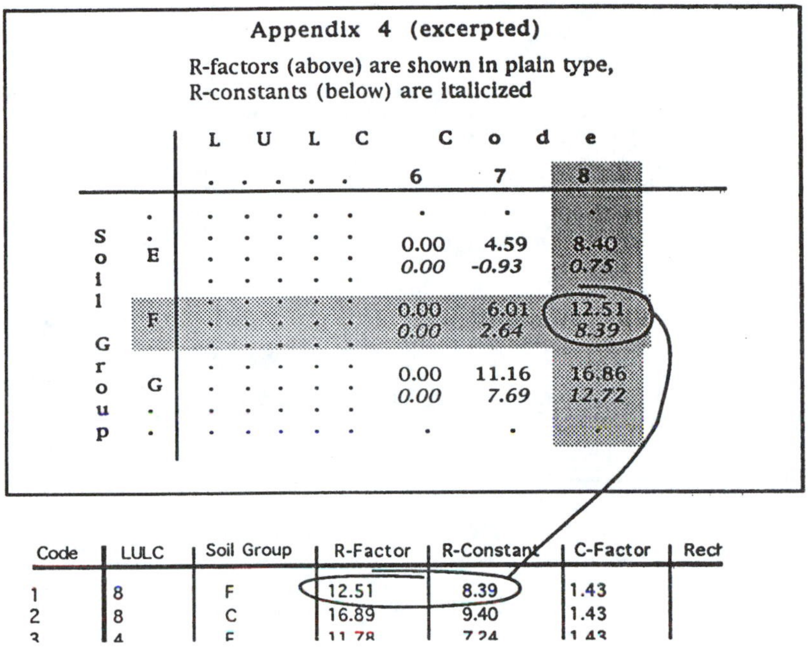

Appendix 4 (excerpted)

R-factors (above) are shown in plain type,
R-constants (below) are italicized

Soil Group	LULC Code	6	7	8
.	. . . . . .	.	.	.
E	. . . . . .	0.00 *0.00*	4.59 *-0.93*	8.40 *0.75*
F	. . . . . .	0.00 *0.00*	6.01 *2.64*	12.51 *8.39*
G	. . . . . .	0.00 *0.00*	11.16 *7.69*	16.86 *12.72*
.	. . . . . .	.	.	.

Code	LULC	Soil Group	R-Factor	R-Constant	C-Factor	Rech
1	8	F	12.51	8.39	1.43	
2	8	C	16.89	9.40	1.43	
3	4	F	11.78	7.24	1.43	

Figure 9. Adding R-factor and R-constant to the spreadsheet.

Using appendix 6, find the climate factor (C-factor) for your municipality and enter it in the spreadsheet. Calculate recharge using the following equation:

$$\text{Recharge} = (\text{R-factor} \times \text{C-factor} \times \text{B-factor}) - \text{R-constant} \quad (3)$$

Round all results to the nearest tenth of an inch (figs. 10 and 23).

Code	LULC	Soil Group	R-Factor	R-Constant	C-Factor	Recharge
1	8	F	12.51	8.39	1.43	14.9
2	8	C	16.89	9.40	1.43	22.0
3	4	F	11.78	7.24	1.43	14.7
4	1	F	4.97	3.05	1.43	6.2
5	4	C	14.75	8.15	1.43	19.3
6	1	C	6.22	3.44	1.43	8.1

Figure 10. Completed spreadsheet with calculated recharge.

Remember, the basin factor is a constant, 1.3. In this example a column is not set up for it. This particular example also has polygons which fall entirely within one municipality. Thus the climate factor is the same for all polygons. A different example might have different climate factors for different pologons.

6. Prepare recharge base map

Each polygon should now have a recharge value expressed in inches per year. Write these values in the polygons on the combination map. The resulting recharge base map includes the combination polygons, their codes, and their recharge values. To distinguish between the codes and the recharge values on the map, make the codes whole numbers, and write the recharge values to the tenths place. In figures 11 and 22, the recharge values are italicized to further distinguish them from the codes.

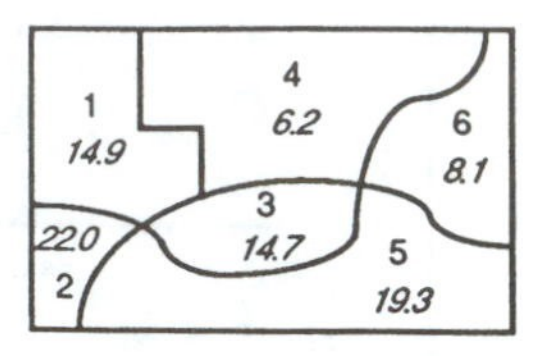

Figure 11. Recharge base map.

The recharge base map is to be used to produce the shaded (ranked) recharge maps described in section III.

7. Mapping by soil unit for more accurate results

More accurate recharge values may be obtained by using soil unit rather than recharge soil group (a group of soil units). R-factors and R-constants for recharge soil groups are generalized values derived from recharge calculated for specific soil units. Results using recharge soil groups may differ by as much as 1 1/2 inches per year from the more accurate values calculated using soil units. Using soil units for recharge calculations is especially applicable for studies of small areas. However, it adds complexity (and accuracy) to larger maps by yielding more polygons. Depending on the size of the study area, this may be a compelling reason to use a GIS. Regardless of the reasons, if a GIS is used, soil units should be incorporated to take advantage of GIS capabilities.

A soil-unit map is made in basically the same way as a soil-group map except for two steps. In step 3b, instead of writing the soil-group code on the mylar template, transfer the first two letters of the soil symbol onto the mylar (the last letter or letter/number combination of the symbol are not needed for this analysis; refer to the county soil-survey book for symbol-to-unit translation). Before proceeding with step 4, shade all soil polygons that contain both a recharge factor and recharge constant of 0.00. Then, in step 5, refer to appendix 5, not appendix 4, for R-factors and R-constants.

8. Recharge estimates for specific percentages of impervious area

For specific development scenarios, it might be desirable to estimate ground-water recharge for a given percentage of impervious cover instead of for the ranges noted in appendix 2. The following calculation can give such estimates for any soil map unit by using a weighted average of the proportion of landscaped open space:

$$\text{Recharge} = (\text{recharge for LULC } 0) \times ((100 - \%\ \text{impervious cover})/100) \qquad (4)$$

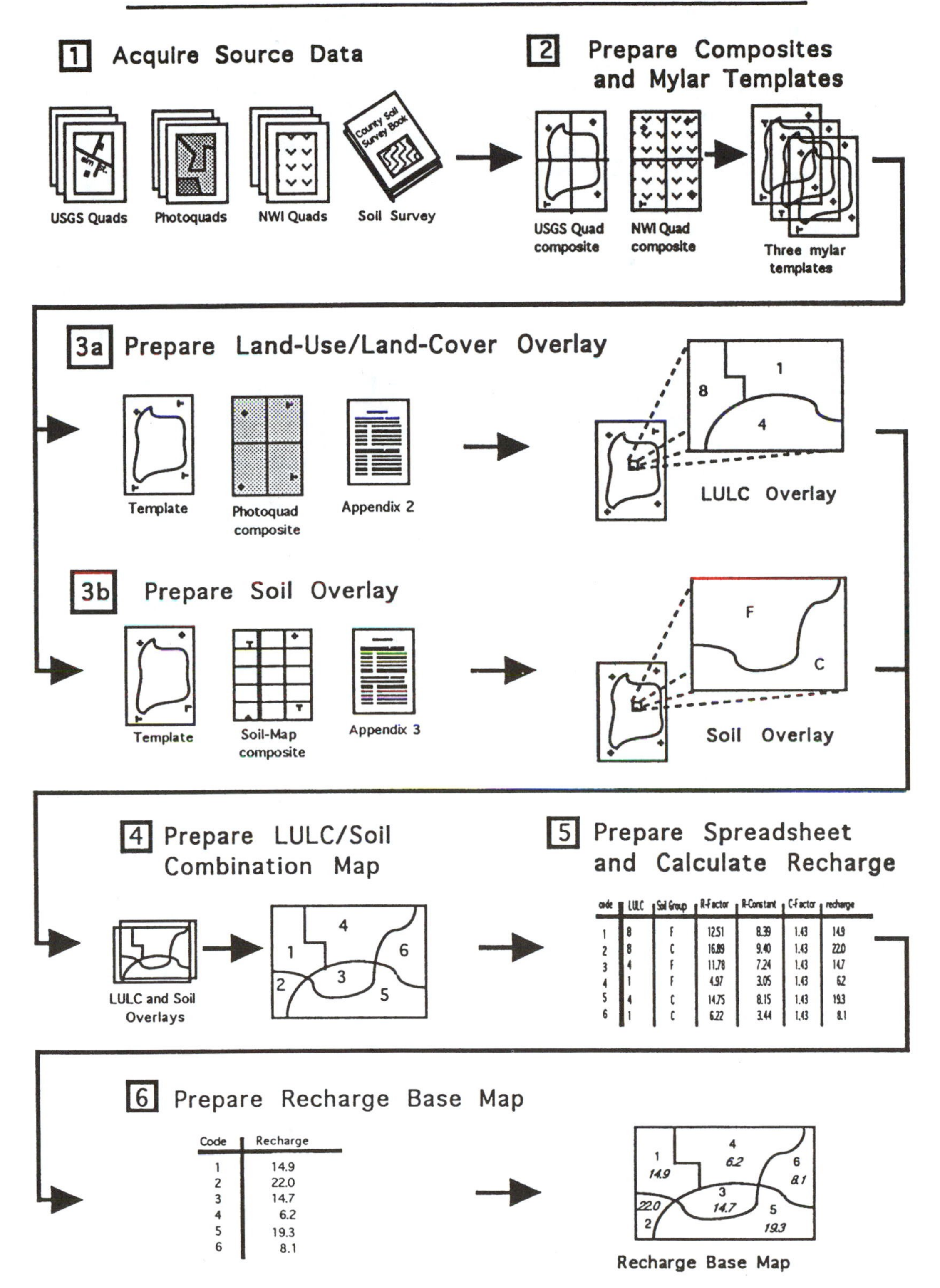
Ground-Water-Recharge Mapping Procedure
1 Acquire Source Data
County Soil Survey Book
USGS Quads
Photoquads
NWI Quads
Soil Survey
2 Prepare Composites and Mylar Templates
USGS Quad composite
NWI Quad composite
Three mylar templates
3a Prepare Land-Use/Land-Cover Overlay
Template
Photoquad composite
Appendix 2
1
8
4
LULC Overlay
3b Prepare Soil Overlay
Template
Soil-Map composite
Appendix 3
F
C
Soil Overlay
4 Prepare LULC/Soil Combination Map
LULC and Soil Overlays
4
1
6
3
2
5
5 Prepare Spreadsheet and Calculate Recharge

code	LULC	Soil Group	R-Factor	R-Constant	C-Factor	recharge
1	8	F	12.51	8.39	1.43	14.9
2	8	C	16.89	9.40	1.43	22.0
3	4	F	11.78	7.24	1.43	14.7
4	1	F	4.97	3.05	1.43	6.2
5	4	C	14.75	8.15	1.43	19.3
6	1	C	6.22	3.44	1.43	8.1

6 Prepare Recharge Base Map

Code	Recharge
1	14.9
2	22.0
3	14.7
4	6.2
5	19.3
6	8.1

1
14.9
4
6.2
6
8.1
3
14.7
22.0
2
5
19.3
Recharge Base Map

APPENDIX E

THE ANSWER TO KEEPING CATCH BASINS FREE OF SILT DURING CONSTRUCTION.

2 STYLES:
HI – FLOW • REGULAR FLOW

ALSO AVAILABLE WITH ABSORBANT PILLOW FOR OIL CONTROL

And It's Simple.

- •REMOVE DRAIN GRATE
- •INSERT SILTSACK™
- •REPLACE GRATE TO HOLD SILTSACK IN POSITION
- •SILTSACK TRAPS SILT
- •REMOVE FILLED SILTSACK (With front-loader or other equipment)
- •CLEAN AND REUSE OR
- •DISCARD & REPLACE

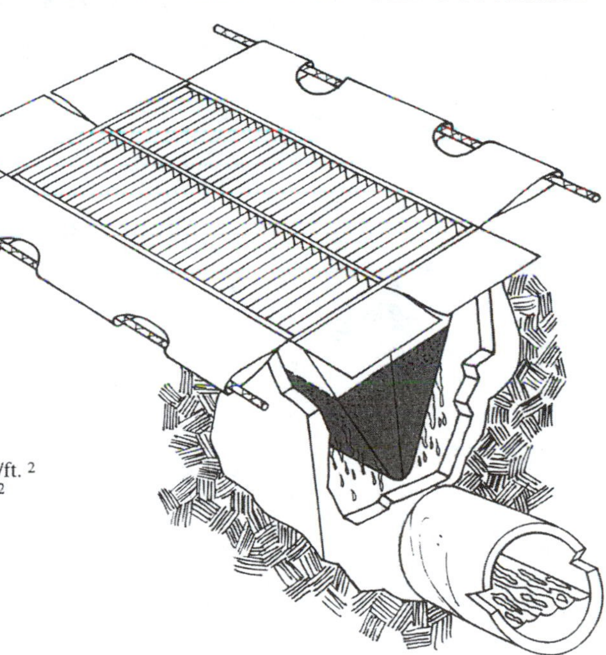

•Sized to fit any size or shape catch basin
•All seams double stitched
•Permeability - Regular flow Siltsack - 40 gal. /min./ft. ^{2}
Hi – flow Siltsack - 200 gal./min./ft.^{2}
•UPS Shippable

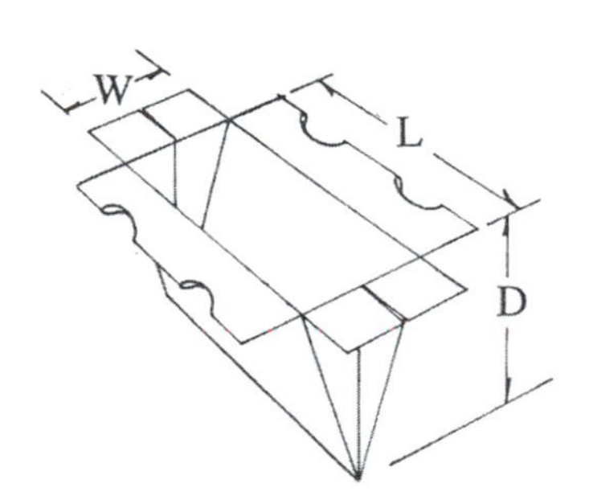

For More Information or to Order, Phone Toll Free

1-800-644-9223

ACF ENVIRONMENTAL • 1801-A Willis Road, Richmond, VA 23237 • U.S. PATENT NO. 5372714

SILTSACK

EASY TO INSTALL EASY TO EMPTY

PROPER INSTALLATION:

Install Siltsack in catch basin, making sure emptying straps are laid flat, outside of basin and held in place by drain grate.

Hold down removal flap pockets and emptying straps by covering with soil.

Properly installed, Siltsack is out of sight and catches silt without the worry of silt fences or straw dams failing.

HOW TO EMPTY SILTSACK:

To prepare for emptying Siltsack, remove soil covering removal flap pockets and insert rebar through pockets.

Remove catch basin cover grate.

Remove Siltsack from catch basin by attaching to both bars and lifting with available equipment.

Move filled Siltsack to dumping area and set on ground.

Remove straps from lifting bars.

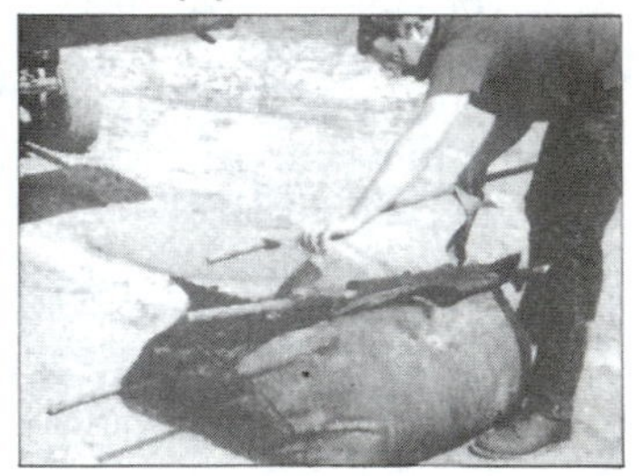

Insert a lifting bar through both emptying straps.

With available equipment, lift Siltsack with emptying straps, which are attached to the bottom of bag.

When raised by emptying straps, bag will turn inside-out and empty Siltsack.

Siltsack is inside-out, empty and ready for reuse or disposal.

PUMPED-SILT CONTROL SYSTEM

USE DIRTBAG ANYWHERE AN ACCUMULATED BODY OF DIRTY WATER MUST BE PUMPED.

Protect the environment effectively and economically! Collect sand, silt and fines. Avoid silting streams, surrounding property, and storm sewers. As more and more emphasis is put on saving our wetlands, regulations are becoming more stringent regarding the pumping of dirty water from holes around construction sites-from foundations, pipe line construction, repairing municipal water/sewer lines, marine construction, utility, highway and site development. The applications are endless where DIRTBAG is the answer!

Installation is easy. First, DIRTBAG is easy to transport to the site. Transporting a simple folded fabric bag beats hauling a load of straw bales! To install , simply unfold and insert up to 4" pump discharge hose into handy sewn-in spout and secure with the attached straps. Pump dirty water into DIRTBAG. The bag collects the silt as the clean water gently filters out from all sides. Consider the alternatives such as straw bale forts which are more cumbersome to transport, build and to clean up afterward. And DIRTBAG poses no threat to the environment when disposed of properly.

Designed and produced from a variety of fabrics to meet engineering specifications for flowrates, strength, and permeability. Stabilized to provide resistance to ultra-violet degradation. Meets municipal, state, and Corps of Engineers specifications. Available in 10' x 15', 12½' x 15', and 15' x 15' sizes. Custom Sizes available.

Once you use a DIRTBAG you'll never do it any other way!

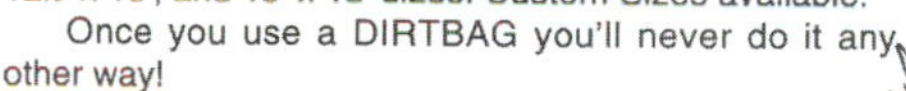

ACF

DIRTBAG® SPECIFICATIONS
CONTROL OF SEDIMENT IN PUMPED WATER

1.0 DESCRIPTION

1.1 THIS WORK SHALL CONSIST OF FURNISHING, PLACING AND REMOVING THE **DIRTBAG** PUMPED SEDIMENT CONTROL DEVICE AS DIRECTED BY THE DESIGN ENGINEER OR AS SHOWN ON THE CONTRACT DRAWINGS. THE DIRTBAG PUMPED-SILT CONTROL SYSTEM IS MARKETED BY:

ACF ENVIRONMENTAL, INC. • 1801 A-WILLIS ROAD • RICHMOND, VA 23237
PHONE: 1-800-644-9223 • FAX: 1-804-271-3074

2.0 MATERIALS

2.1 DIRTBAG

2.1.1 THE DIRTBAG SHALL BE A NONWOVEN BAG WHICH IS SEWN WITH A DOUBLE NEEDLE MACHINE USING A HIGH STRENGTH THREAD.

2.1.2 THE DIRTBAG SEAMS SHALL HAVE AN AVERAGE WIDE WIDTH STRENGTH PER ASTM D-4884 AS FOLLOWS.

DIRTBAG STYLE	TEST METHOD	TEST RESULT
DIRTBAG 53	ASTM D-4884	60 LB/IN
DIRTBAG 55	ASTM D-4884	100 LB/IN

2.1.3 THE DIRTBAG WILL HAVE AN OPENING LARGE ENOUGH TO ACCOMMODATE A FOUR (4) INCH DISCHARGE HOSE WITH ATTACHED STRAP TO TIE OFF THE HOSE TO PREVENT THE PUMPED WATER FROM ESCAPING FROM THE DIRTBAG WITHOUT BEING FILTERED.

2.1.4 THE GEOTEXTILE FABRIC SHALL BE NONWOVEN FABRIC WITH THE FOLLOWING PROPERTIES:

Properties	Test Method	Units	Nonwoven 53	Nonwoven 55
Weight	ASTM D-3776	Oz/yd	8	10
Grab Tensile	ASTM D-4632	Lbs.	203	250
Puncture	ASTM D-4833	Lbs.	130	165
Flow Rate	ASTM D-4491	Gal./Min/Ft^2	80	70
Permitivity	ASTM D-4991	Sec.-1	1.5	1.3
Mullen Burst	ASTM D-3786	Lbs.in^2	400	550
UV Resistant	ASTM D-4355	%	70	70
AOS % Retained	ASTM D-4751	%	100	100

ALL PROPERTIES ARE MINIMUM AVERAGE ROLL VALUE EXCEPT THE WEIGHT OF THE FABRIC WHICH IS GIVEN FOR INFORMATION ONLY.

3.0 CONSTRUCTION SEQUENCE

3.1 GENERAL

3.1.1 INSTALL THE DIRTBAG ON A SLIGHT SLOPE. IT SHOULD BE PLACED SO THE INCOMING WATER FLOWS INTO THE BAG AND

WILL FLOW THROUGH THE DIRTBAG AND THEN FLOW OFF THE SITE WITHOUT CREATING MORE EROSION. THE NECK OF THE DIRTBAG SHOULD BE TIED OFF TIGHTLY TO STOP THE WATER FROM FLOWING OUT OF THE DIRTBAG WITHOUT GOING THROUGH THE WALLS OF THE BAG. TO INCREASE THE SURFACE AREA BEING USED, THE DIRTBAG MAY BE PLACED ON A GRAVEL BED TO ALLOW WATER TO FLOW IN ALL DIRECTIONS. LIFTING STRAPS OR ROPES, SHOULD BE PLACED UNDER THE DIRTBAG TO FACILITATE REMOVAL WHEN FULL OF SEDIMENT.

3.1.2 THE DIRTBAG IS CONSIDERED FULL AND SHOULD BE DISPOSED WHEN IT IS IMPRACTICAL FOR THE BAG TO FILTER THE SEDIMENT OUT AT A REASONABLE FLOW RATE AND SHOULD BE REPLACED WITH A NEW DIRTBAG.

3.1.3 DISPOSAL MAY BE ACCOMPLISHED AS DIRECTED BY THE DESIGN ENGINEER. IF THE SITE ALLOWS, THE DIRTBAG MAY BE

BURIED ON SITE AND SEEDED, VISIBLE FABRIC REMOVED AND SEEDED OR REMOVED FROM SITE TO A PROPER DISPOSAL AREA.

4.0 BASIS OF PAYMENT

4.1 THE PAYMENT FOR ANY DIRTBAG USED DURING CONSTRUCTION IS TO BE INCLUDED IN THE BID OF OVERALL EROSION AND SEDIMENT CONTROL PLAN UNLESS A UNIT PRICE IS REQUESTED.

References

1. American Farmland Trust and Northern Illinois University, *Living on the Edge: The Costs and Risks of Scatter Development,* 1998.
2. Marsh, G. P., *Man and Nature,* Scribner, New York, 1864.
3. McHarg, I. L., *Design with Nature,* Garden City Press, Garden City, NY, 1969.
4. Arendt, R., *Conservation Design for Subdivisions,* Island Press, Washington, D.C., 1996.
5. Lyle, J. T., *Regenerative Design for Sustainable Development,* John Wiley & Sons, New York, 1994.
6. Gurwitt, R., Villages on the freeway, *Governing Mag.,* November 1992.
7. Armstrong, L. J., *Contribution of Heavy Metals to Stormwater from Automotive Disc Brake Pad Wear,* Santa Clara Valley Nonpoint Source Pollution Control, Oakland, CA, October 12, 1994.
8. Drummond, J., cited by Smith, J., in Sustainable cities: Concepts and strategies for eco-city development, ero-home media, Los Angeles, CA, *Nation's Bus.,* September 1991.
9. McConnell, H. H., and Lewis, J., Add salt to taste, *Environment*, November 1972.
10. Kunstler, J. H., Home from nowhere, *Atlantic Mon.,* 278, No. 3, 43, 1996.
11. Florida Design Initiative, *e design Online,* Florida Energy Office, Florida A&M University, Tallahassee, FL, 1998.
12. Newman, M. C. and McIntosh, A. W., *Metal Ecotoxicology, Concepts and Applications,* Lewis Publishers, Chelsea, MI, 1991.
13. Hagman, D. G., and Juergensmeyer, J. C., *Urban Planning and Land Development Control Law,* 2nd ed., West Publishing, St. Paul, MN, 1986.
14. Udall, S. L., *The Quiet Crisis,* Holt Rhinehart & Winston, New York, 1963.
15. Leopold, A., The conversation ethic, *J. For.,* October 1933.
16. U.S. Environmental Protection Agency, Environmental Monitoring and Assessment Program–Overview, EPA/600/9-90/001, January 1990.
17. Solnit, A., Reed, C., Glassford, P., and Erley, D., *The Job of the Practicing Planner,* American Planning Association, Chicago, IL, 1988, 17.
18. Yannacone, V. J., Jr., and Cohen, B. S., *Environmental Rights and Remedies,* Vol. 2, Lawyer's Cooperative Publishing, New York, 1972.
19. Center for Watershed Protection, *Rapid Watershed Planning Handbook — A Comprehensive Guide for Managing Urbanizing Watersheds,* Ellicott City, MD, October 1998.
20. Oreskes, N., Schraeder-Frecette, K., and Belitz, K., Verification, validation, and confirmation of numerical models in the earth sciences, *Science Mag.*, 263, February 4, 1994. American Association for the Advancement of Science.
21. Naeem, S., personal commentary from an online open Forum on Ecoevaluation sponsored by the International Society for Ecological Economics, May 22, 1997.
22. Daily, G. C., Alexander, S., Ehrlich, P. R., Goulder, L., Lubchenco, J., Matson, P. A., Mooney, H. A., Postel, S., Schneider, S. H., Tilman, D., and Woodwell, G. M., *Ecosystem Services: Benefits Supplied to Human Societies by Natural Ecosystems,* Issues in Ecology Series, No. 2, Ecological Society of America, Spring 1997.
23. Costanza, R., d'Arge, R., de Groot, R., Farber, S., Grasso, M., Hannon, B., Limburg, K., Naeem, S., O'Neil, R. V., Paruelo, J., Raskin, R. G., Sutton, P., and vanden Belt, M., The value of the world's ecosystem services and natural capital, *Nature (London)*, 387, 253, 1997.
24. Gluckman, D., Growth management offers new view of property rights, *Fla. Environ.,* November 14, 1989.
25. Robinson, N. A., updated by Reilly, K., *Environmental Regulation of Real Property,* Law Journal Seminary Press, New York, 1994.
26. Harding, J. S., Benfield, E. F., Bolstad, P. U., Helfman, G. S., and Jones, E. B. D., III, Stream biodiversity: the ghost of land use past, *Proc. Natl. Acad. Sci. U.S.A.*, 95, 14843, 1998.
27. U.S. Environmental Protection Agency, Wetlands Fact Sheets, EPA 843-F-95-001, February 1995.

28. Graczyk, T. K., Fayer, R., Trout, J. M., Lewis, E. J., Farley, C. A., Sulaiman, I., and Lai, A. A., *Giardia* sp. cysts and infectious cryptosporidium parvum oocysts in the feces of migratory Canada geese (*Branta canadensis*), *Appl. Environ. Microbiol.,* 64, 2736, 1998.
29. Adams, L. W., and Dove, L. E., *Wildlife Reserves and Corridors in the Urban Environment — A Guide to Ecological Landscape Planning and Resource Conservation,* National Institute for Urban Wildlife, Columbia, MD, 1989.
30. Aldrich, J. W., and Coffin, R. W., Breeding bird populations from forests to suburbia after thirty-seven years, *AM Birds,* 34, 3, 1980.
31. O'Brien, S. J., Genetic erosion — a global dilemma, *Natl. Geographic Mag.,* 181, No. 4, 136, 1992.
32. Packer, C., Captives in the wild, *Natl. Geographic Mag,* 181, No. 4, 121, April 1992.
33. Maxted, J. R., Dickey, E. L., and Mitchell, G. M., The Water Quality Effects of Channelization in Coastal Plain Streams of Delaware, Division of Water Resources, Delaware Department of Natural Resources and Environmental Control, Dover, DE, 1995, 1.
34. Dooley, J. H., Comprehensive Chemistry of Select Greensand from the New Jersey Coastal Plain, New Jersey Geological Survey, Trenton, NJ, 1998.
35. Friedman, D. B., *Connecting People to the Soil: Mimicking Nature in Watershed Management,* Ocean County Soil Conservation District, NJ, March 26, 1999.
36. New Jersey Department of Environmental Protection, A Method for Evaluating Ground Water Recharge in New Jersey, New Jersey Geological Survey, 1993.
37. Doyle, D., Sustainable development: growth without losing ground, *J. Soil Water Conserv.,* January–February 8, 1991.
38. Miller, G. T., Jr., *Living in the Environment,* 6th ed., Wadsworth Publishing, Belmont, CA, 1990.
39. Galli, J., Thermal Impacts Associated with Urbanization and Stormwater Management Best Management Practices, Metropolitan Washington Council of Governments, Maryland Department of Environment, Washington, D.C., 1990.
40. Stankowski, S. J., Population Density as an Indirect Indicator of Urban and Suburban Land Surface Modifications, U.S. Geological Survey Professional Paper, 800-B:B219–B224, 1972.
41. Arnold, C. L., Jr., and Gibbons, J. L., Impervious surface coverage — the emergence of a key environmental indicator, *J. Am. Plann. Assoc.,* 62, No. 2, 1996.
42. Klein, R., Urbanization and stream quality impairment, *Water Resour. Bull.,* (American Water Resources Association), 15, No. 4, 1979.
43. Griffin, D. M., Analysis of non-point pollution export from small catchments, *J. Water Pollut. Control Fed.,* 52, No. 4, 780, 1980.
44. Schueler, T. R., Controlling Urban Runoff: A Practical Manual for Planning and Designing Urban BMP's, Publication No. 87703, Metropolitan Washington Council of Governments, 1987.
45. Todd, D. A., Impact of land use and nonpoint source loads on lake quality, *J. Environ. Eng.,* 115, No. 3, 633, 1989.
46. Schueler, T. R., Mitigating the adverse impacts of urbanization on streams: a comprehensive strategy for local governments, in Watershed Restoration Handbook, Publication #92701, Metropolitan Washington Council of Governments, 1992.
47. Booth, D. B., and Reinfelt, L. E., Consequences of urbanization on aquatic ecosystems — measured effects, degradation thresholds, and corrective strategies in Proc. Watershed 93 Conf., Alexandria, VA, March 1993, 545.
48. Schueler, T. R., The importance of imperviousness, *Watershed Prot. Tech.,* 1, No. 3, 100, 1994.
49. U.S. Environmental Protection Agency, Rapid Bioassessment Protocols for Use in Streams and Rivers: Benthic Macroinvertebrates and Fish, EPA/440/4-89/001, 1989.
50. Dickman, M., Brindle, I., and Benson, M., Evidence of teratogens in sediments of the Niagra River watershed as reflected by chironomid (Diptera: Chironomidae) deformities, *J. Great Lakes Res.,* 18, No. 3, 467, 1992.
51. Warwick, W. F., Morphological deformities in Chironomidae (Diptera) larvae from the Lac St. Louis and Laprairie basins of the St. Lawrence River, *J. Great Lakes Res.,* 16, No. 2, 185, 1990.
52. Diggins, T. P., and Stewart, K. M., Deformities of aquatic larval midges (Chironomidae: Diptera) in the sediments of the Buffalo River, New York, *J. Great Lakes Res.,* 19(4), 648–659, 1993.
53. van Urk, G. F., Kerkum, C. M., and Smit, H., Life cycle patterns, density and frequency of deformities in *Chironomous* larvae (Diptera: Chironomidae) over a contaminated sediment gradient, *Can. J. Fish Aquat. Sci.,* 49, 2291, 1992.

54. Ross, S. T., Matthews, W. J., and Echelle, A. E., Persistence of stream fish assemblages: effects of environmental change, *Am. Nat.,* 126, 24, 1985.
55. Matthews, W. J., Fish faunal structure in an Ozark stream: stability, persistence and a catastrophic flood, *Copeia*, 388, 1986.
56. Ovellet, M., Bonin, J., Rodrigue, J., Des Granges, J. L., and Lair, S., Hindlimb deformities (ectromelia ectrodactyly) in free living anurans from agricultural habitats, *J. Wildl. Dis.*, 33, 95, 1997.
57. Metropolitan Washington Council of Governments, Riparian Buffer Strategies for Urban Watersheds, Publication No. 95703, Washington, D.C., 1995.
58. Metropolitan Seattle Water Pollution Control Department, Biofiltration Swale Performance, Recommendations and Design Considerations, 1992.
59. Petersohn, W. T., and Correll, D. L., Nutrient dynamic in an agricultural watershed: observations on the role of the riparian forest, *Ecology,* 65, 1466, 1984.
60. Wong, S. L., and McCuen, R. H., The design of vegetative buffer strips for runoff and sediment control, in *Stormwater Management for Coastal Areas,* Department of Civil Engineering, University of Maryland, College Park, 1982.
61. Trimble, G. R., and Sartz, R. S., How far from a stream should a logging road be located? *J. For.,* 55, 339, 1957.
62. Heninger and Ray Engineering Associates, Vegetated Buffer Zones, prepared for the Southwest Florida Management District, 1991.
63. Schueler, T., Site Planning for Urban Stream Protection, Metropolitan Washington Council of Governments, Center for Watershed Protection, Publication No. 95708, December 1995.
64. New Jersey Department of Environmental Protection, Report to Governor Thomas H. Kean, the New Jersey Legislature and the Board of Public Utilities — Evaluation and Recommendations Concerning Buffer Zones around Public Water Supply Reservoirs, Rutgers University, New Brunswick, NJ, 1989.
65. Forest Service, U.S. Department of Agriculture, Riparian Forest Buffers — Function and Design for Protection and Enhancement of Water Resources, Document No. NA-PR-07-91, 1991.
66. Brinson, M. M., Bradshaw, H. S., Holmes, R. N., and Elkins, J. B., Jr., Litterfall, stemflow and throughfall in an alluvial swamp forest, *Ecology*, 61, 827, 1980.
67. Correll, D. L., The Nutrient Composition of Soils of Three Single Land Use Rhode River Watersheds, Environmental Data Summary for the Rhode River Ecosystem (1970–1978), Vol. A: Long Term Physical/Chemical Data, Chesapeake Bay Center for Environmental Studies, Edgewater, MD, 1982.
68. Lowrance, R., Todd, R., Fail, J., Jr., Hendrickson, O., Jr., Leonard, R., and Asmussen, L., Riparian Forests as nutrient filters in agricultural watersheds, *BioScience*, 34, No. 6, 374, 1984.
69. Spurr, S. H., and Barnes, B. V., in *Forest Ecology*, 3rd ed., John Wiley & Sons, 1980, 211.
70. Sykes, K. J., Perkey, A. W., and Palone, R. S., Crop Tree Management in Riparian Zones, Forest Service, U.S. Department of Agriculture, Morgantown, WV, no date.
71. Dillaha, T. A., III, Sherrard, J. H., and Lee, D., Long-term effectiveness of vegetative filter strips, *Water Environ. Technol.,* 418, November 1989.
72. Maxted, J. R. and Shaver, E., The use of detention basins to mitigate stormwater impacts to aquatic life, presented at a Natl. Conf. Urban Retrofit Opportunities for Water Resour. Prot. Urban Areas, Chicago, IL, February 9–12, 1998.
73. Bell, Warren B., Appropriate BMP technologies for ultra-urban applications, presented at Regional Conf. on Minimizing Erosion, Sediment and Stormwater Impacts: Prot. and Enhancement of Aquatic Resour. in the 21st Century, State of Delaware, September 13–18, 1998.
74. Desbonnet, A., Pogue, P., Lee, V., and Wolff, N., *Vegetated Buffers in the Coastal Zone: A Summary and Bibliography,* Coastal Resources Center, University of Rhode Island, Kingston, RI, 1994.
75. Hooper, S. T., Distribution of Songbirds in Riparian Forests of Central Maine, M.S. thesis, University of Maine, Orono, 1991.
76. Tilghman, N. G., Characteristics of urban woodlands affecting breeding bird diversity and abundance, *Landscape Urban Plann.*, 14, 481, 1987.
77. U. S. Environmental Protection Agency, Wildlife *Exposure Factors Handbook,* Vol. 1 of 2, EPA/600/R-93/187a, Washington, D.C., December 1993.
78. Rudolph, C. D., and Dickson, J., Streamside zone width and amphibian and reptile abundance, *Southwest. Nat.,* 35, No. 4, 472, 1990.

79. Croonquist, M. J., and Brooks, R. P., Effects of habitat disturbance on bird communities in riparian corridors, *J. Soil Water Conserv.,* January–February 65, 1993.
80. Keller, C., Chandler, M. E., Robbins, S., and Hatfield, J. S., Avian communities in riparian forests of different widths in Maryland and Delaware, *Wetlands,* 13, No. 2 Special Issue, Society of Wetland Scientists, June 137, 1993.
81. Dickson, J. G., and Huntley, J. C., Riparian zones and wildlife in southern forests: the problem and squirrel relationships, Proceedings: Managing Southern Forests for Wildlife and Fish, General Technical Report S O 65, Forest Service U.S. Department of Agriculture, January 1987.
82. Wackernagel, M., Onisto, L., Callejas Linares, A., Lopez, I. S., Falfan, J., Garcia, M., Suarez Geerrero, A. I., Suarez Guerrero, M. G., Ecological Footprints of Nations: How Much Nature Do They Use? How Much Nature Do They Have? Commissioned by the Earth Council for the Rio+5 Forum, International Council for Local Environmental Initiatives, 1997.
83. Tiner, R. W., Keys to Landscape Position and Landform Descriptors for U. S. Wetlands (Operational Draft), Northeast Region, U.S. Fish and Wildlife Service, Hadley, MA, 1997.
84. Tiner R., Schaller, S., Peterson, D., Snider, K., Ruhlman, K., and Swords, J., Wetland Characterization Study and Preliminary Assessment of Wetland Functions for the Casco Bay Watershed, Southern Maine, Northeast Region, National Wetlands Inventory, Ecological Services, U.S. Fish and Wildlife Service, Hadley, MA, NWI Report, 1999.
85. Weaver, R., Ultimate terms in contemporary rhetoric, in *The Ethics of Rhetoric,* Henry Regnery, Chicago, 1953, 211.
86. Harding, J. S., Benfield, E. F., Bolstad, P. U., Helfman, G. S., and Jones, E. B. D., III, Stream biodiversity: the ghost of land use past, *Proc. Natl. Acad. Sci. U.S.A.*, 95, 14843, 1998.
87. Verry, E. S., Effects of forestry practices on physical and chemical resources, Conf. Proc., at the Water's Edge: The Science of Riparian For., BU-6637-S, Minnesota Extension Service, University of Minnesota, January 1996.
88. Dwyer, J. P., Wallace, D., and Larsen, D. R., Value of woody river corridors in levee protection along the Missouri River in 1993, *J. Am. Water Resourc. Assoc.,* 33, No. 2, April 1997 (American Water Res. Assoc.).
89. Woodward, S. E., and Rock, C. A., Control of residential stormwater by natural buffer strips, *Lake Reserv. Manage.*, 11, No. 1, 37, 1995.
90. Doll, A., Lindsey, G., and Albani, R., Stormwater utilities: key components and issues, in Adv. Urban Wet Weather Reduction Conf., sponsored by Water Environment Federation, Cleveland, OH, June 28–July 1, 1998.
91. Burk, J. D., Hurst, G. A., Smith, D. R., Leopold, B. D., and Dickson, J. G., Wild turkey use of streamside management zones in loblolly pine plantations, *Proc. 6th Natl. Wild Turkey Symp.*, 6:84–89, 1990.
92. Korbinger, N. P., Evaluation and Management of Highway Runoff Water Quality, U.S. Department of Transportation, Publication No. FHWA-96-032, June, 1984.

Index

Note: Notations appended to page numbers represent the following: f refers to a figure, n refers to a footnote, t refers to a table.

A

Adversarial relationship with the environment, 26–27
Aerial photography for mapping, 59–60
Affordable housing in municipal master plan, 176
Agins v. City of Tiburon, 54
Agricultural conservation plan zone, 152
Ambler Realty Co. v. City of Euclid, 54
American chestnut trees (*Castanea dentata*), 89
amphibian deformations, 121, 122
Anti-hunting efforts, 80–81
Aquatic insect populations, *see* Benthic macroinvertebrates
Aquatic life preservation areas, 134
Arsenic contamination, 91
Automobiles contamination from, 3
Automobiles, contamination from, 1, 2

B

Beaches
  contamination by floatable debris, 16
  drifter study, 16
  property use restrictions, 55
Beachfront Management Act (1988), 55
Beachfront regulations, 136
Benthic macroinvertebrates
  advantages of as indicator, 127t
  diversity as indicator of system health, 117–118, 124, 125, 126
  as indicator of system health, 130–131, 132, 143
Beryllium contamination from greensand, 93
Best management practices (BMPs)
  ameliorative, 58
  during construction process, 179
  erosion and sediment control, 180
  fabric filter fences, 179–180
  monitoring by professionals, 181
  required for development, 139
Bird species shifts, 83
BMP, *see* Best management practices (BMPs)
Brook trout (*Salvelinus fontinalis*), 103
Buffer zones, 134, 162
  donations of, 149n
  effectiveness of, 136–138
  grassland, 140
  for noise issues, 171
  purchase of vulnerable areas, 149
  sizing, 136–138, 138–139, 165
  tailoring to site, 135–136
  tree species in, 143
  vegetative filter strips (VFSs), 143–144
  wooded, 140, 142–143

C

Caliche, 91
California Coastal Commission v. Nolan, 55
Carson, Rachel, 25
Castanea dentata (American chestnut tree), 89
CCC (Civilian Conservation Corps), 24
CERCLA (Comprehensive Environmental Response Compensation and Liability Act), 64
Chestnut trees, 89
Chromium contamination from greensand, 93
City Beautiful Movement, 21–22
City of Euclid v. Ambler Realty Co., 54
City of New York v. Penn Central Transportation Co., 54
City of Tiburon v. Agins, 54
Civilian Conservation Corps (CCC), 24
Clean lakes program, 160
Clean Water Act (1972, 1977) (CWA), 26, 64, 79
  pollutant discharge illegal under, 103
Compaction of soil, 93–94
Comprehensive Environmental Response Compensation and Liability Act (CERCLA), 64
Comprehensive plans, beginnings of, 22–23
Comprehensive Shoreland Protection Act (1994), 136
Confirmation theory, 39
Conservation easements, 12
Conservation plan used in New Jersey, 23
Conservation programs developed by Franklin D. Roosevelt, 24
Contaminated sites, 107
Contamination plume maps in master plans, 155–156
Core reserves, 164
Cost-benefit calculations of ecosystem contributions, 51
Coyote, 81
Curare, 86

D

Daylighting of streams, 154
Daylighting streams, 157f
DeBenedictis v. Keystone Bituminous Coal Association, 54
Deer tick, 82

Deforestation, 3, 86, 103
Deicing salt, 1
Detention basins, 33
Developers' use of experts, 33
Digitalis, 86
Digital orthophotos, 59–60
Disjointed incrementalism, 22, 26
Displacement of wildlife, 80
Dolan v. City of Tigard, 55–56
Dredging, 16, 160
Drifter study, 16

E

Earth Day, 25, 33
Easements, 12, 144
Ecological footprints of countries, 9, 11–12t
Ecological infrastructure, 48
  continuum nature of, 33
  cost-benefit of, 51
  inventories, 61–62
  recognition of, 51
  surface waterways, 63–74
  water resources, 63
Ecologist, first, 24
Ecology defined, 6
Economic expectations of landowners, 4
Economic value of natural resources, 47–49
Ecosystem, *see also* Waterway ecosystem
  addressed as a whole, 31
  collapse, 12
  continua, 8
  definition, 6
  global, 16
  preservation of, 51
  recycling of components, 8
Ecosystem services, 49t
  economic value of, 47–49
  global values, 50t
Ecotones, 8
Ectopistes migratorius (Passenger pigeon), 96
EIS (Environmental impact statement), 3, 79–80
Eminent domain, 54
Endangered species, 96
  habitat conservation plans (HCPs) in protecting, 163
  and wetlands, 79
Endangered Species Act (1973) ESA), 163
  habitat conservation plans (HCPs), 163
Endocrine disruptors in addressing wildlife overpopulation, 84
Environmental impact statement (EIS), 3, 79–80
Environmental indicators, 109, 114, 117
Environmental problems as land use and planning problems, 16
Environmental protection
  ethic development, 24–26
  legislation, 25
EPA, *see* U.S. Environmental Protection Agency (USEPA)
Epidemics and land use, 21
Equivalent residential or runoff unit (ERU), 154
Erosion control for open space, 85
ERU (Equivalent residential or runoff unit), 154
ESA (Endangered Species Act (1973)), 163
Euthanasia of white-tailed deer, 85
Eutrophication of lakes, 75, 76f
Exposure indicators, 109, 114, 117
Extinction of species, 96

F

Fabric filter fences improved materials, 179–180
Farm conservation plan, 152
Farmland
  complaints by new rural residents, 108
  and groundwater contamination, 108
  as open space, 85
  open space preservation, 107–108
  as pollutant source, 143
  preferential taxation rates, 152–153
  wildlife habitat preservation, 107–108
Fault zones, 95
Fish advisories, 29f, 121
Fish kills, 26
Fish populations as indicator of system health, 118–119
Flood Control Act (1938,1960), 97–98
Flooding, 97–98, 98–99
Floodplains, 96–99
  regulations, 55–56
Foliated metamorphic rock, 95
Food web, 8
Fragipans, 91
Future generations' burden, 31

G

GDP (Gross domestic product), 47
Genetic bottlenecks, 83–84
Geographic Information System (GIS), 107
  format for data storage, 31
  in inventory of infrastructure, 152
  maps in municipal master plan, 109
Geologic factors in planning, 94–96
GIS, *see* Geographic Information System (GIS)
GNP (Gross national product), 47
Government agencies failure in environmental efforts, 26
Grading of land, 3
Gravity sewer line collection systems, 106–107
Greensand as fertilizer, 93
Gross domestic product (GDP), 47
Gross national product (GNP), 47
Groundwater, 75, 77–78
  aquifer maps, 156
  contaminated of, 72
  contamination of aquifers and, 155
  contamination plume mapping, 155–156
  contribution to springs and seeps, 72
  depletion of, 9, 15
  exfiltration, 72
  leachate infiltration, 94
  pollutant infiltration, 77

as potable water source, 77
recharge, 31, 77
economic value of, 51–52
effect of impermeable surfaces, 103
prediction, 160
system health measured by potable wells, 155
Groundwater table, 77

H

Haas v. Udell, 29n
Habitat conservation plans (HCPs), 163, 164
Habitat fragmentation, 83–84
HCP (Habitat conservation plan), 163, 164
Heat island effect, 86, 103
Heavy metal contamination of soil, 91, 93
HGM (Hydrogeomorphic features), 165
Highways
effects of, 1
poor planning of, 22
related pollutants, 2t
Historical perspective in municipal master plan, 61
Holistic management, 57–58
Hybridized zoning, 29
Hydrogeomorphic (HGM) features, 165
Hydrological unit code (HUC), 67, 69, 72

I

Immunocontraception for white-tailed deer, 84–85
Impermeable surfaces, 102–103; *see also* Storm drain systems
contribution to flooding, 55–56
covering productive soils, 94
equivalent residential or runoff unit (ERU) basis for user fees, 154
limitations on, 148–149
quantity related to water quality, 171
replacing permeable surfaces, 3
and runoff, 64
user fees based on size of, 154
and watershed health, 103
Inbreeding caused by habitat fragmentation, 83–84
Indicators, *see also* Environmental indicators; Exposure indicators; Response indicators; Stressor indicators
types of, 57
Infrastructure health, 109
Infrastructure, man-made and natural, 61–62

K

Keystone Bituminous Coal Association v. DeBenedictis, 54

L

Lacey Act (1900), 80
Lake management districts, 160
Lakes, 74–75, 160
Landscape chemicals, 3
Landscape fragmentation effects on wildlife, 162
Landscapes, intrusion into undisturbed, 13
Land use planning
and case law, 22
history of, 21–23
importance of preventive measures, 33
primary considerations, 52
Leachate drainage into groundwater, 94
Limestone caves, 94–95
Long Beach Township v. Riggs, 29n
Lucas v. South Carolina Coastal Commission, 55
Lyme disease, 82, 163

M

Mahon v. Pennsylvania Coal, 54
Massachusetts Rivers Protection Act (1996), 136
Master plans, first use of, 22–23
Methyl *tert*-butyl ether, 3
MMP, *see* Municipal master plan (MMP)
Mobile phone towers scenic considerations, 176
Models, 33–40
closed system assumptions not valid, 35–36
as estimators, 34–35
heuristic, 35
in public policy, 39–40
sensitivity analysis, 40
MTBE, *see* Methyl *tert*-butyl ether
Municipal Land Use Law (New Jersey), 23
Municipal master plan (MMP), 29, 155–156; *see also* Buffer zones; Ecological infrastructure
affordable housing in, 176
components of, 30t
consequences of inadequate, 30
developers' cost-benefit decisions, 31, 169–170
development of plans for, 31–32, 52, 57–58
development process considerations, 31–32
ecologically based, 13, 29
farmland in, 107–108
floodplain planning, 99
geologic factors, 94–96
historical perspective, 61
impaired stream areas, 149–150
incorporation of wetlands information, 165–167
indicators used in developing, 57
inventory of resources, 61–62
inventory of unique habitats, 99
mapping for, 59–60
opposition by vested interest groups, 53
potable water intakes included in, 108
preservation of agricultural soils, 94
preventing degradation of surface waterways, 134–136
preventive planning, 58
and private property rights, 53–56
and problem species, 163
in protection of intact ecological infrastructure, 52
public participation in process, 58
purpose, 1

residents' contributions, 30–31
scientific data and, 57
soil mapping in, 91
storm drain systems in, 100–101
transgenerational mindset, 31, 32
tree specimens in, 89
unique habitats in inventory, 99
vegetative inventory in, 89
watershed protection strategy, 158, 160
wildlife problems addressed in, 164
zoning improvements, 176
zoning subordinate to, 29
Municipal tax map, 59

N

National Environmental Monitoring Improvement Act (1984), 26
National Environmental Policy Act (1969), 25
National Flood Insurance Program (NFIP), 98–99
National Pollutant Discharge Elimination System (NPDES), 103
permits for wastewater discharge, 105
permit violations, 105
point source dischargers, 114
National wetlands inventory maps (NWI), 165–167
Natural Resource Conservation Service (NRCS), 67
Natural resources
inventories, 61–62
un-sustainable consumption, 9, 12
NFIP (National Flood Insurance Program), 98–99
Nickel contamination from greensand, 93
No-growth strategies, 12
Noise complaints, 171
Nolan v. California Coastal Commission, 55
NPDES, *see* National Pollutant Discharge Elimination System (NPDES)
NRCS (Natural Resource Conservation Service), 67
NWI (National wetlands inventory maps), 165–167

O

Open space
acquisitions, 12
management plans for, 85
Open systems, 38
models invalid for, 36

P

Parcel base map, 59–60
Passenger pigeon (*Ectopistes migratorius*), 96
Penn Central Transportation Co. v. City of New York, 54
Pennsylvania Coal v. Mahon, 54
Planning processes disaggregated, 34
Point source discharges, 103, 105–107
Pollinators economic value of, 48
Ponds, 74–75
Potable water intakes, 108
Potable water wells and groundwater depletion, 15
Preseault v. United States, 55
Preventive planning, 58
Private property rights, 53–56
Property parcel map, 59–60

Q

Quinine, 86

R

Raccoon and habitat change, 81
Raptor nesting habitat, 89
RCRA (Resource Conservation and Recovery Act), 64
Recharge, *see* under Groundwater
Recreation use of wetlands, 79
Regulatory takings, 54–56
Research data government development of, 13
Research results, *see also* Scientific data
Resource Conservation and Recovery Act (RCRA), 64
Response indicators, 109, 117–119, 121, 123–128
human oriented, 171
Ridgeline/panorama protection, 176
Riggs v. Long Beach Township, 29n
Riparian corridors, *see* Buffer zones
River basin based planning, 57–58
Roadways, *see* Highways
Rollback taxes, 149n
Roosevelt, Franklin D. and conservation programs, 24
Runoff
capture by soil, 94
due to compaction, 93
and impermeable surfaces, 64
regulating, 137
storm drain systems, 100

S

Salvelinus fontinalis (Brook trout), 103
Sand filtration structures, 153, 154f, 155f, 156f, 157f
Scientific data
accessibility of, 33
availability, 109
contribution to planning process, 33–34
in local planning, 34
quantity of, 33
SCS (Soil Conservation Service), 24
Sediments contaminated, 16
Sensitivity analysis of models, 40
Sewage pump stations, 106–107
Sewer system planning, 22
Shoreland regulations, 136
Silent Spring, 25
Sinkholes, 95
SMZ (Stream management zones), 134–136, 165
Soil
bulk density measurements, 93–94
effects of air exposure by excavation, 91

formation, 90–91
glauconite as contamination source, 93
horizons, 91
leaching, 91
in master plan, 168–171
precipitation capture, 94
preservation of agricultural, 94, 168–171
Soil Conservation Service (SCS), 24
Soil conservation surveys, 91
South Carolina Beachfront Management Act (1988), 55
South Carolina Coastal Commission v. Lucas, 55
Standard City Planning Enabling Act (1928), 22
Standard State Zoning Enabling Act (1924, 1926), 22, 23
Steep slopes, 96
Stewardship, 1, 23
Storm drains as wildlife highways, 81
Storm drain systems
ecological effects, 99–100
flooding and, 100
pollutant accumulation, 99–100
runoff quantity, 100
water quality effects, 100
Storm water runoff
applied as sheet flow, 146–148, 165
effect of bypassing buffers, 144–145
Storm water utility, 154
Strahler method, 74
Stream
diversity as indicator of system health, 104t
habitat quality, 65f, 66f
heat island effect, 86
hierarchy, 74
Stream management zones (SMZs), 134–136, 165
Stressor indicators, 109, 114
Suburbia, ills of, 1
Superfund Act (Comprehensive Environmental Response Compensation and Liability Act), 64
Surface water pollutants, 114
Surface waterways, *see also* Benthic macroinvertebrates
daylighting, 154
determining degradation causes, 151t
lake management districts, 160
preventive planning for, 134–136
rehabilitation considerations, 151–154

T

Tennessee Valley Authority (TVA), 24
"Think globally, act locally", 18
Threatened species, *see* Endangered species
Total maximum daily load program (TMDL), 105–106
Toxic Substances Control Act (TSCA), 64
Traffic, *see* Highways
Transgenerational mindset, 32
Tree species water tolerance, 8, 145t, 146t, 147t, 148t
Trout waterway protection, 31, 103, 134, 165
TSCA (Toxic Substances Control Act), 64
TVA (Tennessee Valley Authority), 24

U

Udell v. Haas, 29n
Undisturbed landscapes, intrusion on, 13
United States v. Preseault, 55
Urban heat island effect, 86, 103
U.S. Army Corps of Engineers jurisdiction over wetlands, 79
USDA Soil Conservation Service (SCS), 67
U.S. Environmental Protection Agency (USEPA)
jurisdiction over wetlands, 79
place-based management recommended, 26

V

Vanadium contamination from greensand, 93
Vegetative filter strips (VFSs), 143–144
Vegetative resources, 86
individual tree specimens, 89
medicines from, 86
urban heat island effect, 86
VFS (Vegetative filter strips), 143–144
Village of Euclid v. Ambler Realty Co., 22

W

Water quality degradation, 151
Water quality monitoring, 105
Watershed based planning, 57–58
Watershed boundaries, 67
Watershed management, 64, 67, 69, 72–74
assessment reports, 72
wildlife habitat, 72
Watershed management areas (WMAs), 72
Watershed protection included in municipal master plan, 158, 160, 171
Waterway contamination, 26
Waterway ecosystem, 15
Western coyote, 81
Wetlands, 165–167
contribution to water quality, 79
definition, 78–79
national wetlands inventory maps (NWI), 165
protection, 79
recreation, 79
replacement proposals, 165
White-tailed deer, 81, 82
automobile-related accidents, 128, 162, 163
displacement of, 34
Lyme disease and, 82, 163
overpopulation, 84–85
as problem species, 163
Wildlife
core reserves for, 164
displacement myth, 80
fragmented landscape, 162
habitat loss, 79–84
intention to protect, 162
legislation, 80

local community activities, 162
mammalian adaptation to habitat changes, 81
in refuges, 82
storm drains as highways, 81
value in planning process, 79
Wildlife habitat
bird species changes caused by changes in, 83
and buffer zones, 162
buffer zones for, 164–165
core reserves for, 164
criteria, 164t
fragmented by development, 82
mitigation measures, 163
myths, 162
riparian corridor as, 162
species shifts caused by changes in, 82–83
Wildlife population displacement, 33–34
Wild trout protection areas, 134
WMA (Watershed management areas), 72
Woodland destruction, 86

Z

Zinc contamination from greensand, 93
Zoning
effects on ecosystem, 16
first applications of, 22
hybridized, 29
mission of, 29
"one lot, one structure", 175
vs. planning, 29
proposals for improving, 175
subordinate to master plan, 29
Zoning maps, 4
deficiencies of, 8
density allocations, 5
Zoning ordinances, 22
ecologically based, 13, 29
Standard State Zoning Enabling Act (1924, 1926), 22